STRAIN AND STRESS
ANALYSIS
BY HOLOGRAPHIC AND
SPECKLE
INTERFEROMETRY

STRAIN AND STRESS ANALYSIS BY HOLOGRAPHIC AND SPECKLE INTERFEROMETRY

Valery P. Shchepinov,
Vladimir S. Pisarev,
Sergey A. Novikov,
Vitaly V. Balalov,
Igor N. Odintsev
and
Mikhail M. Bondarenko

JOHN WILEY & SONS
Chichester · New York · Brisbane · Toronto · Singapore

National   01243 779777
International (+44) 1243 779777

Other Wiley Editorial Offices

John Wiley & Sons, Inc., 605 Third Avenue,
New York, NY 10158–0012, USA

Jacaranda Wiley Ltd, 33 Park Road, Milton,
Queensland 4064, Australia

John Wiley & Sons (Canada) Ltd, 22 Worcester Road,
Rexdale, Ontario M9W 1L1, Canada

John Wiley & Sons (Asia) Pte Ltd, 2 Clementi Loop #02-01,
Jin Xing Distripark, Singapore 0512

Library of Congress Cataloging-in-Publication Data

Strain and stress analysis by holographic and speckle interferometry /
    edited by Valery P. Shchepinov, Vladimir S. Pisarev.
        p. cm.
    Includes bibliographical references and index.
    ISBN 0-471-96077-2 (hardcover : alk. paper)
    1. Deformations (Mechanics). 2. Strains and stresses.
    3. Holographic interferometry. 4. Speckle metrology.
    5. Structural analysis (Engineering) I. Shchepinov, V. P. (Valerii Pavlovich),
    1947– . II. Pisarev, Vladimir S.
    TA417.6.S74 1996   95-22521
    624.1'71–dc20   CIP

British Library Cataloguing in Publication Data
A catalogue record for this book is available from the British Library

ISBN 0 471 96077 2

Typeset in 10/12pt Times by Dobbie Typesetting Ltd, Tavistock, Devon
Printed and bound in Great Britain by Biddles Ltd, Guildford, Surrey
This book is printed on acid-free paper responsibly manufactured from sustainable forestation,
for which at least two trees are planted for each one used for paper production.

CONTENTS

PREFACE

Holographic and speckle interferometry are currently the most advanced optical methods employed in experimental mechanics which implement laser radiation. A remarkable feature of these methods is their capability to perform contactless and high-sensitivity measurements of parameters inherent in solid mechanics on optically rough surfaces of real structures. These parameters are displacements and displacement derivatives along the spatial coordinates. The whole-field character of such measurements is of great importance in many mechanical applications.

This book introduces and discusses the development and application of holographic and speckle interferometry employed in the investigation of structure and material deformation and fracture processes. The exceptional features and advantages of these methods are discussed and compared with conventional methods of experimental mechanics. Much attention is given to revealing the capabilities of holographic and speckle interferometry in solving topical mechanical problems. Among these are problems related to stress concentration near cut-outs and cracks in curved shells, investigation of microplastic strains, elasto-plastic local strain history in the low-cyclic fatigue range, fatigue crack initiation and propagation, and determination of contact and residual stresses.

The book contains mainly the results of investigations of researchers and post-graduates from the Physics of Strength Department at the Moscow Physical Engineering Institute. Most of the experimental work was carried out in the holographic laboratories of this department and Central Aerohydro-dynamics Institute (TsAGI) over the last ten years. This book develops the approach to quantitative fringe pattern interpretation for an accurate determination of displacement components contained in the monograph by Yu. I. Ostrovsky, V. P. Shchepinov and V. V. Yakovlev, *Holographic Interferometry in Experimental Mechanics*, Springer Series in Optical Science, 60, Springer-Verlag, Berlin, 1991.

For a correct solution of many topical strength problems by means of holographic and speckle interferometry, the researcher should have some previous knowledge of the physical principles of optical measurements, fringe pattern recording and interpretation techniques, approaches to design of experiment, etc. In addition, there is a series of problems associated with the choice of a reliable method to transform detected displacement, slope and curvature values to strains and stresses. The main reason for such problems is due to the high sensitivity of the holographic and speckle interferometry methods with respect to both strain-induced and rigid-body displacements of structures to be studied.

This book is divided into two parts. The first part (Chapters 1–4) introduces the physical and metrological aspects of holographic and speckle interferometry from the deformation analysis viewpoint. Much attention is given to questions on how to obtain a high-quality fringe pattern corresponding to deformation conditions of the object under study. The second problem is ensuring the maximum possible accuracy of the displacement component determination. This part also discusses the non-conventional method for choosing optimal parameters of interferometer systems, original techniques of specklegram recording based on four-multiple exposing, and a new approach to slope and curvature measurements by means of compensation holographic interferometry.

The second part initially discusses the important problem of transition from displacement or slope values measured on a single (external) face of a thin-walled structure to strain and stress distributions through the object thickness (Chapter 5). The first step in this procedure, dealing with strain determination on the external face, can be performed through the use of geometrical relations only. On the other hand, calculation of the strains and stresses in the normal to the surface direction through the object thickness is connected with the use of hypotheses concerning the deformation character of a thin-walled structure. A detailed analysis of the influence of both these steps on the accuracy of the final results is presented. A series of current mechanical problems are used as examples for this analysis. One concerns the determination of strain and stress concentration in plates and curved shells with large cut-outs, taking into account real boundary conditions and a few hole interactions.

Original approaches to determination of parameters of fracture mechanics for through and partially through cracks in curved shells based on the precise determination of displacement fields and weight functions are developed in Chapter 6.

One of the most important aspects of holographic and speckle interferometry implementation in the field of solid mechanics is the efficiency and reliability of these methods to determine and control mechanical properties and mechanical behaviour of a material at different stages of the deformation process. Chapter 7 introduces new holographic compensation interferometric techniques for accurate measurement of the elasticity modulus, Poisson's ratio and microplasticity parameters. These techniques have demonstrated their viability in many cases, providing more information than traditional experimental methods.

New results are presented on the evolution of local elasto-plastic strains in the contact interaction zone and the change of the material local mechanical properties for the low-cyclic fatigue case, including processes of the fatigue crack initiation and propagation (Chapter 8).

The non-conventional approach to reliable determination of residual stresses by combining the hole-drilling technique and holographic interferometry is established in Chapter 9.

New techniques are discussed for contact surface and contact pressure determination in the case of rough surface mechanical interaction (Chapter 10). The methods of correlation holographic interferometry and correlation speckle photography used for these purposes are based on the measurement of the contrast change of interference fringe patterns with carrier.

The last chapter of the book illustrates how the methods of holographic and speckle interferometry can be implemented for solving some characteristic problems in the field of strength analysis of different structures.

The book is illustrated with numerous interferograms. This was done to demonstrate the various fringe patterns which must be quantitatively interpreted to obtain strain and stress values with the required accuracy. Note that none of these interferograms were prepared as illustrations only. All fringe patterns are primary sources of the corresponding displacement and strain or stress distribution contained in this book.

All the results presented show that holographic and speckle interferometry can be used as an effective means for development in various scientific and applied subjects in the fields of solid and fracture mechanics.

This book is mainly aimed at researchers and engineers specialized in the field of strength of various structures and materials. It can also serve as a guide to post-graduates and students in understanding the fundamentals of holographic and speckle interferometry of opaque bodies and the way these methods can be applied in experimental mechanics.

The main objective of this book is to expand the field of research in experimental mechanics using the presented methods and results of investigations. The authors hope this will eventually lead to the wide implementation of holographic and speckle interferometric techniques in new experimental strain analysis for the future.

One last comment should be made here. Within the context of this book we do not discuss problems dealing with the automation of both recording and interpretation of interference fringe patterns. However, this topic is presented in more detail in the following monographs: R. S. Sirohi (ed.), *Speckle Metrology*, Series: Optical engineering, Marcel Dekker, New York, 1993, and P. K. Rastogy (ed.), *Holographic Interferometry*, Springer series in optical science, 68, Springer-Verlag, Berlin, 1994. These books also describe the current status of research on the developement and implementation of holographic and speckle interferometry in the fields of experimental mechanics which lie outside the scope of our book.

We gratefully acknowledge the kind cooperation of our colleagues V. S. Aistov, Dr V. I. Gorodnichenko and A. Y. Shcikanov in the experimental work and discussions. We would like to thank Dr T. K. Begeev and Dr V. I. Grishin who made most of the finite element calculations used in this book. The English text was edited by Dr V. S. Pisarev. Special thanks are given to Mrs Carol McHugh who in the course of preparation of the English text greatly contributed to the clarity and intelligibility of the presentation. The authors are most grateful to those who have made this publication possible: Mr F. Petricca, Mr G. Petricca and Mrs R. Alati for their active and constructive support and Miss M. A. Petricca for her careful contribution to the detailed drawings.

Moscow, Russia,   V. P. Shchepinov,
Zug, Switzerland   V. S. Pisarev.
                                                                   March 1995

LIST OF AUTHORS

Vitaly V. Balalov, State Academy of Service, 141220 Cherkizovo, Russia

Mikhail M. Bondarenko, Bauman State Technical University, 107005 Moscow, Russia

Sergey A. Novikov, Prochnost Co. Ltd, 109428 Moscow, Russia

Igor N. Odintsev, The Russian Minatom Engineering Center of Atomic Equipment Strength Reliability & Life Time, 101000 Moscow, Russia

Vladimir S. Pisarev, Central Aerohydrodynamics Institute, 140160 Zhukovsky, Russia

Valery P. Shchepinov, Physical Engineering Institute, 115409 Moscow, Russia

The manuscript was prepared by the authors in cooperation with ALITECO AG, Alpenstrasse 14, 6300 Zug, Switzerland

1 INTRODUCTION TO OPTICAL HOLOGRAPHY AND THE SPECKLE EFFECT

Holography is a method of wave recording and reconstruction based on recording interference fringe patterns which result from coherent superposition of the wave scattered from an object surface and a reference wave derived from the same source used for the illumination of the object. The resulting intensity distribution that can be contained, for instance, on a photoplate is called a hologram. If the hologram is illuminated with the original reference wave it reconstructs the object wave, i.e. it reproduces the same amplitude and phase distribution as that created by the object wave during recording.

The possibility of holographic imaging was first demonstrated by Gabor [1]. Further developments in holography related to the off-axis reference beam technique were achieved by Leith and Upatnieks [2]. Denisyuk proposed the technique of producing reflection holograms [3].

A brief consideration of the principles of hologram recording and the main properties of holograms is presented in this chapter. The basic physical principles of holography needed for holographic interferometry application in the field of experimental mechanics of deformable bodies will be covered. A deeper understanding of optical holography can be acquired by reading monographs by Butters [4], Viénot et al. [5], Hariharan [6], Collier et al. [7], Abramson [8], Caulfield [9], Ostrovsky [10], and Francon [11].

When a body with an optically rough surface is illuminated by a coherent laser light, a random interference pattern consisting of so-called speckles can be observed. On the one hand, the speckles represent an optical noise that can reduce the quality of the holographic interference fringe pattern to be interpreted in terms which are interesting from a mechanical point of view. On the other hand, the effect of the speckles can be effectively used for various metrological applications. For the latter reason a brief description of the optical background which is necessary for an understanding of metrological

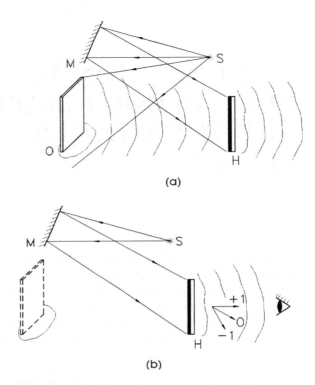

(a)

(b)

Figure 1.1. Schematic representation of hologram formation (a), and wavefront reconstruction (b)

applications of different speckle interferometric techniques is given in the third section of this chapter. More detailed information concerning speckle effects and related phenomena can be found, for instance, in work by Francon [12], Dainty [13].

1.1 Basic equations of holography and hologram classification

Figure 1.1(a) is a schematic diagram of the off-axis arrangement for hologram formation developed for the first time by Leith and Upatnieks [14]. The light wave emerging from the point source S illuminates the object surface O. The photoplate H receives the light wave scattered by the object surface (object wave). A part of the illuminating wave (reference wave) is reflected by the mirror M onto the same photoplate H immediately. Within the score of the

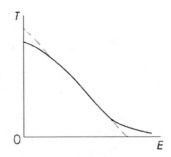

Figure 1.2. Typical dependence between amplitude transmittance T and exposure E of photographic emulsion

scalar beam treatment, the complex amplitudes of the object and reference waves on the hologram plane can be expressed as

$$A_0 = a_0 \exp(-i\phi), \quad A_r = a_r \exp(-i\varphi) \tag{1.1}$$

where a_0 and a_r, ϕ and φ are the amplitudes and phases of the object and the reference waves, respectively.

If the object and reference waves are coherent, they are capable of forming an interference fringe pattern in a recording medium covering the photoplate with the following resultant intensity distribution:

$$I = |A_0 + A_r|^2 = (A_0 + A_r)(A_0^* + A_r^*) =$$
$$a_0^2 + a_r^2 + A_0 A_r^* + A_0^* A_r \tag{1.2}$$

where $*$ denotes a complex conjugation. The fringe pattern thus recorded is a hologram. Below we will consider the case of pure amplitude recording of this intensity distribution only, i.e. the case when the recording medium changes its amplitude transmittance due to illumination. In order to describe the hologram recording process on the photoplate, it is convenient to represent the properties of photographic emulsion in the form of dependence between the amplitude transmittance of the photographic layer T and exposure $E(E = It$, where I is the intensity distribution of the light directed at the photoplate and t is the exposure duration). This dependence is presented schematically in Figure 1.2.

The central segment of the plot of amplitude transmittance versus exposure represents, as seen from Figure 1.2, a linear function that can be accurately expressed by the relation

$$T = b_0 + b_1 E \tag{1.3}$$

where the coefficient b_1 defines the inclination of the straight section (for negative recording to which Figure 1.2 corresponds, $b_1 < 0$; in the case of positive recording, $b_1 > 0$). The amplitude transmittance of a hologram now

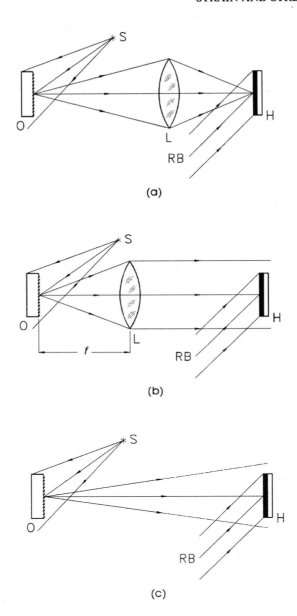

Figure 1.3. Generation of holograms of various types: image plane (a), Fraunhofer (b), Fresnel (c), Fourier (d), Denisyuk (e). The following notations are used in the diagrams: S is the illuminating point source, O is the object, L is the lens, H is the hologram, RB is the reference wave, OB is the object wave

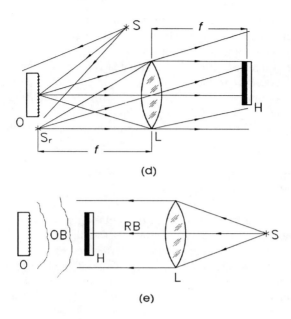

(d)

(e)

can be obtained by substituting into equation (1.3) the intensity distribution from equation (1.2)

$$T = b_0 + b_1 t(a_0^2 + a_r^2) + b_1 t A_0 A_r^* + b_1 t A_0^* A_r$$

The distribution of the complex amplitudes, immediately behind the hologram plane, when the photoplate is illuminated with the reference wave A_r (see Figure 1.1(b)) is given by

$$TA_r = b_0 + b_1 t(a_0^2 + a_r^2)A_r + b_1 t a_r^2 A_0 + b_1 t a_r^2 A_0^* \tag{1.4}$$

The first term on the right-hand side of expression (1.4) is directly proportional to the complex amplitude of the reference wave and corresponds to a zero-order wave. The second term of this equation coinciding within the real constant factor with the complex amplitude of the object wave describes the object wave reconstructed by the hologram (positive first-order wave). This wave produces a virtual 3-D object image located at the position where the object was placed during the hologram recording. The third term in expression (1.4) differs from the wave conjugated to the object wave by the complex factor. This term corresponds to the negative first-order reconstructed wave that forms a distorted real image of the object. In order to obtain an undistorted real object image, the hologram has to be illuminated with a wave

A_r^* conjugate to the reference wave, i.e. the wave with a wavefront curvature equal in magnitude and opposite in sign to that of the reference wave, but propagating in the opposite direction.

The angles of propagation of all three waves reconstructed simultaneously with the hologram are influenced by the angle between the object and reference beams. In the optical arrangement of Leith and Upatnieks for hologram recording shown in Figure 1.1(a) the inclination of the reference wave with respect to the object wave and photoplate plane can be chosen so that the mutual interference of zero and first-order waves behind the hologram will be avoided.

The equations (1.2) and (1.4) that describe the hologram recording and wavefront reconstruction respectively are usually called the basic equations of holography.

There are two approaches to hologram classification which can be adopted. The first entails the method of formation of the object and reference waves. The second is based on the manner in which interference structures are recorded.

If the object image O is focused in the plane of the hologram H by means of a lens L, the amplitude and phase distributions on the photoplate will coincide with those in the object surface (see Figure 1.3(a)). This is the image plane hologram. It is important that in this case each point of the object surface corresponds to a small area on the hologram, thus resulting in a local recording. This means that spectral composition and the wavefront configuration of the reconstructing wave are no longer important to reproduce the object image correctly.

Another type of hologram can be recorded when the photoplate H is placed at an infinite distance from the object, i.e. in the Fraunhofer diffraction region. This is called the Fraunhofer hologram. In the case considered each point of the object sends a parallel light beam onto the hologram. In order to reconstruct the hologram, the object should be located at an adequate distance from the plate, or at the lens focus (see Figure 1.3(b)).

Fresnel holograms are frequently used in various applications and are considered the most widespread. They are formed when the recording medium is placed in the near-field diffraction region (Fresnel diffraction zone)—see Figure 1.3(c). By increasing the distance between the object and hologram, the Fresnel hologram is transformed into a Fraunhofer hologram. When this distance is reduced to zero, an image plane hologram will be recorded.

If both the object and the point source of the reference wave are located at infinity, the amplitude distribution of each wave in the plane of the hologram represents the Fourier transform of the amplitude distribution of the object and the reference source, respectively. Such a hologram is called the Fourier transform hologram. These holograms are usually produced by placing the object O and the reference wave source S in the focal plane of a lens (see Figure 1.3(d)).

A common characteristic of all the above-considered optical arrangements for hologram recording is that both the object and reference beam reach the hologram plate from the same side which is usually covered with the recording medium. Figure 1.3(e) illustrates the situation when the reference and object beams strike the photosensitive layer on opposite sides. The maximum possible angle between them in this case is equal to 180°. This so-called opposed-beam configuration was first developed by Denisyuk [3]. A particular feature of this set-up is that it is possible to reconstruct the light waves recorded on the hologram by both laser and normal white light illumination.

The spatial frequency v or period d of the interference pattern depend on the angle between reference and object beam α in the following way:

$$v = \frac{1}{d} = 2\frac{\sin(\alpha/2)}{\lambda} \tag{1.5}$$

where λ is the wavelength. Note that the tangential to the surface of constant intensity, at each point of the volume of reference and object beam overlapping, bisects the angle α, i.e. coincides with the bisector of the interior angle between the wave vectors of the interfering waves.

In the off-axis optical set-up of Leith and Upatnieks the coherent reference beam is usually formed separately. The spatial frequency of the interference fringe pattern caused by hologram recording in this case corresponding to equation (1.5) is usually the value of order $1000\,\text{mm}^{-1}$. Therefore, recording such holograms requires the use of photographic material capable of relatively high spatial resolution.

If a hologram is recorded by using the opposed-beam configuration (such a hologram is usually called a reflection hologram), the angle between the object and reference beam is close to 180°. In this case the spatial frequency of the interference structure tends to its maximum value $2/\lambda$. Therefore, the requirements for spatial resolution of the photosensitive medium used for reflection hologram recording are higher than those in the case of off-axis recording.

The interference pattern can be recorded in a photosensitive material in either of the following ways:

- In the form of variations in the transmittance, or reflection index and holograms, in this case called amplitude holograms.
- In the form of variations in the thickness or refractive index, these holograms are called phase holograms.

In practice, phase and amplitude modulation usually occur simultaneously. For example, a conventional photographic plate responds to an interference pattern with variations in blackening, refractive index and relief. After bleaching, the hologram will correspond only to phase modulation.

1.2 Principal properties of holograms

The main characteristic of a hologram compared to a normal photograph consists of the following. A conventional photograph is capable of recording the intensity distribution of the incident light waves only. A hologram contains information about both the intensity distribution of the wavefront scattered by the object surface and the phase distribution of the object wave relative to the phase of the reference wave. The variation in the amplitude of light waves is recorded on a hologram as the variation in contrast with the interference pattern, while the information during the phase is preserved in the shape and frequency of the interference fringes.

Amplitude holograms are usually recorded on negative photographic material. The properties of such a hologram are identical to those of a positive hologram, i.e. the bright (dark) spots of an object correspond to bright (dark) spots in the reconstructed image. This results from the fact that, as mentioned above, the information on the object wave amplitude is contained only in the contrast of the interference fringe pattern. This contrast distribution is independent of whether a positive or a negative recording procedure is implemented. The change of a negative process on a positive process or vice versa leads to the change of the phase of the reconstructed object wave by π. This cannot be detected visually, but can manifest itself in some holographic interferometric experiments.

When the light emerging from each point of the object illuminates the whole surface area of the hologram during recording, each small part of the hologram is capable of reconstructing the whole image of the object. Of course, the smaller the section of the hologram being used for the reconstruction, the smaller the light wave set carrying information of the object. As the size of a hologram section decreases, the quality of the reconstructed image deteriorates.

When the image-plane hologram is recorded, each point of the object surface illuminates only a small area of the corresponding point on the hologram surface. Therefore, a fragment of such a hologram will be able to reconstruct only the part of the object corresponding to it. The whole brightness range which can be reproduced by a photographic plate does not exceed, as a rule, one or two orders of magnitude. Real objects, however, are characterized by much larger variations in brightness. A hologram acting as an optical element has focusing properties, thus the whole energy of the reference wave incident on the hologram surface is redistributed into the object wave. In this way the hologram is capable of reproducing up to five or six decades of image brightness.

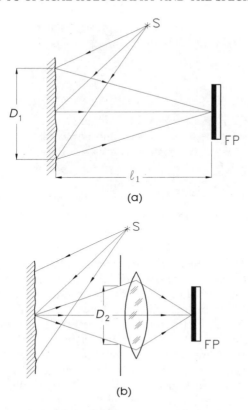

Figure 1.4. Optical arrangement for recording of objective (a) and subjective (b) speckle pattern

When the accurate copy of the reference beam used for hologram recording reconstructs the object wave, the virtual image coincides in shape and position with the actual object. If the reference wave source is displaced from its original position, the reconstructed virtual image will be deformed and distorted compared to the real object.

The limit of the spatial resolution capability of the Fresnel hologram is determined by light diffraction on its aperture and can be calculated in the same way as for a conventional optical system. In accordance with Rayleigh's criterion, the angular resolution δ_β of a circular hologram of diameter D is

$$\delta_\beta = \frac{1.22\lambda}{D} \tag{1.6}$$

where λ is the wavelength. However, in practice reaching the limiting resolution given by equation (1.6) is hindered for several reasons. One of these is the

Figure 1.5. Typical image of subjective speckle pattern

so-called speckle phenomenon. The physical aspects of a speckle formation will be considered in the next section. The second reason for the limited resolution (1.6) reduction consists of the fact that in most optical configurations the limited dimension of the hologram is connected with the resolution of the recording medium, since a growth in the hologram size results in an increase of the reference-to-object beam angle and, hence, of the spatial frequency of the interference pattern.

The most important characteristic of a hologram is the diffraction efficiency which defines the brightness of the reconstructed image. It represents the ratio of the light flux in the reconstructed wave to that of the incident on the hologram surface. The diffraction efficiency depends on the type of hologram and the properties of the recording medium used, as well as the recording conditions.

If the exposures at the peaks of an interference pattern are substantially beyond the linear segment of the amplitude transmittance versus exposure dependence (Figure 1.2), the hologram recording procedure becomes non-linear. A linear recorded hologram can be represented as a diffraction grating with a sinusoidal amplitude transmittance distribution which is not capable of producing higher than a first-order diffraction. In non-linear recording, a hologram also represents a periodic grating which can result in the appearance of diffracted waves of higher orders. The effects of non-linearity on the first-order image is background halation, distortion of relative intensities coming from different points of the object, and, in some cases, the appearance of false images.

1.3   Properties of individual speckles and speckle patterns

The speckle effect is one of the main phenomena resulting from the coherent illumination interaction with the optically rough surface. By speckle effect, we mean the phenomenon of the interference of mutually coherent waves having accidentally varying phases. The result of such an interference is the distribution of maxima and minima of light intensity which is stationary in time but random in space. The individual intensity peaks representing the bright or dark spots are called speckles. The grouping of these speckles is usually called the speckle pattern. The term optically rough surface means the surface that has a microrelief, the height of individual peaks of which is comparable to or higher than the wavelength of the light used for the surface illumination. This condition is valid, as a rule, for surfaces of real engineering structural elements when a visible laser radiation is implemented for illumination.

A surface roughness is the essential characteristic of each structural element and depends on the manufacturing technology used such as cutting, casting, rolling and so on. Physically, a surface roughness represents a group of microhills and microhollows. In most cases these elements of a microrelief are randomly arranged. This representation of a surface microrelief allows us to interpret it mathematically as a realization of an ergodic and stationary random process depending on the spatial coordinates on an object surface.

When a rough surface is illuminated with coherent light, the information on a surface microrelief is transmitted into a scattered wave (speckle field) and then into a surface image formed by some optical system. Thus, the random dependence of the surface roughness from the spatial coordinates is transformed into the set of dark and bright spots both in the image plane (image speckle pattern) and in any intermediate plane located between the illuminated surface and the image plane.

A speckle structure obtained in different ways can be classified as 'objective' and 'subjective'. Figures 1.4(a) and 1.4(b) schematically show the optical arrangement for recording of the 'objective' and 'subjective' speckle pattern, respectively. The main distinction between the two above-mentioned speckle patterns consists of the fact that a small fragment of the 'objective' fringe pattern contains information on the whole illuminated part of the object surface, whereas a small section of the 'subjective' fringe pattern corresponds to a definite part of the object surface, the size of which is equal to the resolution of the recording optical system. Therefore, in order to record the 'subjective' speckle pattern, the optical recording system would not be capable of resolving the individual elements of the surface microrelief. A typical speckle pattern obtained with the optical set-up shown in Figure 1.4(b) is presented in Figure 1.5. This figure shows the magnified image of the section of the optically rough surface uniformly illuminated by laser light. The image of the same surface obtained with white illumination, i.e. without the speckle effect, represents the homogeneous background. In the following we will mainly use the speckle structures formed with different lens systems and, therefore, the term 'subjective' will be omitted.

A three-dimensional speckle pattern is characterized by a great space anisotropy. Analysis has revealed that from the form of an individual speckle, two directions can be selected in 3-D space: the direction in which most of the wave set scattered by the object surface is propagated, and the orthogonal direction. Individual speckles have an elliptical shape. The larger axis of each such ellipse coincides with the main direction of the wave scattering. The average transversal size S_τ of an individual 'objective' speckle in the cross-section located on the distance l_1 from the optically rough surface illuminated with coherent light (see Figure 1.4(b)) is given by

$$S_\tau \sim \frac{\lambda l_1}{D_1} \tag{1.7}$$

where λ is the wavelength and D_1 is the maximum dimension of the surface area under illumination. The analogous speckle size taken along the main direction of the scattered light propagation can be estimated as

$$S_n \sim \frac{4\lambda l_1^2}{D_1^2}. \tag{1.8}$$

Consider now the spatial dimensions of individual speckles formed with a lens system. The average speckle size $\overline{S}_\tau$ in the above-mentioned transversal direction is

$$\overline{S}_\tau \simeq 1.22\lambda F \frac{M+1}{M} \tag{1.9}$$

where $F = f/D$ is the numerical aperture of the recording lens; f is the focal length of the lens; D is the diameter of the lens input pupil; and M is the transverse magnification factor. Obviously, the minimum value of the average speckle size (1.9) is reached when $M \to \infty$. The latter means that the object surface coincides with the focal plane of the lens and, hence, the object surface image is focused at infinity. Thus, the minimum speckle size is given by

$$\overline{S}_\tau = 1.22\lambda F.$$

The analogous parameter for the optical system when magnification M is equal to unity is

$$\overline{S}_\tau = 2.44\lambda F. \tag{1.10}$$

One of the most important characteristics of speckle patterns which represent a realization of a random process is the probability density function $\rho(I)$ of the intensity distribution I in an arbitrary point of space.

Taking into account that the phase shifts of the light waves scattered by an optically rough surface are distributed in the range of $-\pi$ to π, we can deduce that the resultant speckle field is a rotational (circular) Gaussian variable. The intensity of the speckle field in such cases is characterized by the following exponential probability density function by Goodman [16]:

$$\rho(l) = \begin{cases} \frac{1}{\bar{I}}\exp - \frac{I}{\bar{I}}, & \text{when } I > 0 \\ 0 \text{ otherwise} \end{cases} \tag{1.11}$$

where $\bar{I}$ is the average intensity value. A special feature of the distribution (1.11) is the fact that random fluctuations of intensity with respect to their average magnitudes are equal to $\bar{I}$.

A contrast of the speckle pattern, defined as the ratio of the standard deviation σ_l to the average intensity value, is equal to unity at any point of the speckle field. This fact is clearly illustrated in the photograph of the speckle pattern presented in Figure 1.5. This exerts a highly negative effect on the resolution of the optical systems when coherent radiation is used to form an image of both opaque and transparent objects. Instead of a homogeneous illumination having almost equal brightness, the surface is covered with a spotty high-contrast pattern which practically eliminates the fine structure of the image. This phenomenon does not depend on the way the image is viewed, i.e. by eye, through the recording lens, or direct recording onto the photoplate.

References

1. Gabor, D. (1948) A new microscopic principle. *Nature* **161**, 777–778.
2. Leith, E. and Upatnieks, J. (1963) Wavefront reconstruction with continues tone objects. *J. Opt. Soc. Amer.* **53**, 1377–1381.
3. Denisyuk, Yu. N. (1962) On the reproduction of the properties of an object in the wavefield of the radiation scattered by it. *Dokl. AN SSSR* **144**, 1275–1276.
4. Butters, J. N. (1971) *Holography and its Technology.* Peter Peregrinus, London.
5. Viénot, J.-C., Smigielsky, P. and Royer H. (1971) *Holographie optique.* Dunod, Paris.
6. Hariharan, P. (1984) *Optical Holography: principles, techniques and applications.* Cambridge University Press, Cambridge.
7. Collier, R. J., Burckhardt, C. B. and Lin, L. H. (1971) *Optical Holography.* Academic Press, New York.
8. Abramson, N. (1981) *The Making and Evaluation of Holograms.* Academic Press, London.
9. Caulfield, H. J. (ed.) (1979) *Handbook of Optical Holography.* Academic Press, New York.
10. Ostrovsky, Yu. I. (1977) *Holography and its Application.* Mir, Leningrad.
11. Francon, M. (1969) *Holographie.* Masson, Paris.

12. Francon, M. (1978) *La granularité laser (speckle) et ces applications en optique.* Masson, Paris.
13. Dainty, J. C. (ed.) (1975) *Laser Speckle and Related Phenomena.* Springer-Verlag, Berlin.
14. Leith, E. and Upatnieks J. (1961) New technique in wavefront reconstruction. *J. Opt. Soc. Amer.* **51**, 1469.
15. Stetson, K. A. (1975) A review of speckle photography and interferometry. *Opt. Engineer.* **14**, 482–489.
16. Goodman, J. W. (1975) Statistical properties of laser speckle patterns. In Dainty J. C. (ed.), *Laser Speckle and Related Phenomena*, pp. 9–75, Springer-Verlag, Berlin.

2 DISPLACEMENT MEASUREMENT BY HOLOGRAPHIC INTERFEROMETRY

The main objective of holographic interferometry is to record, observe, and obtain quantitative interpretation of interference patterns, at least one has been recorded onto, and reconstructed by, a hologram. In contrast to conventional interferometry, holographic interferometry is capable of performing interferometric measurement of the displacement components of an optically rough object surface. This leads to the fact that holographic interferometry has gained widespread acceptance in experimental mechanics of deformable bodies. Various interference fringe patterns, allowing us to determine surface displacement fields under both static and dynamic loading of an investigated object, can be constructed through the use of holographic interferometry.

In a general case, fringe patterns obtained by means of holographic interferometric techniques contain information on all three displacement components of the surface of a strained body. Therefore, the problem of quantitative interpretation of such interferograms is complex. A rich variety of approaches to processing data derived from holographic interferograms of opaque bodies have been developed, but the method based on absolute fringe orders identification has acquired the widest recognition for quantitative interpretation of interference fringe patterns resulting from holographic recording of different states of the strained object. That is why the absolute fringe order counting technique of holographic interference fringe patterns interpretation is considered only in this chapter. The potential of this technique in the field of strain investigation can be considerably extended through the use of reflection holograms.

An accurate determination of the displacement vector of a point on the surface of a strained body has to be solved before practical implementation of different holographic interferometric techniques.

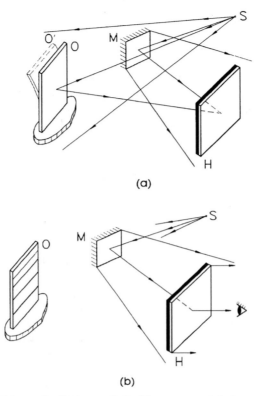

(a)

(b)

Figure 2.1. Schematic diagram of doubly exposured hologram of a deformed object recording (a), and observation of reconstructed fringe pattern (b)

An experimental set-up which is capable of ensuring a choice of parameters of an interferometer optical system proceeding from the C-optimality criterion and, thus, to minimize the errors made in determination of the displacement vector components, is presented in this chapter.

In order to illustrate how optimal systems of holographic interferometers can be practically implemented, the most characteristic examples of displacement components determination of objects of different shapes and strained states is described.

Note that methods of reliable quantitative interpretation of strain-induced interference fringe patterns are the main subject of this chapter. A more detailed description of the formation of various fringe patterns in holographic interferometry can be found, for instance, in the following books by Erf [1], Vest [2], Schumann and Dubas [3], Ostrovsky et al. [4], Abramson [5], Jones and Wykes [6], Schumann et al. [7], Ostrovsky et al. [8], and Rastogi [9].

Figure 2.2. Holographic interferogram of the bending of a cantilever beam obtained by two-exposure method

2.1 Formation of holographic fringe patterns

The two-exposure method is the most widely used approach in quantitative determination of 3-D surface displacement fields of a strained body which employs holographic interferogram recording. Figure 2.1 shows the Leith–Upatnieks optical system for hologram formation. Object O (which is a cantilever beam rigidly clamped along one edge) is illuminated by a laser light emerging from a point source S. A photographic plate H receives two light waves simultaneously from the object (object wave) and from the reference source (reference wave). A part of the illuminating wave reflected by means of a mirror M onto a photographic plate H serves as a reference beam.

The main principle of the two-exposure technique consists of carrying out consecutive exposures, on the same photoplate H, of two holograms of the object in different states (an original O and a changed O′), for example before and after deformation. In Figure 2.1(a) the state O′ corresponds to a beam bending.

When a double–exposure hologram is illuminated with a facsimile of the reference wave, both waves scattered by the surface of the body in two of its states O and O′ will be reconstructed simultaneously. Their interference produces an image of the surface of the body superimposed with an interference fringe pattern which contains information on the changes the body has undergone between the two exposures. In the case presented in Figure 2.1(b) the interference fringes describe the deflection field of the plate surface. The doubly exposed holographic interferogram of a rigidly clamped (along one short edge) cantilever beam under bending is shown in Figure 2.2.

A more detailed analysis of a fringe pattern formation in the two-exposure method can be performed by using the fundamentals of optical holography. The complex amplitude of the light wave reflected from the surface of a body in its initial state in the hologram plane A_1 has the following form:

$$A_1 = a_0 \exp(-i\phi) \tag{2.1}$$

where a_0 and ϕ are the amplitude and phase of the object wave, respectively.

After an object surface transition to new states, caused, for instance, by the object deformation, the complex amplitude in the hologram plane A_2 can be written as

$$A_2 = a_0 \exp[-i(\phi + \delta)] \tag{2.2}$$

where δ is the strain-induced phase change of the object wave. The deformation of the object is assumed to be so small that it affects only the phase of the object wave. In addition, it is necessary to acknowledge the important assumption that the changes in the microrelief of the surface between the exposures may be neglected. Research experience proves that the latter assumption is valid in the elastic strain range and in a large part of the elasto-plastic strain range for most materials used in structures manufacturing.

The expression for the complex amplitude of the reference wave in the hologram plane A_r can be written in analogy to that for the object wave as

$$A_r = a_r \exp(-i\varphi) \tag{2.3}$$

where a_r and φ are the amplitude and phase of the reference wave, respectively.

The light intensity distribution in the plane of the recording medium, e.g. a photographic plate, after the first exposure will be

$$I_1 = |A_1 + A_r|^2 \tag{2.4}$$

and the corresponding distribution after the second exposure can be written in analogy to equation (2.4):

$$I_2 = |A_2 + A_r|^2. \tag{2.5}$$

The total exposure E can be obtained by summarizing equations (2.4) and (2.5):

$$E = \{|A_1 + A_r|^2 + |A_2 + A_r|^2\}\tau \tag{2.6}$$

where τ is the duration of each from two exposures.

Let's assume that the photographic emulsion is processed in such a way that its amplitude transmittance T versus exposure E dependence has a linear character and can be expressed by equation (1.3). When the double-exposure hologram is illuminated with a replica of the reference wave, a new wave will form at the rear side of the hologram, whose complex amplitude A is proportional to the transmittance

$$A = Ta_r \exp(-i\varphi). \tag{2.7}$$

Substituting relations (1.3), (2.6), (2.1)–(2.3) into expression (2.7) we can obtain, after simple transformations,

$$\begin{aligned}
A = &[b_0 + 2b_1\tau(a_0^2 + a_r^2)]a_r \exp(-i\varphi) \\
&+ b_1\tau a_0 a_r^2\{\exp(-i\phi) + \exp[-i(\phi + \delta)]\} \\
&+ b_1\tau a_0 a_r^2\{\exp(i\phi) + \exp[i(\phi + \delta)]\}\exp(-i2\varphi).
\end{aligned} \tag{2.8}$$

The first term in expression (2.8) represents, within a constant factor of $b_0 + 2b_1\tau(a_0^2 + a_r^2)$, the amplitude of the reference wave traversing the hologram, and corresponds to a zero-order wave. The second term describes, within a real factor of $b_1\tau a_0 a_r^2$ two object waves forming virtual images. The third term describes two distorted real images.

The intensity distribution I_e in the virtual image (which is the subject of main interest) will correspond to the square of the second term in equation (2.8):

$$\begin{aligned}
I_e &= b_1^2\tau^2 a_0^2 a_r^4 |\exp(-i\phi) + \exp[-i(\phi + \delta)]|^2 \\
&= 4b_1^2\tau^2 a_0^2 a_r^4 \cos^2\left(\frac{\delta}{2}\right) \propto I_0 \cos^2\left(\frac{\delta}{2}\right)
\end{aligned} \tag{2.9}$$

where $I_0 = a_0^2$ is the intensity of the image of the object in the original state. Expression (2.9) shows that the intensity of the reconstructed image is modulated by the cosine squared function.

The contrast or visibility γ_H of the fringes in an interference pattern of any nature is determined as

$$\gamma_H = \frac{I_{max} - I_{min}}{I_{max} + I_{min}}. \tag{2.10}$$

In the case of two-exposure hologram recording, as it follows from expression (2.9), the fringe contrast γ_H (2.10) is equal to unity. Thus the two-exposure method allows us to record the holographic interference pattern with the highest possible fringe contrast for bodies with diffusely reflected surfaces.

An important feature of the double-exposure hologram is the possibility to observe light waves shifted in time. In other words, the light waves, forming a holographic interferogram, pass the same space trajectories, but in different time.

As noted above (see Section 1.3), an image reconstruction by means of a double-exposure hologram is covered by the so-called speckle structure which contains information about surface microrelief. It should be pointed out that the average number of dark speckles situated within a total area of dark fringes is more than the average number of bright speckles contained within a total area of bright fringes.

More than two object waves can be recorded on a single hologram simultaneously. The case when the additional phase difference acquired by each subsequent object wave to be constant and equal to δ is of great practical importance from an experimental strain analysis viewpoint. The constant phase difference increment in the case of a linear elastic deformable body can be reached by applying equal loading steps between exposures only.

In the case concerned, the complex amplitudes of the objects waves in the hologram plane can be expressed in analogy to the two-exposure method:

$$A_1 = a_0 \exp(-i\phi)$$
$$A_2 = a_0 \exp[-i(\phi + \delta)]$$
$$A_3 = a_0 \exp[-i(\phi + 2\delta)]$$
$$\cdots\cdots$$
$$A_m = a_0 \exp\{-i[\phi + (m - 1)\delta]\}$$

where a_0 and ϕ are the amplitude and phase of each object wave; A_1 is the complex amplitude of the object wave recorded during the first exposure; and A_m is the complex amplitude of the object wave recorded during the mth exposure.

The hologram obtained in such way is usually called the multiple-exposure hologram. The technique of consecutive recording, on a single hologram, of several light waves scattered by the surface of a strained body is known as the multi-exposure method.

In an analogy to the two-exposure method, using the complex amplitude of the reference wave given in equation (2.3), assuming that the duration of each of the m exposures is equal to τ and the recording is a linear process, we can use equations (1.3) and (2.7) to find the intensity of the virtual image I_{em} for the multi-exposure method:

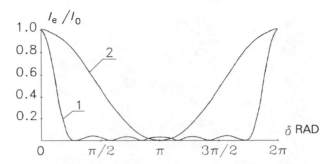

Figure 2.3. Intensity distribution of virtual image reconstructed by doubly-exposed hologram (2) and multiply-exposed hologram (1)

$$I_{em} \propto I_0 \frac{\sin^2\left(m\frac{\delta}{2}\right)}{\sin^2\left(\frac{\delta}{2}\right)} \qquad (2.11)$$

where $I_0 = a_0^2$ is the intensity distribution on the surface of a stationary object.

Expression (2.11) shows that the intensity modulation of the image in multi-exposure hologram recording is a complex function of a phase difference δ.

The virtual-image intensity distribution of the image reconstructed with a multiple-exposure hologram for $m = 7$ is presented in Figure 2.3 by curve 1. Curve 2 in this figure describes the intensity distribution for the double-exposure method. Figure 2.3 shows that when the number of exposures increases, the width of the principal intensity maxima decreases. At the same time, the intermediate maxima between them are of a substantially lower intensity.

To further illustrate the multi-exposure method, Figure 2.4(a) shows a fringe pattern obtained by means of reflection hologram reconstruction, which corresponds to the surface displacement component fields of a circular cylindrical shell with a circular cut-out loaded with tensile forces. This holographic interferogram was obtained for $m = 5$, i.e. for four loading steps of force $P = 180$ N. For comparison, Figure 2.4(b) shows a fringe pattern recorded by the double-exposure technique for the same object and a load increment of $P = 700$ N. The fringe pattern in a multiple-exposure hologram clearly reveals the differences in the principal and intermediate intensity peaks.

It should be noted that for some practical cases the multi-exposure method application allows a considerable increase of the loads applied to the body to be investigated and, hence, leads to an increase of the displacement measurement range for constant interferometer sensitivity.

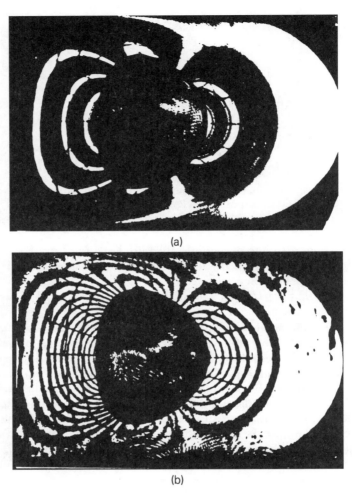

(a)

(b)

Figure 2.4. Holographic interferogram of a cylindrical shell under tension obtained by multi-exposure (a) and two-exposure method (b)

It has been shown by Iwata and Nagata [10] that recording a multiple-exposure hologram with high-quality interference patterns characterized by sharp bright fringes means that displacement component values, corresponding to equal loading increments at the m consecutive loading steps, vary from one another by less than

$$\frac{\lambda}{2(m-1)} \tag{2.12}$$

where λ is the laser light wavelength.

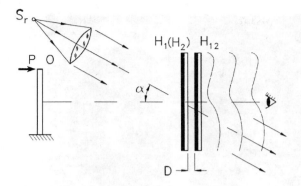

Figure 2.5. Schematic diagram of optical set up of hologram duplication

This particular attribute of multiple-exposure holograms can be used as the most effective application in the field of experimental strain analysis in two ways. The first application is directed at the design and subsequent testing of the special loading apparatus or devices used for a high accurate loading of specimens and structures, a strain or stress analysis of which has to be performed by means of different holographic and speckle interferometric techniques. The recording of high-quality multiple-exposure holograms in the full loading range to be inspected demonstrates the reliable performance of the loading device during optical interferometric measurements.

A negative result of the multiple-exposure hologram recording procedure means that either the loading apparatus has unstable operational character-istics or the deformation process to be studied in an elastic strain range has a non-linearity. The latter is the second multi-exposure approach for structure strain analysis.

The object waves reflected from the surface of a strained body in its two states can also be recorded on separate holograms and the interference fringes observed in their simultaneous reconstruction.

This approach to fringe pattern forming developed by Abramson [11] for certain mechanical problems is called sandwich-holographic interferometry. The main feature of this approach is its capability to record the consecutive single-exposure hologram set, corresponding to different object surface states, any pair of which can be afterwards superimposed to receive an interference fringe pattern. Apparently, the most effective application of the sandwich-holographic interferometry method is in the investigation of different types of residual strains, since the process cannot be repeated on the same object.

The consecutive duplication of two holograms on a single photoplate can be used for interferometric comparison of the light waves which have been recorded on different holograms by Odinstev et al. [12]. One of the possible

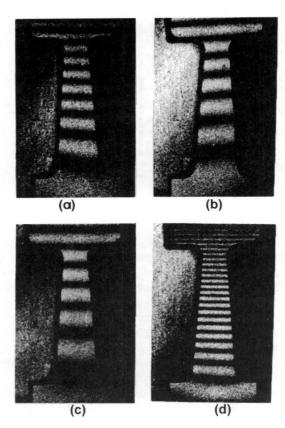

Figure 2.6. Holographic interferograms describing microcreep deformation process obtained by hologram duplication technique H_{12} (a), H_{23} (b), H_{34} (c) and H_{14} (d)

optical system set-ups for hologram duplication is shown in Figure 2.5. The hologram H_1 (or H_2) is placed in the same position as that during recording. For this purpose, a special kinematic fixture for the exact placement of the hologram to the position in which it was recorded can be used. Photoplate H_{12}, used for the original hologram duplication, and one of the holograms to be copied (H_1 or H_2) are placed near one another as shown in Figure 2.5 inside the device for sandwich-hologram recording in such a way that their emulsions face the object position under study.

After illumination of the above-mentioned holograms H_1 (or H_2) and photoplate H_{12} with the reference wave in the arrangement depicted in Figure 2.5 (S_r is a point reference beam source, O is the object to be studied), a reconstructed wave with original hologram H_1 (or H_2) forming a virtual image of the object O will serve as the object wave during H_{12} hologram recording. The reference wave for the hologram H_1 (or H_2), emitted by a point source S_r and transmitted through the hologram H_1 (or H_2) then serves as a reference wave during the H_{12} hologram recording procedure.

In order to reduce the influence of glass inhomogeneity and differences in the thickness of the photoplates H_1 or H_2 on the phase distribution of the light waves recorded in the hologram H_{12}, the angle α between the object and reference beams (see Figure 2.5) should be as small as possible. The validity of this condition can be realized by increasing the distance between the hologram and the object under study. The distance D between the hologram H_1 (or H_2) and hologram H_{12} should be kept within the minimum possible level.

An investigation of the microcreep deformation process is dealt with here as a detailed illustration of a practical application of the hologram duplication technique. The specimen used was a cantilever beam, having an equal-strength segment, made of M1 copper and loaded by a constant concentrated force. The original reflection holograms of the specimen were recorded at equal time intervals of 20 min starting from zero. Holographic interferograms reconstructed with double duplicated holograms H_{12}, H_{23}, H_{34} and H_{14} are shown in Figures 2.6(a), (b), (c) and (d), respectively. The fringe patterns presented describe a kinetic of the microcreep process of the beam in the course of one hour and contains information about total residual strains accumulated during this time.

The real-time holographic methods involve the interference of waves, one of which is reconstructed by means of the hologram, of an object recorded in its initial state while the other is scattered directly by the strained object (Stetson and Powell [13]). To observe the interference fringe pattern by means of the real-time technique, a holographic interferogram of the initial object surface state should be recorded on a photoplate mounted in a special kinematic fixture. The design of this fixture permits the removal of a photographic plate from the optical set-up and repositioning it within a fraction of a fringe spacing [14]. Under these conditions, the plate can be photographically processed outside the holographic set-up.

After photographic processing and returning the photographic plate to the location of the exposure we can observe the real object and its holographic virtual image simultaneously. If observation of the object through the hologram is continued while the object is gradually loaded, interference fringes will become evident and will change their shape and frequency in accordance with the changes in the object. The fringe pattern observed with this real-time technique characterizes the dynamics of the displacement of points on the surface of the object.

For a positive recording process (i.e. for $b_1 > 0$ in expression (1.3)), it can be shown that the intensity of the object image is modulated by cosinusoidal fringes, just as it takes place in the case of the double-exposure method.

The contrast of fringes in a real-time holographic interference pattern depends on the intensity ratio of the interfering waves. If the optimum hologram recording conditions are satisfied, maximum fringe visibility can be reached.

A practical realization of the real-time method is limited by the stringent requirement that the hologram has to be returned to the initial position within one tenth of the wavelength of the laser light used. Thus one has to match, within a fraction of the spatial period, the pattern formed by the object and reference beams prior to loading with that recorded on the hologram. The absence of fringes on the surface of the unstrained object is evidence of the exact hologram repositioning after photographic processing.

A relatively wide application of the real-time technique is found in the field of non-destructive testing and vibration modes analysis. In this approach silver halide emulsions [15], photothermoplastics and photorefractive crystals [16] are the most frequently used recording materials. It should also be noted that the real-time method can be used for quantitative displacement component measurement in the single case when the fringe pattern allows a direct mechanical interpretation of how it takes place in the case of a plate or beam bending.

In the above-mentioned methods of holographic interference fringe formation (two- and multi-exposure, duplication and real-time) the holograms are recorded under conditions where the object is in a steady state. Naturally, it is interesting to discuss the capability of hologram recording during deformation or motion of the object to be studied. Powell and Stetson [17] were the first who solved this problem in the case of harmonic oscillation of a body.

We shall consider here only such motions or deformations where the so-called time-function can be isolated [18]. In this case the changing in time of the phase of the object wave in the hologram plane, $\delta(t)$, can be represented as the following product:

$$\delta(t) = \delta_m f(t) \tag{2.13}$$

where δ_m is the phase difference, corresponding to a maximum displacement of the points of the object surface, and $f(t)$ is some function of time.

We shall assume the function $f(t)$ to be the same for all points on the object surface. Then for complex amplitude of the object wave $A(t)$ in the hologram plane at some instant in time, we can write

$$A(t) = a_0 \exp[-i\delta_m \cdot f(t)] \tag{2.14}$$

where a_0 is the complex amplitude of the object wave. We shall assume also that the exposure time of the hologram is equal τ. The complex amplitude of the reconstructed light wave will be proportional to the time-averaged

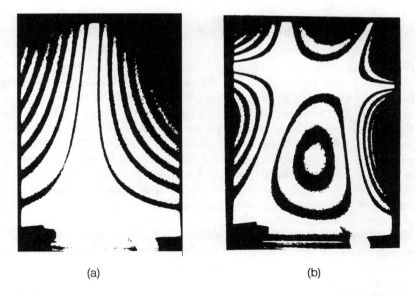

(a)   (b)

Figure 2.7. Interferogram of compressor turbine blade obtained by time-average holography: bending resonance mode (a) and bending-torsional resonance mode (b)

amplitude of the object wave $A(t)$. Taking into account expression (2.14), we can obtain the intensity of the virtual image in this case:

$$I_{ea} \propto \left| \frac{1}{\tau} \int_0^\tau A(t)\,dt \right|^2 = a_0^2 \left| \frac{1}{\tau} \int_0^\tau \exp[-i\delta_m \cdot f(t)]\,dt \right|^2. \qquad (2.15)$$

Expression (2.15) describes in the most general form all above-presented cases of generation for holographic fringe patterns.

The latter technique of recording a hologram during deformation or motion of the object has been called the time-average method. We shall consider the case where the phase of the object wave changes in accordance with a harmonic law with a time-function of the form

$$f(t) = \sin(\omega t) \qquad (2.16)$$

where ω is the circular frequency of the oscillation. Substitution of the oscillation law (2.16) in relation (2.15) results in the following intensity distribution of a virtual image, taking into account the evident condition $\tau \gg \omega^{-1}$:

$$I_{ea} \propto a_0^2 \left| \lim_{\tau \to \infty} \frac{1}{\tau} \int_0^\tau \exp(-i\delta_m \cdot \sin\omega t) \mathrm{d}t \right|^2 = I_0 J_0^2(\delta_m) \qquad (2.17)$$

where $I_0 = a_0^2$ is the intensity distribution of the stationary object; $J_0(\delta_m)$ is the zero-order Bessel function of the first kind. In this case the meaning δ_m corresponds to the vibration amplitude of the object.

Relation (2.17) shows that the image of the vibrating body surface reconstructed with a time-average hologram will be covered by a system of interference fringes. In the case concerned, the intensity of bright fringes will dramatically decrease with the growth of amplitude δ_m. The brightest fringe position on a fringe pattern obtained by means of reconstruction of a time-average hologram coincides with parts of the object surface, where the condition $\delta_m = 0$ is valid. These points of the vibrating object surface are usually called 'node lines'.

For more detailed illustrations of the time-average technique, two holographic reconstructions of the compressor turbine blade corresponding to different resonance vibration modes are shown in Figure 2.7. The brightest node lines are clearly seen in both photographs. The decrease in brightness of the bright fringes (the main difference between time-average and double-exposure interferograms) can also be observed in the photographs presented in Figure 2.7.

Note that we shall use the term 'time-average method' in all cases where an exposure is made of a moving or strained object. The intensity of a reconstructed image corresponding to different methods for obtaining holographic fringe patterns can be described in a general form in terms of the so-called characteristic fringe function $M(\delta)$ which was introduced by Stetson [18]:

$$I_e \propto I_0 |M(\delta)|^2 \qquad (2.18)$$

Subsequent to equation (2.15), the fringe function can be represented in the following form:

$$M(\delta) = \frac{1}{\tau} \int_0^\tau \exp[-i\delta_m f(t)] \mathrm{d}t \qquad (2.19)$$

Thus, in holographic interferometry of diffusely reflecting objects the intensity distribution of a reconstructed image is modulated by the square of the fringe function.

2.2 Interpretation of holographic interferograms

The quantitative determination of object surface displacement fields by interference fringe patterns is one of the main tasks in holographic

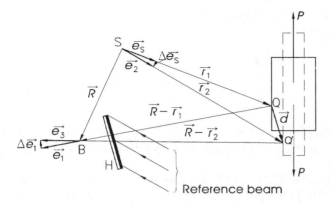

Figure 2.8. Schematic diagram of interferometer arrangement for derivation of the relation between phase difference and displacement vector of the surface point

interferometry of diffusely reflected bodies. To solve this problem it is necessary to establish the relations between the fringe parameters and the surface displacement components of a strained object.

One possible way of reaching this objective involves the application of the so-called geometrical model developed by Aleksandrov and Bonch-Bruevich [19], Ennos [20] and Sollid [21]. In this model, a rough surface is taken to consist of mutually incoherent point scatters. Therefore, only the wave emerging from the same points (which will henceforth be called 'corresponding' or 'identical') on the surface in two of its different states will be able to interfere. It should be noted that the validity of the geometrical approach has been confirmed by a large set of various experimental data [2, 5, 8, 9].

A diagram of the conditions of illumination and observation of the corresponding points Q and Q' on a surface before and after its deformation is shown in Figure 2.8. The virtual images reconstructed through use of a double-exposure hologram can be considered as two simultaneously existing real surfaces of the object under study. In the case involved, the phase difference of light waves results from the displacement of point Q on the object surface under defined directions of the illumination and observation.

The phase difference δ between the two interfering rays can be written as

$$\delta = \frac{2\pi}{\lambda} (SQ'B - SQB) \qquad (2.20)$$

where λ is the wavelength of laser light; SQ'B is the optical path from the light source S through the point Q' to point B; SQB is the analogous optical path for point Q.

Using the nomenclature in Figure 2.8 we can present the above-mentioned optical paths in the following way:

$$SQ'B = \vec{e}_2 \vec{r}_2 + \vec{e}_3 (\vec{R} - \vec{r}_2) \tag{2.21}$$

$$SQB = \vec{e}_S \vec{r}_1 + \vec{e}_1 (\vec{R} - \vec{r}_1) \tag{2.22}$$

where $\vec{r}_1$, $\vec{r}_2$, and $\vec{R}$ are vectors of the points Q, Q' and B respectively; $\vec{e}_S$, $\vec{e}_1$ and $\vec{e}_2$, $\vec{e}_3$ are the unit vectors of illumination and observation of points Q and Q', respectively.

Substituting equations (2.21) and (2.22) into equation (2.20), and taking into account the following presentation of the illumination and observation unit vector of the point Q',

$$\vec{e}_2 = \vec{e}_S + \Delta \vec{e}_S, \quad \vec{e}_3 = \vec{e}_1 + \Delta \vec{e}_1$$

we can obtain, after some transformations, the intermediate expression for the phase difference

$$\delta = \frac{2\pi}{\lambda} [(\vec{e}_1 - \vec{e}_S)(\vec{r}_2 - \vec{r}_1) + \Delta \vec{e}_S \vec{r}_2 + \Delta \vec{e}_1 (\vec{R} - \vec{r}_2)]. \tag{2.23}$$

The vector difference $\vec{r}_2 - \vec{r}_1$ represents the displacement vector $\vec{d}$ to be determined. Usually, in optical systems of the holographic interferometers, the distance from the light source to the object is much greater that the displacement value $|\vec{d}|$:

$$|\vec{d}| \gg |\vec{r}_1| \simeq |\vec{r}_2|$$

and, therefore, for the relative vector directions we can safely assume

$$\Delta \vec{e}_S \perp \vec{r}_2, \quad \Delta \vec{e}_1 \perp (\vec{R} - \vec{r}_2).$$

If this assumption is valid, expression (2.33) acquires the following final form:

$$\delta = \frac{2\pi}{\lambda} (\vec{e}_1 - \vec{e}_S) \vec{d}. \tag{2.24}$$

Bright interference fringes will be observed when

$$\delta = 2\pi n \qquad n = 0, 1, 2, \ldots \tag{2.25}$$

and dark ones when

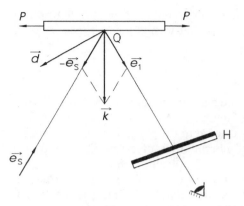

Figure 2.9. The sensitivity vector of a holographic interferometer in the case of fringe pattern interpretation in terms of absolute orders

$$\delta = 2\pi(n - \frac{1}{2}) \qquad n = 1, 2, \ldots \qquad (2.26)$$

where n is the absolute order of a bright (dark) fringe at the surface point under consideration.

Substitution of relations (2.25) and (2.26) into (2.24) leads to the following equation for the bright fringes:

$$(\vec{e}_1 - \vec{e}_S)\vec{d} = n\lambda, \qquad n = 0, 1, 2, \ldots \qquad (2.27)$$

and for the dark ones

$$(\vec{e}_1 - \vec{e}_S)\vec{d} = \left(n - \frac{1}{2}\right)\lambda \qquad n = 1, 2, \ldots \qquad (2.28)$$

Equation (2.27), or (2.30), which found the connection between the displacement vector of a surface point, the parameters of the interferometer system used (directions of illumination and observation), and the absolute fringe order at this point on the pattern, is usually called the principal relation of holographic interferometry. It should be noted that below, without a loss of generality, only the equation for bright fringes (2.27) will be used.

For the following comparative metrological analysis of different optical systems it is convenient to introduce the concept of the sensitivity vector $\vec{k}$ as the difference between the unit vectors of observation and illumination [19, 20]:

$$\vec{k} = \vec{e}_1 - \vec{e}_S. \qquad (2.29)$$

Using this notation, the principal relation of holographic interferometry can be converted to a form, which in many cases may be more convenient for the analysis

$$\vec{k}\,\vec{d} = n\lambda. \tag{2.30}$$

Let us consider the diagram of illumination and observation of some surface point Q shown in Figure 2.9. Without a loss of generality, one can assume that the displacement vector $\vec{d}$ of the point of interest lies in the plane defined by the unit vector $\vec{e}_S$ of illumination of point Q, and the unit vector $\vec{e}_1$ of the observation of point Q through the hologram. As Figure 2.9 shows, the sensitivity vector $\vec{k}$ is directed along the bisector of the angle between the directions of unit vectors $\vec{e}_S$ and $\vec{e}_1$. Thus a single fringe pattern obtained for one fixed observation direction provides the possibility of determining the projection of the displacement vector $\vec{d}$ onto the sensitivity vector $\vec{k}$ in accordance with equation (2.30). This means that object deformation generally does not have a direct mechanical interpretation, as occurs, for instance, in the moiré methods.

The following factors can affect the fringe pattern configuration:

• The variation over the object surface to be studied of both magnitude and direction of the sensitivity vector
• The changing over the object surface of both magnitude and direction of the displacement vector.

These circumstances introduce serious difficulties into quantitative interpretation of holographic interferograms of diffusely reflecting objects, which is necessary for displacement component determination.

Holographic interference fringes caused by displacement and deformation of a non-transparent object with a diffusely reflecting surface are localized in the general case. During reconstruction of any type of holographic interferogram described above, interference fringes may be observed either behind or in front of the surface image, and only in some particular cases on the surface itself. Thus, the correct projection of localized fringe patterns has to be ensured before using the geometrical model for a quantitative displacement determination. A method for this projection is based on the use of the so-called observer-projection theorem established by Stetson [18]. It should be noted that a procedure of linking localized fringes to the object surface under study may frequently be accompanied by additional difficulties of fringe pattern interpretation.

The extent of the region where fringes are localized depends on the size of the aperture of the viewing system. As its size decreases, the extent of the localization region increases, with the fringes appearing at the surface of the object. We should remember, however, that the aperture size reduction is always accompanied by a growth of the average speckle size (1.9) in an image of a diffusely reflected surface reconstructed with coherent light. Thus, the decreasing process of the aperture size leads to a corresponding degradation in fringe resolution. When the average speckle size becomes equal to the average fringe spacing, the contrast interference pattern disappears. This prompts us to question the diffraction efficiency (or reflection ability) of the holograms which must reconstruct the interference patterns with high-frequency fringes in the localization region. This type of circumstance frequently occurs in areas of strain concentration.

As it follows from the principal relation of holographic interferometry (2.30), the use of one illumination direction and one observation direction allows us to determine, at any point on the object surface, the projection of the displacement vector on the sensitivity vector direction only. However, to obtain the complete quantitative description of an arbitrary vector at any point, we must determine three vector projections on three non-complanar directions. In other words, it is necessary to record interference fringe patterns from three different directions. If it is possible to interpret three interferograms obtained by means of absolute fringe orders assignment, the following system of linear algebraic equations can be obtained:

$$\begin{aligned} \vec{k}_1 \vec{d} &= n_1 \lambda \\ \vec{k}_2 \vec{d} &= n_2 \lambda \\ \vec{k}_3 \vec{d} &= n_3 \lambda \end{aligned}$$ (2.31)

where n_1, n_2, n_3 are the absolute fringe orders at the point of interest determined from the corresponding interferograms; $\vec{k}_1$, $\vec{k}_2$, $\vec{k}_3$ are the sensitivity vectors defined by the chosen configuration of illumination and three viewing directions; and λ is the laser light wavelength.

The solution of coupled equations (2.31) in the case of non-complanar sensitivity vectors provides three components of displacement vector $\vec{d}$. The possibility of the absolute fringe orders determination is an essential condition with regard to the applicability of the approach to the quantitative interpretation of holographic fringe patterns. That is why this method is sometimes called the absolute fringe orders counting method or the zero-motion fringe method. This approach to interferogram interpretation was first proposed by Ennos [20] and generalized by Sollid [21].

In most practical cases it is convenient to represent the coupled equations (2.31) through the vector components in the Cartesian coordinate system (x_1, x_2, x_3), which is defined on the object surface to be investigated. The system of linear algebraic equations for the displacement vector components d_1, d_2, d_3 determination at each point of interest has the form

$$K\vec{d} = \lambda N \tag{2.32}$$

where $\vec{d} = [d_1, d_2, d_3]^T$ is the displacement vector to be determined; $N = [n_1, n_2, n_3]^T$ is the absolute fringe order vector; and K is 3×3 matrix of sensitivity of holographic interferometer, whose elements are equal to projections of sensitivity vectors on the corresponding coordinate axes.

The numbers or absolute orders of interference fringes can be determined by counting them from a fringe of any order on condition that this fringe will be reliably identified on all three interferograms. Relation (2.9) shows that in the case of two-exposure recording for fixed points on the surface where $\delta = 0$, the corresponding intensity is at a maximum. These fringes are usually called zero-motion fringes. Thus, in two exposure patterns the zero-order fringe will always be bright. It is evident that the zero-motion fringe is the most convenient point of the fringe counting origin. As follows from the principal relation (2.32), a zero phase difference corresponding to the zero-order fringe is reached either when the displacement vector is zero ($\vec{d} = 0$) or when it is orthogonal to the sensitivity vector ($\vec{d} \perp \vec{k}$).

Now we consider the major techniques used to find the zero-motion fringe position if it is situated on an observed fringe pattern. The simplest case is when there are regions on the object surface which are known not to have undergone displacement caused by loading between or during exposure. In the interferogram shown in Figure 2.2 the zero-motion fringe lies near the clamped end of the beam.

The main importance for zero-motion fringe identification consists, from a practical point of view, of a method based on a combined application of the two-exposure and time-average techniques, which was first proposed by Köpf [22]. The diagram of a hologram exposure, necessary to accomplish this approach, is shown in Figure 2.10(a). After the first exposure of duration τ_1 the initial state of the object surface was recorded on the hologram. The corresponding object wave phase is denoted by ϕ_0. Subsequent steps consisted of a time-averaged hologram recording during time τ_2 and simultaneous object loading in such a way that the rate of phase change remained constant. Therefore, the object wave phase at a current time t is

$$\phi_0 + \left(\frac{\delta}{\tau_2}\right)t$$

where δ is the maximum phase difference between the first and second exposures.

Finally, the deformed object in a stationary state is subjected to a third exposure of duration τ_1 and at object wave phase $\phi_0 + \delta$. In order to determine the intensity distribution of the reconstructed virtual image by means of the above-described three-exposure technique, it is necessary to represent the

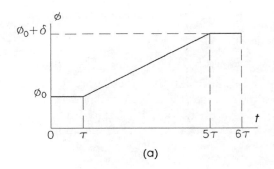

(a)

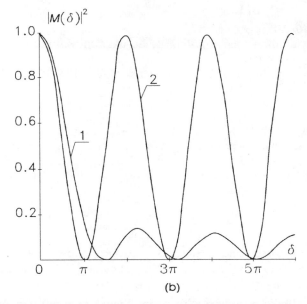

(b)

Figure 2.10. (a) Dependence of the phase ϕ of the object wave against exposure time for recording of hologram with bright zero-order fringe, and (b) the form of function $|M(\delta)|^2$ in the case of three-exposure (1) and two-exposure (2) hologram recording

integral (2.19) as a sum of three integrals to find the squared characteristic fringe function:

$$|M(\delta)|^2 = \left[2\tau_1 \cos\left(\frac{\delta}{2}\right) + \tau_2 \frac{\sin\frac{\delta}{2}}{\frac{\delta}{2}} \right]^2. \tag{2.33}$$

For $\tau_2 = 0$ this expression gives the fringe function of the two-exposure method (2.9), and for $\tau_1 = 0$, that of the time average technique with the phase being varied at a constant rate [23].

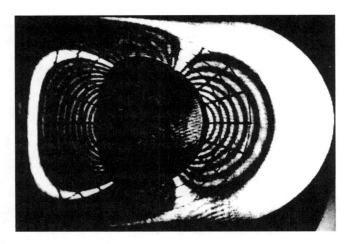

Figure 2.11. A holographic interferogram of a cylindrical shell subjected to tension with bright zero-motion fringe

It should be noted that when the ratio $\tau_2/2\tau_1$ increases, the intensity of the zero-motion fringes will increase relative to the other bright fringes. At the same time, the roots of function (2.33) and of the function $\cos^2\delta$ will become slightly different. However, these differences may be neglected for

$$\frac{\tau_2}{2\tau_1} \simeq 2.$$

The function (2.33) corresponding to ratio $\tau_2/2\tau_1 = 2$ is shown in Figure 2.10(b) by curve 1. Curve 2 in this figure describes the squared characteristic fringe function of the double-exposure method. A comparison of these two functions clearly shows an almost complete coincidence of its roots, and that the fringe pattern corresponding to expression (2.33) has a more intense zero-motion fringe than other bright fringes. That is why the zero-order fringe can be easily identified on three-exposure interferograms. This fact is of great importance for a reliable determination of the absolute fringe orders.

A holographic interferogram of a circular cylindrical shell with a circular cut-out subjected to tension, which was obtained by means of combined application of the two-exposure and time average techniques, is shown in Figure 2.11. The fringe pattern clearly shows two different locations of the zero-motion fringes.

The same view of the two-exposure interferogram obtained for the same tensile load increment as the above-mentioned three-exposure interferogram is shown in Figure 2.4(b).

An important special characteristic of the three-exposure method consists of its ability to discover not only regions free of deformation (for example, the right part on the photograph in Figure 2.11 where the shell is clamped) but also areas of the object surface where the sensitivity vector is orthogonal to the displacement vector. The latter is clearly illustrated by the bright fringe on the left-hand side of Figure 2.11.

If zero-motion fringes are not displayed in an interferogram, the fringe orders can be determined through the use of elastic strings connecting the object under investigation [24]. The zero-motion fringe in this case is located at the point of the string fixed to the rigid body. The interferograms presented in Figure 2.20 show such a string made of rubber.

The result of the displacement vector determination at each surface point by means of the linear equation system (2.32) will also depend on the choice of the fringe orders sign for all observation directions. This is connected to the fact that the fringe orders can be read out on the positive or negative side from zero-motion fringes. It is important to note that the assignment of the fringe sign in holographic interferometry usually abides some conditional rules which must be the same for all viewing directions. In real practical cases, the application of these rules to fringe interpretation may be relatively evident or quite complex. In instances, photographs of fringe patterns may turn out to be sufficient for a correct sign assignment whereas under some conditions, an analysis of the fringe pattern variation with varied viewing directions within a hologram aperture is required.

As follows from expression (2.18), which relates the phase difference caused by the object deformation with the intensity distribution in a fringe pattern, the solution of (2.31) can yield only the magnitude of the displacement vector to within the sign. To determine the sign of the displacement vector, i.e. its direction, one must use *a priori* information about the character of deformation of the object, for example, the direction of tensile force for the interferogram in Figure 2.11, or employ other experimental methods for displacement components determination at some reference points of the object surface.

The quantitative interpretation of holographic interferograms by means of absolute fringe orders counting can be carried out in both a single- and a multi-hologram interferometer set-up [25–29]. In the first case the fringe patterns required for the displacement component determination are viewed from the three directions within the single hologram aperture. It should, however, be noted that the sensitivity of this interferometer with respect to the in-plane displacement components is usually not sufficient for its reliable determination.

In the multi-hologram technique, for each of three observation directions an individual hologram is provided. By using this approach, one can obtain practically any desired geometry of the sensitivity vectors.

The interferometers based on reflection hologram recording can also be considered as a single hologram system. But it is necessary to keep in mind that the opposed beam optical arrangement, in contrast to the Leith–Upatnieks optical system, offers broad possibilities for the displacement measurements. Due to the small distance between the reflection hologram and the object surface under study, we can obtain, first, a wide range of changes in the viewing angles. This property is necessary to minimize the errors of the in-plane displacement component determination. The second special feature of the reflection hologram approach is its ability to change continuously angle of any observation direction. The latter procedure should be frequently used for the correct assignment of the absolute fringe number signs.

Other techniques can be used in addition to the above-considered method of quantitative interpretation of holographic fringe patterns. Approaches based on relative fringe order counting [19], fringe order difference determination [6], fringe-vector method of Pryputniewicz [23], fringe visibility analysis, and others, are known as those having application to strain-induced displacement determination. However, the absolute fringe order counting technique is the most widely used and most powerful method in the field of experimental strain analysis by means of holographic interferometry. For this reason we have used only absolute fringe order terms in the analysis presented in this chapter.

2.3 Displacement determination error analysis and design of holographic experiment

A practical implementation of the absolute fringe order technique in measuring the three components of the displacement vector involves the solution of a system of linear equations (2.32). The coefficients of the sensitivity matrix K on the left-hand side of equation system (2.32) (the geometrical parameters of the interferometer optical system) and the vector N on the right-hand side of the equation (the absolute fringe orders at the point under study corresponding to all three viewing directions) are known only within certain errors. One can express the error in the vector N also in the form of a vector ΔN, and that in the elements of the matrix K, by a matrix ΔK.

The errors in the displacement vector $\vec{d}$, which can be expressed through the vector $\vec{d} = [\Delta d_1, \Delta d_2, \Delta d_3]$, are related to the vector ΔN and matrix ΔK, and this relation is defined by the actual form of the sensitivity matrix K [30, 31].

The optimal choice of the parameters of the interferometer optical system ($\vec{e}_S, \vec{e}_1, \vec{e}_2, \vec{e}_3$) to minimize the displacement measurement errors $\Delta \vec{d}_i$, ($i = 1, 2, 3$) in accordance with some optimality criterion is the main problem of the experimental design [32]. A review of different approaches of the design of the experiment applied to strain analysis of deformable bodies by means of holographic interferometric techniques is presented in reference [8].

To solve the problem of the design of the experiment, the perturbations of the sensitivity matrix K and the absolute fringe order vector N of equation system (2.32) must be connected with the displacement component measurement errors Δd_i ($i = 1, 2, 3$). To find the required relation, two approaches can be used: statistical and non-statistical.

The first may serve as a basis for a comparative analysis of optical configurations [2, 8, 33–36]. However, due to a complex mathematical formulation, it cannot provide a direct way to select the optical system ensuring the lowest possible error in displacement component measurement.

The non-statistical analysis of the displacement measurement errors, and the so-called criterion of C-optimality resulting from it, is, in our opinion, the best approach to ensure the most accurate determination of the displacement vector and its components. This approach has an effective application for metrological analysis of interferometer optical systems and optimal choice of its parameters [37–46].

The presence in equation (2.32) of an *a priori* known sensitivity matrix K as well as the linear relation between the measured N and unknown $\vec{d}$ parameters is the basis of the following analysis. In this case the upper bound on the error made in calculating the displacement vector $\Delta \vec{d}$ can be estimated in the following form [47, 48]:

$$\frac{\| \Delta \vec{d} \|}{\| \vec{d} \|} \leqslant \frac{\| K \| \| K^{-1} \|}{1 - \| K \| \| K^{-1} \| \left(\frac{\|\Delta K\|}{\|K\|} \right)} \left(\frac{\| \Delta N \|}{\| N \|} + \frac{\| \Delta K \|}{\| K \|} \right) \tag{2.34}$$

where $\| \cdot \|$ are the vector and matrix norms compatible with one another.

The elements of the matrix ΔK are related to the errors of the directly measured linear or angular parameters of an interferometer system through the following expression [31]:

$$\Delta K_{ij} = \sum_{m=1}^{l} \frac{\partial k_{ij}}{\partial \theta_m} \, d\theta_m, \quad (i, j, = 1, 2, 3)$$

where θ_m denotes the generalized parameters of the optical system entering the elements of the matrix K; $d\theta_m$ is the measurement error for the mth parameter; and l is the number of the generalized parameters.

The value $\| K \| \| K^{-1} \|$, which is used in inequality (2.34), is usually denoted cond K and called the condition number of a system of linear algebraic equations. The condition number of a linear equation system depends only on the kind of the sensitivity matrix K, i.e. on the geometry of an interferometer's optical system. Thus equation (2.34) allows a preliminary estimation of the errors in the absolute value of the displacement vector. As it is shown in inequality (2.34), the non-statistical approach to experimental design consists of choosing a system of equations (2.32) with a minimum condition number. The corresponding criterion, in accordance with which the sensitivity matrix K

of a holographic interferometer (with possible minimum condition number) should be chosen, is called the C-optimality criterion.

The value of cond K depends on the type of the matrix norm used. In practice, spectral condition numbers are the most convenient since linear transformations ensuring maintenance of other matrix norms can be carried out by a less general matrix class than the class of orthogonal matrices. In accordance with the definition of the spectral condition number it follows that [47]

$$\text{cond } K = \| K \| \ \| K^{-1} \| = \frac{\mu_1}{\mu_2} \geqslant 1 \qquad (2.35)$$

where μ_1 and μ_2 are the maximum and minimum singular numbers of the matrix K which are equal to the maximum and minimum square roots of eigenvalues of matrix $(KK^{T})^{-1}$. As follows from (2.35), the theoretical possible minimum condition number is unity:

$$\text{cond } K = 1. \qquad (2.36)$$

Experimental designs which can be described with sensitivity matrix K and meet condition (2.36) are called C-optimal designs.

In most practical cases, the objective of the holographic experiment is to measure all three displacement components. To make use of the C-optimality criterion to minimize errors made in displacement component determination Δd_i ($i=1, 2, 3$) a simpler variant of estimation (2.34) must be taken. As known, part of the relative error (2.34) from measurements of the geometrical parameters of the optical system (i.e. of the sensitivity matrix $\| \Delta K \| / \| K \|$) is small in comparison with unity and considerably smaller than the relative error from the absolute fringe order determination $\| \Delta N \| / \| N \|$ [33, 35, 37]. Thus, in most cases inequality (2.34) can be represented in the following form:

$$\frac{\| \Delta \vec{d} \|}{\| \vec{d} \|} \leqslant \text{cond } K \frac{\| \Delta N \|}{\| N \|}. \qquad (2.37)$$

The errors in the individual components $|\Delta d_i|$ ($i=1,2,3$) can be estimated by means of inequality [49]:

$$|\Delta d_i| \leqslant \gamma_0 \sum_{j=1}^{3} |K_{ij}^{-1}| \qquad i = 1, 2, 3 \qquad (2.38)$$

where $\gamma_0 = \beta_N + \beta_K \Sigma_{i=1}^{3} |d_i|$; β_N and β_K are the maximum elements of the vector ΔN and matrix ΔK, respectively.

Although factor γ_0 is expressed through the components of the displacement $\vec{d}$ vector, inequality (2.38) is related to the absolute errors $|\Delta d_i|$ ($i=1, 2, 3$) and allows preliminary estimation of the individual components d_i ($i = 1$, 2, 3). Taking equation (2.38) into account, one can express the norm $\| \Delta \vec{d} \|$ in terms of any component of the displacement vector, e.g. Δd_1:

$$|\Delta d_i| = |\Delta d_1| G_i, \| \Delta \vec{d} \| = |\Delta d_1| \| G \| \tag{2.39}$$

where

$$G_i = \left[\sum_{j=1}^{3} |K_{ij}^{-1}| \right] \left[\sum_{j=1}^{3} |K_{1j}^{-1}| \right]^{-1} \qquad i = 1, \ 2, \ 3.$$

Combining equation (2.39) and (2.37) yields the desired estimate for the errors in the displacement components:

$$\frac{|\Delta d_i|}{\| \vec{d} \|} \leqslant \text{cond } K \frac{G_i}{\| G \|} \frac{\| \Delta N \|}{\| N \|} \qquad i = 1, \ 2, \ 3. \tag{2.40}$$

For the following analysis it should be noted that when the absolute fringe orders are used to interpret an interferogram, the interferometer sensitivity with respect to the component normal to the surface of the object is known to exceed (sometimes substantially) that of the in-plane components. In fact, as seen, for instance, in equation (2.46), the sensitivity to the normal component is always greater than unity and that of in-plane components is always less than unity. As a result of this anisotropy in the sensitivity, it is impossible to construct the optimum design cond $K = 1$. Error estimates made with inequalities (2.37) and (2.40) become very conservative and, hence, cannot be recommended for practical use.

In searching for ways to estimate minimum errors in such cases one can use the dependence of the condition number on the scale of the right-hand side of equation (2.32). A linear transformation of the variables $\vec{d} = M\vec{d}_m$ yields an equivalent system which can be more conveniently used to analyse the measurement accuracy:

$$K_m \vec{d}_m = \lambda N \tag{2.41}$$

where $K_m = KM$ and $\vec{d}_m = M^{-1}\vec{d}$ are the scaled sensitivity matrix and displacement vector, respectively, and M is a 3×3 diagonal matrix. For the optimum condition which provides a minimization of errors made in the measurement of displacement components with an interferometer specified by equation (2.41), one can obtain

$$\text{cond } K_m = 1 \tag{2.42}$$

where cond $K_m = \| K_m \| \| K_m^{-1} \|$. The most required adequate condition to meet equation (2.42) is the orthogonality of the matrix K_m. Subsequently, the most required adequate condition of the matrix orthogonality is the orthogonality of the vectors constructed from the columns (rows) of the matrix K_m and equality of their dimensions to unity.

In the case of the K_m matrix orthogonality, combining equations (2.40) and (2.42) provides the following equation:

$$\frac{|\Delta d_{im}|}{\| \vec{d}_m \|} \leqslant \frac{1}{\sqrt{3}} \frac{\| \Delta N \|}{\| N \|} \qquad i = 1, 2, 3 \qquad (2.43)$$

where Δd_{im} ($i = 1, 2, 3$) are the displacement measurement errors corresponding to the scaled equation system (2.41).

The displacement components errors reduced to the original scale $|\Delta d_i|$ ($i = 1, 2, 3$) can be determined from the inequality

$$|\Delta d_i| \leqslant M(i)|\Delta d_{mi}| \qquad i = 1, 2, 3 \qquad (2.44)$$

where $M(i)$ is the element of matrix M, describing the anisotropy in the interferometer optical system sensitivity.

Inequality (2.44) and the linear algebraic equation system (2.41), where matrix K_m is reduced to the original form, allows us to represent the errors Δd_i ($i = 1, 2, 3$) as a direct function of the Δn_j ($j = 1, 2, 3$) values which are the errors made in the absolute fringe measurement for the jth viewing direction. Let's assume that these errors are equal for all three observations, i.e. $\Delta n_j = \Delta n$ ($j = 1, 2, 3$). Under such a condition the combination of inequalities (2.43) and (2.44) results in the following final estimation:

$$|\Delta d_i| \leqslant \lambda \Delta n M(i) \qquad i = 1, 2, 3. \qquad (2.45)$$

To obtain equation (2.45) it is necessary to take into account that a linear transformation carried out by any orthogonal matrix does not change the length of an arbitrary vector, i.e.

$$\| K_m \vec{d}_m \| = \| \vec{d}_m \| = \lambda \| N \|.$$

Thus the design of the interferometer optical system ensuring the minimization of the displacement component measurement errors in accordance with equation (2.44) consists of choosing the sensitivity matrix K in an original scale, which can be scaled by some matrix M to result in the validity of condition (2.42).

Now we shall consider the detailed description of the optimization procedure for a widely used interferometer optical system consisting of one illumination direction (unit illumination vector $\vec{e}_S$) and three observation directions (unit observations vectors $\vec{e}_j$; above and below $j = 1, 2, 3$)—see Figure 2.12. The Cartesian coordinate system origin O is usually placed at the geometrical centre

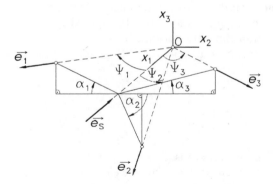

Figure 2.12. Schematic diagram of intreferometer and parameters to be optimized

of the region to be investigated. The coordinate axis x_1 coincides with the normal to the object surface under study at point O. To simplify the final results, without loss of generality, one can assume that angles ψ_1 and ψ_2 corresponding to unit vectors $\vec{e}_1$ and $\vec{e}_2$ are equal to one another.

The corresponding sensitivity matrix K of the interferometer concerned at point O has the following form:

$$
\begin{aligned}
K = \{K_{ij}\} = \\
i, j = 1, 2, 3
\end{aligned}
\begin{bmatrix}
1 + \cos\psi_1; & -\sin\psi_1\cos\alpha_1; & \sin\psi_1\sin\alpha_1 \\
1 + \cos\psi_1; & \sin\psi_1\cos\alpha_2; & -\sin\psi_1\sin\alpha_2 \\
1 + \cos\psi_3; & \sin\psi_3\cos\alpha_3; & \sin\psi_3\sin\alpha_3
\end{bmatrix}.
\tag{2.46}
$$

The matrix elements K_{ij} are equal to the difference in projection between the observation and the illumination unit vectors (sensitivity vector $\vec{e}_j - \vec{e}_S$)) on the coordinate axis x_i ($i = 1, 2, 3$). The values of the matrix K (2.46) elements K_{ij} directly depend on the angle values of α_j and ψ_j (see Figure 2.12). The choice of the correct interpretation of α_j and ψ_j parameters to minimize the errors of calculated displacement error components Δd_i ($i = 1, 2, 3$) and to ensure the stability of the equation (2.32) solution procedure, is the main subject of the following analysis.

Note that the elements of the first column of equation (2.46) are always greater than unity, and those of the second and third columns less than unity. This represents a mathematical consequence of the above-mentioned fact that the use of the absolute fringe order counting technique for interferogram interpretation results in an anisotropy of the interferogram sensitivity with respect to the normal to the object surface and tangential displacement components. In such case, as shown previously, the matrix K scaling procedure must be taken to optimize an interferometer optical system.

The evident relationship between matrix K elements, K_{ij} which result from equation (2.46) provide the possibility to define the scaled matrix K_m configuration:

$$K_m = \begin{bmatrix} K_{11}^m & -K_{12}^m & K_{12}^m \, \text{tg} \, \alpha_1 \\ K_{11}^m & K_{22}^m & -K_{22}^m \, \text{tg} \, \alpha_2 \\ K_{31}^m & K_{32}^m & K_{32}^m \, \text{tg} \, \alpha_3 \end{bmatrix}. \tag{2.47}$$

Now to establish the relations between the interferometer optical system parameters, it is convenient to represent the elements of the matrix K_m (2.47) as functions depending on the same parameter. If, for instance, we introduce into consideration the angle ω so that $(K_{11}^m)^2 = \sin^2 \omega$, and meet the conclusion that the length of the vector formed by the first column of equation (2.47) must be equal to unity, the matrix K_m can be expressed in the following form:

$$K_m = \begin{bmatrix} \sin\omega & -\cos\omega \, \cos\alpha_1 & \cos\omega \, \sin\alpha_1 \\ \sin\omega & \cos\omega \, \cos\alpha_2 & -\cos\omega \, \sin\alpha_2 \\ \sqrt{\cos2\omega} & \sqrt{2} \sin\omega \, \cos\alpha_3 & \sqrt{2} \sin\omega \, \sin\alpha_3 \end{bmatrix} \tag{2.48}$$

where $0 < \omega < 45°$ because, for the interferometer system involved (see Figure 2.12), the sensitivity with respect to the normal displacement component for all three viewing directions $\vec{e}_j$, which is described by elements in the first column of matrix (2.48), cannot be equal to zero.

The required orthogonality of the matrix (2.48) rows yields the following relations:

$$\cos(\alpha_2 - \alpha_1) = \text{tg}^2 \, \omega, \tag{2.49}$$

$$\cos(\alpha_1 + \alpha_2) = -\cos(\alpha_2 + \alpha_3) = \sqrt{\cos2\omega}/\sqrt{2}\cos\omega. \tag{2.50}$$

Dependencies (2.49) and (2.50), taking into account that $0 < \omega < 45°$, allow us to obtain the condition of the α_j-angles connection:

$$\cos\left[\frac{(\alpha_1 + \alpha_2 + 2\alpha_3)}{2}\right] = 0$$

or, in another form,

$$\alpha_1 + \alpha_2 + 2\alpha_3 = \pi, \ 0 \leqslant \alpha_j \leqslant \frac{\pi}{2} \qquad j = 1, \ 2, \ 3. \tag{2.51}$$

If the relations (2.49)–(2.51) are valid, the requirement of matrix (2.48) columns orthogonality and equality of its lengths to unity are always met.

Therefore the conditions represented by relations (2.49)–(2.51) are conditions needed to reduce the matrix K_m (2.48) to the orthogonal form and which allow us to use three parameters ω, α_1, and α_2 only in subsequent steps of the optimization procedure.

Now to construct the optimal experimental design (2.42), we must find the scaling matrix M and represent its elements through the angles α_j and ψ_j. As shown in reference [50], using the diagonal matrix in the following form leads to the desired objective:

$$M = \begin{bmatrix} a^{-1} & 0 & 0 \\ 0 & b^{-1} & 0 \\ 0 & 0 & b^{-1} \end{bmatrix}. \tag{2.52}$$

Inserting matrices (2.46), (2.48) and (2.52) into relation $K = M^{-1} K_m$ and the subsequent requirement of equality of the left- and right-hand corresponding elements of the first two rows in the obtained matrix equation, yields the factors a^{-1} and b^{-1}:

$$a^{-1} = \frac{\sqrt{\cos\theta}}{\sqrt{2}\,(1 + \cos\psi_1)\cos\left(\frac{\theta}{2}\right)}, \tag{2.53}$$

$$b^{-1} = \frac{1}{\sqrt{2}\,\sin\psi_1\,\cos\left(\frac{\theta}{2}\right)}. \tag{2.54}$$

In expressions (2.53) and (2.54) the functions $\sin\omega$ and $\cos\omega$ are represented through the parameter $\theta = \alpha_2 - \alpha_1$ by means of relation (2.49).

Expressions (2.53) and (2.54) allow us to carry out a term-by-term comparison of the corresponding elements of the third row in matrix equation $K = M^{-1} K_m$, which leads to the following equation system:

$$\begin{aligned} 1 + \cos\psi_3 &= (1 + \cos\psi_1)\sqrt{\cos 2\omega}\,(\sin\omega)^{-1} \\ \sin\omega_3 &= \sqrt{2}\,\sin\omega_1\,\mathrm{tg}\,\omega. \end{aligned} \tag{2.55}$$

Equation system (2.55) establishes the dependence between the observation angles ψ_1 and ψ_3, which, by using expressions (2.49) and (2.50), can be represented in the following form:

$$\mathrm{tg}\left(\frac{\psi_3}{2}\right) = \cos\theta\,\mathrm{tg}\left(\frac{\psi_1}{2}\right)\left(\sin\frac{\theta}{2}\right)^{-1} \tag{2.56}$$

where $0 < \theta < 90°$.

To permit simultaneous observation of point O on the surface under study from all three viewing directions, the unit vectors $\vec{e}_j$ ($j = 1, 2, 3$) should be

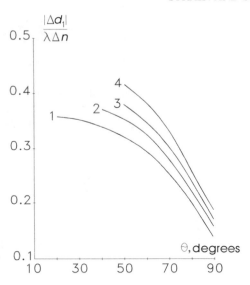

Figure 2.13. Dependences of normalized error $\Delta d_1/\lambda\Delta n$ made in the normal displacement component determination against parameter θ for different values of the observation angle ψ: $\psi = 20$ (1), 40 (2), 50 (3) and 60 (4) degrees

confined within a solid angle of 2π, implying that $\psi_1 < 90°$ and $\psi_3 < 90°$. Combining these conditions with equation (2.56), one can obtain the ranges for the viewing angles, namely

$$0 < \psi_q < 90°$$

$$0 < \psi_r < \text{arctg}\left\{\left[\frac{\sin\left(\frac{\theta}{2}\right)}{\cos\theta}\right]^m\right\}$$

(2.57)

where $m = 1$, $q = 3$, $r = 1$, if $0 < \theta \leqslant 60°$
$\quad\quad m = -1$, $q = 1$, $r = 3$, if $60° < \theta < 90°$.

Thus, relations (2.51), (2.53), (2.54), (2.56) and (2.57) describe a general case of C-optimal experimental design for the determination of the displacement vector components for point 0, constructed for a holographic interferometer optical system (see Figure 2.12).

The errors made in determining the normal Δd_1 and tangential Δd_2 and Δd_3 displacement components, which correspond to optimal interferometer design, can be expressed by equation (2.45):

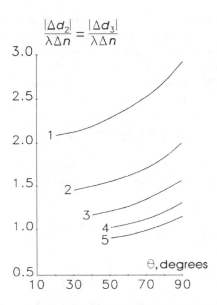

Figure 2.14. Dependences of normalized error $\Delta d_2/\lambda\Delta n = \Delta d_3/\lambda\Delta n$ made in the tangential displacement component determination against parameter θ for different values of the observation angle ψ: $\psi = 20$ (1), 30 (2), 40 (3), 50 (4) and 60 (5) degrees

$$|\Delta d_1| \leqslant a^{-1}\lambda\Delta n, \tag{2.58}$$

$$|\Delta d_2| = |\Delta d_3| \leqslant b^{-1}\lambda\Delta n \tag{2.59}$$

where the factors a^{-1} and b^{-1} are determined in accordance with equations (2.53) and (2.54).

The dependencies between normalized error of normal displacement component determination Δd_1 and parameter θ for different observation angle ψ_1 values are presented in Figure 2.13. Analogous dependencies for the error of tangential displacement components Δd_2 and Δd_3 are shown in Figure 2.14. These relations can serve as nomograms for a selection of optimal interferometer designs with required sensitivity.

When holograms with plane wave illumination are recorded and the fringe patterns obtained are observed by means of a telecentric viewing optical system, the sensitivity vector is constant over the entire object surface being investigated. In this case estimations (2.58) and (2.59) are satisfied for all illuminated points.

In a general case of a spherical illumination wavefront and observation from fixed points, the sensitivity vector varies as one moves from one surface point to another. For such optical systems a C-optimal design of the form (2.42) can be constructed for the geometric centre of the surface section under study. The optimality condition will be satisfied with some small deviations in the vicinity of point O within the constraints imposed on the parameters of the interferometer setup. According to equations (2.40) and (2.45) the component measurement errors near this point can be estimated by the expression

$$|\Delta d_i| \leqslant \sqrt{3}\, \lambda\, \Delta n\, M(i) \operatorname{cond} K_m \frac{G_i^m}{\| G^m \|} \frac{\| d_m \|}{\| N \|} \qquad i = 1,\, 2,\, 3 \qquad (2.60)$$

where Δn is the measurement error on the absolute fringe orders which is assumed to be equal for all viewing directions and G^m is the vector with the components

$$G_i^m = \left(\sum_{j=1}^{2} \left| (K_{ij}^m)^{-1} \right| \right) \left(\sum_{j=1}^{3} \left| (K_{1j}^m)^{-1} \right| \right)^{-1} \qquad i = 1,\, 2,\, 3$$

and diagonal elements of matrix M are determined from equations (2.53) and (2.54) for the origin at O.

Meeting the optimality condition (2.42) means that at fixed viewing angles ψ_1 and ψ_3 the interferometer is capable of measuring the displacement vector components with a minimum error, as defined by the C-criterion. To construct the optimal interferometer system, it is necessary to choose the angle difference $\theta = \alpha_2 - \alpha_1$, determining the range of the observation angles ψ_1 and ψ_3 variation and then to fix the values of the angles α_1, α_2 and α_3 according to (2.51). The selection of the angles ψ_1 and ψ_3 (2.56), within the allowable range (2.57) is governed, as a rule, by the desire to reach the maximum possible sensitivity of measurement for the in-plane displacement components. In practice, however, the freedom of this choice is limited by the requirement that the part of the curved object surface under study must be simultaneously observed from three directions. It should be noted that the possible variation in α_j-angles (2.51) is quite wide and practically eliminates all limitations resulting from the need for simultaneous observation of the surface part under study.

2.4 Optical systems of holographic interferometers for displacement components determination

The method of experimental design based on the implementation of the orthogonal matrices class to select the correct optical interferometer system parameters was dealt with in the previous section. This method enables us to construct an infinite number of optimal configurations which meet the

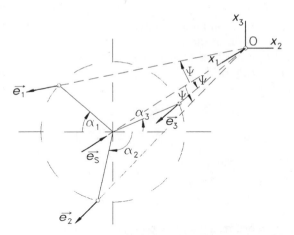

Figure 2.15. Schematic diagram of interferometer setup for three displacement components measurement when all three observation angles ψ are equal one other

C-optimality criterion. However, as in any experimental method, we can always select the interferometer system which ensures the most effective application of holographic interferometry techniques for displacement component determination.

The interferometer optical system in which all three viewing angles are equal to one another, i.e. $\psi_i = \psi$ ($i = 1, 2, 3$), is of particular interest for simultaneous measurement of three displacement components, especially if reflection hologram recording is implemented. Equation (2.56), required for the definition of the parameter θ in the case concerned, has the following form:

$$\cos\theta\left[\sin\frac{\theta}{2}\right]^{-1} = 1,$$

and the corresponding value of θ will be

$$\theta = 60°. \tag{2.61}$$

The interferometer system described by equation (2.61) represents a cone with apex at origin O of the coordinate system (x_1, x_2, x_3) and illumination direction $\vec{e}_S$ coinciding with the cone's axis (see Figure 2.15). In this case, three viewing directions lie on the generatrices of the cone and the angles between the unit observation vector projections onto the x_2 x_3 plane are equal to $2\pi/3$ radians. Note that the x_1-axis is usually assumed to coincide with the normal surface direction at the point O.

The sensitivity matrix of the interferometer involved can be obtained from equation (2.46), taking into account condition (2.61) and setting $\psi_i = \psi$ ($i = 1, 2, 3$). To fix the space orientation of the cone formed by observation unit vectors, it is necessary to define one of three angles α_j, for instance to take $\alpha_1 = 0$. Then expressions (2.51) and (2.61) yield

$$\alpha_2 = \alpha_3 = \frac{\pi}{3}.$$

The corresponding sensitivity matrix can be written as

$$K = \begin{bmatrix} 1 + \cos\psi & -\sin\psi & 0 \\ 1 + \cos\psi & (\frac{1}{2})\sin\psi & -\frac{\sqrt{3}}{2}\sin\psi \\ 1 + \cos\psi & \frac{1}{2}\sin\psi & \frac{\sqrt{3}}{2}\sin\psi \end{bmatrix}.$$

The scaling matrix M in accordance with equations (2.53) and (2.54) has the form

$$M = \begin{bmatrix} [\sqrt{3}(1 + \cos\psi)]^{-1} & 0 & 0 \\ 0 & \sqrt{2}(\sqrt{3}\sin\psi)^{-1} & 0 \\ 0 & 0 & \sqrt{2}(\sqrt{3}\sin\psi)^{-1} \end{bmatrix}. \qquad (2.62)$$

The form of a scaling sensitivity matrix K_m for the optimal optical system concerned is determined from equation $K_m = KM$, which leads to the following result:

$$K_m = \begin{bmatrix} \frac{1}{\sqrt{3}} & \frac{-\sqrt{2}}{\sqrt{3}} & 0 \\ \frac{1}{\sqrt{3}} & \frac{1}{\sqrt{6}} & -\frac{1}{\sqrt{2}} \\ \frac{1}{\sqrt{3}} & \frac{1}{\sqrt{6}} & \frac{1}{\sqrt{2}} \end{bmatrix}.$$

Substituting the diagonal elements of matrix (2.62) into inequality (2.45) yields the maximum possible errors which may be made in the displacement components measurement with the above-described optimal interferometer system:

$$|\Delta d_1| \leqslant \frac{\lambda \Delta n}{\sqrt{3}(1 + \cos\psi)}, \qquad (2.63)$$

$$|\Delta d_2| = |\Delta d_3| \leqslant \frac{\sqrt{2}\lambda \Delta n}{\sqrt{3}\sin\psi}. \qquad (2.64)$$

The optical system of the holographic interferometer, in which three observation directions form a cone with the axis coinciding with the illumination direction, has a very important attribute. Namely, all optical systems which are obtained from any original system by means of the cone's rotation through any angle around the x_1-axis (illumination direction) will meet the C-optimality condition. Thus, additional information can be derived and the measurement accuracy of the displacement components improved for the constant interferometer sensitivity which is defined by angle ψ.

This procedure can be realized by recording the reconstructed fringe patterns corresponding to the photographic plate rotated around the x_1-axis in the $x_2\,x_3$ plane through an angle $\alpha = 2\pi/3m$ radians (m is an integer). These patterns provide the possibility to obtain m systems of equations, each corresponding to an optimum interferometer optical system. By solving these systems, m variants of displacement components values d_i ($i = 1, 2, 3$) are provided. The average values of the obtained displacement components can be expressed by a standard procedure:

$$\bar{d}_i = \left(\frac{\sum\limits_{p=1}^{m} d_i^p}{m}\right) \qquad i = 1, 2, 3. \tag{2.65}$$

Assuming that the errors in the fringe order Δn_j ($j = 1, 2, 3$) are known and for all three viewing directions equal to Δn which is a random quantity with a Gaussian distribution and a zero mean, one can obtain the normal dispersion of the averaged displacement components

$$\gamma_i^2 = \left(\frac{|\Delta d_i|}{\sqrt{m}}\right)^2 \qquad i = 1, 2, 3. \tag{2.66}$$

The difference between the square root of normal dispersion (2.66) and the errors made in the single measurement (2.63) and (2.64) clarifies the degree of a possible improvement of the displacement components measurement accuracy by means of the above-described approach. Expression (2.66) proves that one can decrease the errors of displacement components in $\sqrt{m}$ times.

The interferometer set-up required to achieve the measurement accuracy improvement procedure can be implemented in the following way [42]. A two-exposure hologram or a hologram obtained by combining a time-average and two-exposure recording procedure in an opposite-beam arrangement is fixed in the $x_2\,x_3$ plane (see Figure 2.16) and illuminated with a collimated laser beam along the x_1-axis. A telecentric observation system is placed in the plane $x_1\,x_2$ at the angle ψ to the x_1-axis. The reconstructed fringe patterns are recorded by a detector in the initial position and in m positions corresponding to m consecutive rotations of the hologram in the $x_2\,x_3$ plane through $2\pi/3m$ radians about the x_1-axis. The number m must be chosen so that after transition from one viewing direction to another observation point, the absolute fringe orders

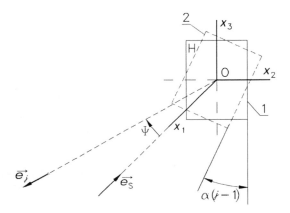

Figure 2.16. Schematic diagram of recording of fringe patterns reconstructed by means of reflection hologram (1 and 2 are two adjacent positions of the hologram)

at most points of interest change to a value by more than Δn from inequalities (2.63) and (2.64).

Practical experience of most characteristic strain-induced fringe patterns proves that the most reasonable maximum value of m is four. Using more than twelve viewing directions does not increase the accuracy although there is a considerable increase in experimental data to be treated. It should be noted that if the holographic interferograms of high fringe quality and density are obtained, it is sufficient to use three or even two equation systems (2.32) at each point of interest for a reliable quantitative fringe interpretation.

Each hologram position corresponding to a consecutive rotation through $\pi/6$ radians around the x_1-axis ($m=4$) allows us to write the row of the sensitivity matrix (2.46):

$$K_j = \left[1 + \cos\psi; \ \sin\psi \, \sin[(j-1)\alpha]; \ -\sin\psi \, \cos[(j-1)\alpha] \right] \qquad (2.67)$$

where $j=1$, 2, 3, . . . 12. Each of the three rows of equation (2.67) corresponding to numbers $j=l$, $j=l+4$, $j=l+8$ allows the sensitivity matrix of the holographic interferometers to meet the C-optimality criterion (2.42).

One of the main advantages of this interferometer configuration over multi-hologram set-ups is that it offers the possibility constantly to observe the process of the fringe order change at any point of interest in the course of the hologram rotation. This is very useful in order to make a reliable assignment of the fringe signs on different interferograms. Moreover, the comparison of the results of the displacement components determination obtained for different meanings of l in most cases is the most reliable criterion for the correctness of fringe orders and sign identification procedures.

Equations (2.63)–(2.66) allows us to obtain preliminary estimations of displacement component determination accuracy, representing the metrological basis of the above-presented technique.

We shall consider a real practical example: $\theta = 60°$, $\psi = 50°$, $\lambda = 0.6328\,\mu m$, $\Delta n = 0.2$ parts of fringe. The last value is the most accurate one for visual absolute fringe order counting in high fringe spacing regions. According to equations (2.63) and (2.64), one can obtain

$$|\Delta d_1| \leqslant 0.045\,\mu m; \qquad |\Delta d_2| = |\Delta d_3| \leqslant 0.13\,\mu m.$$

To decrease errors by half, as follows from equation (2.66), four optimal interferometer systems ($m = 4$, 12 fringe patterns) must be used:

$$\gamma_1 = 0.5|\Delta d_1| \leqslant 0.023\,\mu m, \tag{2.68}$$

$$\gamma_2 = \gamma_3 = 0.5|\Delta d_2| \leqslant 0.065\,\mu m. \tag{2.69}$$

Estimations (2.68) and (2.69) clarify the real practical range of errors in displacement components determination which can be achieved for visual absolute fringe order counting.

Note that an improvement of the measurement accuracy can be accomplished by the above-described technique only, based on the use of more than one C-optimal interferometer system, and thus leads to estimations (2.66). Overdetermining equation system (2.32) with C-optimal sensitivity matrix K by adding other observation directions (i.e. using $m \times 3$ sensitivity matrix K, where $m > 3$), as proposed in reference [25], cannot improve the errors of the displacement components determinations (2.63) and (2.64) [43].

The general approach to the quantitative interpretation of holographic interferograms, which implies the use of three different fringe patterns (as a minimum), is connected with a considerable volume of the input data (absolute fringe orders) to be treated. Although the presence of a zero-motion fringe on the fringe patterns allows us to use the interpretation procedures (relatively simple and suitable from an automation point of view), in many practical cases the question arises whether it is possible to reduce the input data array and how this reduction will influence measurement accuracy.

The positive answer to the first part of this question depends on the preliminary information about the character of deformation of the object to be studied. In particular, to solve any axis-symmetrical problem of the deformable body mechanics under symmetrical loading, it is necessary to determine only two displacement components, usually normal and tangential to the surface under investigation.

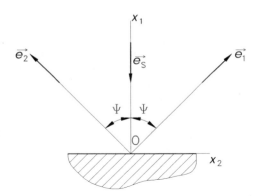

Figure 2.17. Schematic diagram of interferometer setup for two displacement component measurements with symmetrical observation directions

For instance, determination of the stress concentration in a plane specimen with a circular open hole under tension requires information about the distribution along a hole edge of one in-plane displacement component in the loading direction only. An in-plane (tangential) displacement component d_2 (also d_3 if necessary) can be determined by means of an analysis of two fringe patterns only. The corresponding interferometer set-up, which allows us to obtain information about two displacement components at each surface point of interest, a normal to an object surface component d_1 and a tangential one d_2, is shown in Figure 2.17. The object surface point O under study is illuminated in a direction specified by the vector $\vec{e}_S$ normal to the surface at point O, and the two viewing directions are symmetrically spaced about it. The x_1-axis of the Cartesian coordinate system (x_1, x_2, x_3) coincides with the illumination direction. The unit observation vectors $\vec{e}_1$ and $\vec{e}_2$ lie in the x_1x_2 plane. The angle ψ between the viewing direction and the normal to the surface determines the interferometer sensitivity.

The corresponding sensitivity matrix K can be derived from the first two elements of rows (2.67) for $j = 4, 8$:

$$K = \begin{bmatrix} 1 + \cos\psi & \sin\psi \\ 1 + \cos\psi & -\sin\psi \end{bmatrix}. \tag{2.70}$$

The scaling matrix M, corresponding to equation (2.70), is expressed as

$$M = \begin{bmatrix} [\sqrt{2}(1 + \cos\psi)]^{-1} & 0 \\ 0 & [\sqrt{2}\sin\psi]^{-1} \end{bmatrix} \tag{2.71}$$

and the orthogonal scaled sensitivity matrix $K_m = KM$ has the form

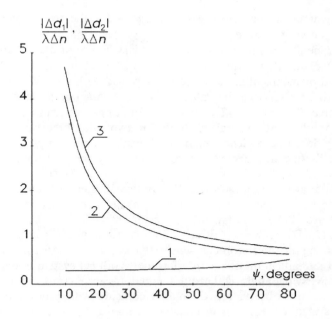

Figure 2.18. Dependences of normalized error $\Delta d_1/\lambda\Delta n$ (1) and $\Delta d_2/\lambda\Delta n$ (2, 3) made in determination of the normal and tangential displacement component determination, respectively, in the case of two (1, 2) and three (1, 3) observation directions

$$K_m = \begin{bmatrix} \frac{1}{\sqrt{2}} & \frac{1}{\sqrt{2}} \\ \frac{1}{\sqrt{2}} & -\frac{1}{\sqrt{2}} \end{bmatrix}.$$

Substituting corresponding diagonal elements of scaling matrix (2.70) into inequalities (2.58) and (2.59) yields the estimations of the displacement component errors:

$$|\Delta d_1| \leqslant \frac{\lambda\Delta n}{\sqrt{2}(1 + \cos\psi)}, \tag{2.72}$$

$$|\Delta d_2| \leqslant \frac{\lambda\Delta n}{\sqrt{2}\sin\psi}. \tag{2.73}$$

A comparison of estimations (2.72) and (2.73) with analogous estimations for the case of three viewing directions is of interest from a metrological standpoint. The normalized dependence between the displacement component measurement errors and observation angle ψ for the two above-mentioned cases are shown in Figure 2.18.

These dependencies prove that a preliminary choice of the direction lying in the $(x_2 x_3)$ plane along which displacement component d_2 (or d_3) are to be determined by means of an interferometer system with sensitivity matrix (2.70) will lead to a decrease in the value of $|\Delta d_2|$ (or $|\Delta d_3|$) in comparison with the case of three observation directions used for fringe pattern interpretation. An increase of observation angle ψ both for two and three observation cases practically does not influence the normal displacement component error value $|\Delta d_1|$, but significantly decreases the in-plane displacement component errors $|\Delta d_2|$ and $|\Delta d_3|$.

It should also be noted that the optical system of an interferometer corresponding to sensitivity matrix (2.70) has no way of carrying out the set of the equal accurate measurement, the results of which can be represented by equations (2.65) and (2.66). Additional information in this case can be obtained after changing the interferometer sensitivity, i.e. observation angle ψ only.

Further simplification of the interferometer optical system may be achieved when the viewing angle ψ becomes equal to zero $\psi = 0$. In this case the illumination and observation directions coincide with one another and the equation system degenerates to the single equation

$$d_1 = n \frac{\lambda}{2}. \tag{2.74}$$

Equation (2.74) allows us to determine the normal to the object surface displacement component d_1 only, but with maximum possible sensitivity $\lambda/2$. Only in this case, the interference fringe patterns observed have a direct mechanical interpretation and represent the lines of the deflection equal level.

The error made in the normal surface displacement component determination can be estimated by the following inequality:

$$|\Delta d_1| \leqslant \frac{\lambda}{2} \Delta n. \tag{2.75}$$

An optical system satisfying the requirement $\psi = 0$ is shown in Figure 2.19. During a hologram recording (Figure 2.19(a)), the laser beam LB which is expanded and collimated by a set of lens L1 and L2 impinges on the beam splitter BS. The plane wave reflected from the beam splitter BS illuminates the object surface O along the direction coinciding with the normal to the object surface. The illumination unit vector $\vec{e}_S$ in this case coincides with the normal surface direction. The wave reflected from the object and passing the splitter is incident on the photographic plate H (object wave). The wave directly passing through the beam splitter is reflected by the mirror M onto the holographic plate (reference wave).

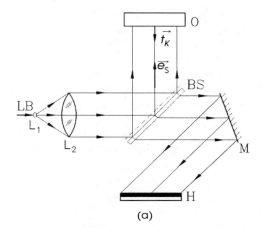

(a)

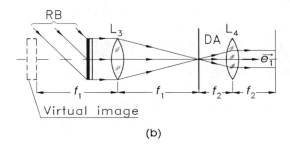

(b)

Figure 2.19. Schematic diagram of interferometer arrangement for determination of the normal to the object surface displacement component: (a) hologram recording, (b) observation of fringe pattern

The reconstructed image is viewed by means of the optical system illustrated in Figure 2.19(b). When the hologram H is illuminated by the plane reference wave RB, the image of the object is viewed on the screen S through a telecentric system composed of lenses L_3 and L_4. The diaphragm DA placed in the lens focus ensures spatial filtering of the object wave, as a result of which the viewing direction defined by the unit vector $\vec{e}_1$ will also be parallel to the x_1-axis everywhere on the object surface.

This type of interferometer system has the remarkable capability of visualizing out-of-plane displacement components in the presence of very large in-plane components sometimes exceeding the former by an order of magnitude [51].

2.5 Some technical aspects of strain-induced displacement measurement of deformed bodies

The deformation of an object can be accompanied by rigid body displacements such as translations, rotations and their combinations. These rigid body displacements also make their own contribution to the fringe pattern formation process.

These combined fringe patterns, which contain information not only about the strains of the object but also about the total surface displacement, are usually not suitable for quantitative interpretation. This represents a major difficulty in holographic interferometry for measurements of strain-induced displacements of the surface of the stressed object. Therefore, the possibility of isolating strain-induced displacements against a background of rigid-body motion determines whether holographic interferometry can be employed in the quantitative study of the deformation. Practical experience has proved that the only reliable way of overcoming this problem is by recording the fringe patterns which result from strain-induced displacement.

To avoid, or considerably reduce, rigid-body displacements, specially designed loading devices can be used, or a fine adjustment made to the standard loading equipment, ensuring the required deformation of an object, while preventing its rigid-body displacements. It is vital that the fine adjustment of the different loading devices enables us to measure the displacements of its various elements (columns, holders, rods, etc.) through holographic tests. The data obtained can be used to bring improvements in the design and operation of the loading devices, i.e. eliminating or reducing undesired displacements.

We shall consider an investigation of the deformation of a circular cylindrical shell subjected to tension to illustrate an approach to strain-induced fringe patterns recording based on the use of specially designed loading devices. It should be noted that the recording of the tensile strain-induced interferograms represents one of the most complex problems from the holographic interferometry standpoint due to, first, the anisotropy of an interferometer sensitivity and, second, the fringe patterns to be obtained are localized far from an object surface.

The object investigated was an aluminium circular cylindrical shell of diameter $D = 60$ mm, length $L = 100$ mm and wall thickness h $= 1.5$ mm. Fresnel holograms are recorded by means of two hologram interferometers, having the sensitivity matrix of form (2.70) and observation angle $\psi = 55°$. Typical fringe patterns on the shell surface obtained for two viewing directions and corresponding to a tensile force action of 3554 N are shown in Figure 2.20.

Distributions of the radial d_1 and axial d_2 displacement components along the shell generatrix are presented in Figure 2.21. The results obtained from the

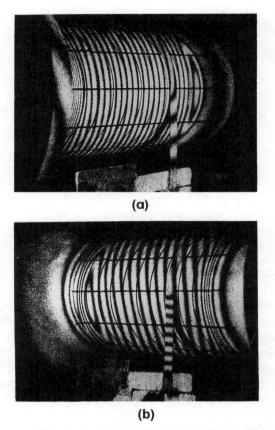

(a)

(b)

Figure 2.20. Holographic interferogram of regular cylindrical shell under tension obtained in two symmetrical directions (a) and (b)

experiment and the corresponding elasticity theory data reveal that a good agreement was reached within errors resulting from: boundary condition modelling, non-coincidences of material properties and material properties used for calculation, load value determination, etc. These special loading devices are used for all experimental investigations of deformation in shells with holes which are described later (see Chapters 5 and 6).

The fixing of a photoplate directly on an object surface and subsequent application of an opposite-beam arrangement for reflection hologram recording is a very effective approach to avoid the negative influence of the rigid-body displacements on formation of the strain-induced fringe patterns. If the object under study is a plane specimen subjected to tension-compression

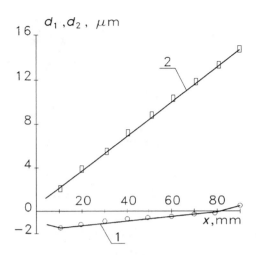

Figure 2.21. Distributions of the radial (1) and axial (2) displacement component along a generatrix of regular cylindrical shell under tension

without bending, the above approach is practically the only way of ensuring correct interpretation of quantitative fringe patterns.

In strain studies of different structures with reflection hologram interferometers the recording medium can be fixed on the object surface under investigation with magnetic or mechanical devices [52]. These devices, however, cannot be employed when investigated objects are loaded with different types of testing machines, since they cannot eliminate the influence of vibrations. These oscillations always arise if a structure element or specimen is loaded, for instance, with a servohydraulic testing machine.

In the case of a thin-walled object with a plane surface, one can fix the photographic plate in place by means of an intermediate transparent optical medium as proposed by Zhylkin and Gerasimov [53]. This approach to fixing a recording medium onto the specimen surface with subsequent reflection hologram recording represents the so-called overlay interferometer, which allows to us minimize the rigid-body influence on strain-induced fringe patterns to be interpreted.

The following development of overlay interferometer was introduced by Gorodnichenko and Pisarev [54]. The developed technique involves some valuable concepts proposed by Boone [55], Neumann and Penn [56]. The effectiveness of this approach can be clearly illustrated by using as an example the problem about tension of an aluminum strip of width $b = 60$ mm and thickness $h = 5$ mm. The specimen was loaded by a servohydraulic closed-loop computer-aided testing machine.

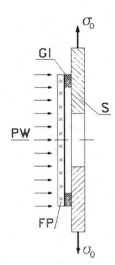

Figure 2.22. Schematic diagram of an overlay holographic interferometer

A diagram of the above-mentioned overlay interferometer set-up is shown in Figure 2.22. In this case a photoplate FP is mounted on a plane specimen S surface by means of special non-transparent glue Gl, covering the small corner parts of the photoplates, so that only air clearance with a depth of a few tens of micrometres exists between the holographic emulsion and the object surface region of interest. It should also be noted that the glue used has a low shear rigidity compared to the metal specimen.

The reliability of the loading unit and hydraulic grips accurately mounted along one line allows us to place the elements of the interferometer optical system (He–Ne laser, mirrors, micro-objective and lens) outside the frame of the testing machine.

A typical interferogram of the stretched strip for loading increment $P = 50$ MPa is shown in Figure 2.23. The interferogram presented corresponds to the linear distribution of the axial displacement component d_2, which is characterized by an inclination angle α_0

$$\alpha_0 = \mathrm{tg}^{-1}\left(\frac{d_2(x_2)}{x_2}\right)$$

where $d_2(x_2)$ and x_2 are the current values of the displacement component and distance starting from the fixed specimen end. To estimate the accuracy of the results obtained from the above-presented experiment, it is appropriate to determine the elasticity modulus value which corresponds to the inclination angle magnitude obtained:

Figure 2.23. Holographic interferogram of the tension of a regular strip obtained by means of overlay interferometer

$$E = \frac{P}{\alpha_0 \, bh}$$

where P is applied tensile load; b is the specimen width; and h is the specimen thickness. The corresponding elasticity modulus value in the case concerned will be

$$E = 72\,000 \pm 2000 \, \text{MPa}$$

which is in good agreement with the data from standard mechanical testing of the specimen material.

The results of this comparison show a high reliability both of the testing machine used for the plane specimen loading and the overlay holographic interferometer with air clearance between the emulsion and the object surface under study. The approach to the investigation of a plane specimen deformation loaded according to plane stress conditions is used in all analogous experiments presented in this book.

To clarify the metrological and methodological possibilities of holographic interferometric techniques applied to deformation study, some characteristic examples, which describe most practical cases, will be considered below.

From the standpoint of holographic interferometric techniques, having, as mentioned above, the anisotropy of the sensitivity with respect to the normal to an object surface and tangential displacement components, these examples can

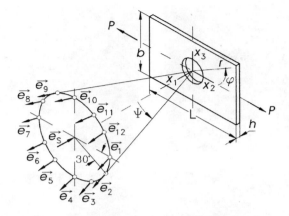

Figure 2.24. Overlay interferometer arrangement referring to the centre of a hole in a plane specimen under tension

be effectively classified using the ratio between the normal displacement component and the maximum from two tangential components. The following situations may be possible in practice:

- The tangential component is greater than the normal one in order of magnitude.
- The normal and tangential components have the same order of magnitude.
- The normal displacement component exceeds the tangential one by more than ten times.

The investigation of a strain/stress concentration in a plane specimen with a central circular open hole subjected to tensile loading represents one of a most interesting examples of the first situation mentioned above. A detailed experimental analysis of the displacement component distributions was carried out for an aluminium strip of width $b = 60$ m, thickness $h = 6$ mm, with a central open hole of diameter $2r_0 = 12$ mm. A diagram of the specimen and notation used is shown in Figure 2.24.

Specimen loading and the procedure of reflection hologram recording using an overlay interferometer was carried out by means of a combined application of the two-exposure and time-average techniques described in Section 2.2, allowing zero-motion fringe visualization. The set of 12 interferograms was reconstructed from each hologram recorded as shown in Figure 2.24.

Typical interferograms, with bright zero-motion fringe corresponding to an observation angle of $\psi = 56°$, necessary to construct two C-optimal optical

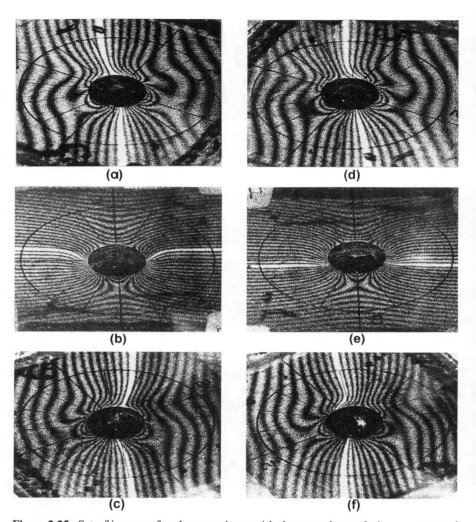

Figure 2.25. Set of images of a plane specimen with the central open hole reconstructed by a single reflection hologram for different viewing directions: $\alpha = 30$ (a), $\alpha = 90$ (b), $\alpha = 150$ (c), $\alpha = 210$ (d), $\alpha = 270$ (e), $\alpha = 330$ (f) degrees

interferometer systems of form (2.61), are shown in Figure 2.25. The reflection hologram, which is the source of the fringe patterns presented above, was obtained for net stress increment $\sigma_0 = 40$ MPa. One equation system of form (2.32) can be obtained by means of the interpretation of hologram viewing for angles α, which are equal to 30°, 150°, and 270° (photographs (a), (c), (e) in

Table 2.1 The absolute fringe orders in 16 points on the circular hole edge in the plane specimen under tension for different observation directions.

α, degrees	1	2	3	4	5	6	7	8	9	10	11	12	13	14	15	16
0	1.5	1.8	1.6	1.4	1.4	2.0	2.5	2.8	2.6	0.5	-2.5	-6.5	-8.3	-6.3	-3.5	-0.5
30	9.7	9.6	7.5	4.6	2.0	0.4	-0.5	-1.4	-2.4	-3.6	-5.5	-6.7	-6.0	-2.5	2.5	7.5
60	16.2	15.0	11.0	6.5	1.5	-1.5	-3.6	-5.3	-6.3	-6.9	-6.7	-6.0	-3.5	2.6	8.7	13.9
90	18.8	17.4	11.5	5.7	0.0	-4.0	-6.0	-7.3	-7.5	-7.4	-6.3	-3.9	-0.4	5.3	11.5	17.5
120	17.4	15.0	9.3	3.7	-2.5	-5.6	-6.4	-6.5	-6.3	-5.3	-3.5	-1.1	1.6	7.0	12.7	16.4
150	11.7	9.2	4.5	-1.3	-5.4	-6.1	-5.5	-3.6	-2.1	-1.1	-0.1	1.3	3.5	6.5	9.6	11.7
180	4.4	1.5	-1.5	-5.5	-7.2	-6.0	-3.0	0.5	3.4	3.6	3.5	3.3	3.4	3.4	3.7	4.5
210	-4.4	-5.6	-7.5	-8.6	-7.7	-4.5	0.5	5.5	8.3	8.4	6.4	3.4	0.5	-1.0	-2.4	-3.5
240	-10.4	-10.9	-10.7	-10.0	-7.4	-2.5	3.4	9.5	11.9	11.3	7.5	2.1	-2.0	-5.8	-8.1	-9.5
270	-13.4	-13.3	-11.6	-9.5	-6.3	-0.7	5.5	11.0	13.2	11.5	6.9	-0.5	-5.1	-9.5	-11.6	-13.3
300	-11.5	-10.7	-8.8	-6.5	-3.0	1.0	6.0	10.5	11.9	9.8	4.5	-2.5	-7.5	-11.4	-11.7	-12.0
330	-6.5	-5.4	-4.4	-2.6	-0.6	1.9	4.7	7.6	8.2	5.7	1.0	-4.5	-8.6	-10.0	-9.4	-7.5

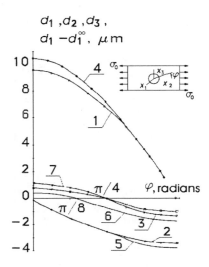

Figure 2.26. Distributions of the displacement components along the circular hole edge in a plane specimen under tension

Figure 2.25). The other equation system results from interferograms obtained for $\alpha = 90°$, $210°$, and $330°$ (fringe patterns (b), (d), (f) in Figure 2.25).

To interpret the interferograms, the absolute fringe orders are determined by direct counting, starting from any zero-motion position, and taking into account that the fringe order reverses its sign as one crosses the zero-motion fringe. It should be kept in mind that the fringe-order signs must be assigned conventionally by the same rule for all viewing directions. The directions of the displacement components would be determined after finishing the interferogram interpretation procedure using the known information about the direction of a tensile force.

The absolute fringe orders were identified at 16 points belonging to the hole edge for each 12 viewing directions used. The data obtained are listed in Table 2.1. This table allows us to construct four equation systems of form (2.32) for all 16 points under study. Each system thus obtained yields three displacement components d_1, d_2, and d_3.

We shall take, for instance, the second equation system ($\alpha = 30°$, $150°$, and $270°$) for the displacement components determination at point 1. The number of the equation system coincides with the number of its first equation. The left-hand side of this system is constructed from the rows of form (2.67) for $j = 2, 6, 10$, and the right-hand side consists of the corresponding data from Table 2.1:

$$(1 + \cos\psi)d_1 + \left(\frac{\sin\psi}{2}\right)d_2 - \left(\frac{\sqrt{3}}{2}\sin\psi\right)d_3 = 9.7\lambda$$

$$(1 + \cos\psi)d_1 + \left(\frac{\sin\psi}{2}\right)d_2 + \left(\frac{\sqrt{3}}{2}\sin\psi\right)d_3 = 11.7\lambda$$

$$(1 + \cos\psi)d_1 + \left(\frac{\sqrt{3}}{2}\sin\psi\right)d_2 = -13.4\lambda$$

(2.76)

where λ is equal to $0.6328\,\mu m$.

If, for instance, we solve all four equation systems corresponding to point 1 (one of them is (2.76)) and then average the results obtained, the averaged displacement component values $\bar{d}_1$, $\bar{d}_2$ and $\bar{d}_3$, would have errors which can be estimated by inequalities (2.66). If the above-mentioned procedure is to be carried out at all 16 edge points of interest, the displacement component distributions would be obtained along the hole boundary. These relations, without taking into account the rigid-body motion of the photoplate between exposures, are shown in Figure 2.26 by curves 1, 2, 3 for components $\bar{d}_2$, $\bar{d}_3$ and $\bar{d}_1$ respectively.

Displacement component distributions along the hole edge obtained by means of a finite element method for elasticity modulus $E = 71000\,MPa$, Poisson's ratio $\mu = 0.34$ and net stress increment $\sigma_0 = 40\,MPa$ are presented in the same figure by curves 4, 5 and 6 for components d_2, d_3, and d_1 respectively. The software package FITING [57] was used for the calculation. The volume mesh consisting of 20-node isoparametric finite elements was selected to simulate the specimen investigated. This mesh took advantage of the symmetry of the problem and loading, and modelled only a one-eighth part of the specimen. For a reliable stress concentration determination, the mesh area between the hole boundary and circle line $r = 20\,mm$ was divided into seven parts in the radial direction, eleven parts in the angular directions and 5 parts through the thickness. The degree of calculation accuracy achieved can be characterized by a coincidence of the stress concentration factor obtained in the middle plane of the specimen and known inquiry data within one per cent [58].

It should be pointed out that experimental displacement component distributions along the hole boundary, which are plotted in Figure 2.26, are found to compare well with the corresponding numerical data obtained by the finite element method.

The comparison of the experimental and numerical data is not the only criterion to estimate the accuracy and reliability of displacement component determination by holographic interferometry technique. Another possible way consists of comparing the data obtained for two slightly different but analogous objects. In our case the specimen analogous to the one presented

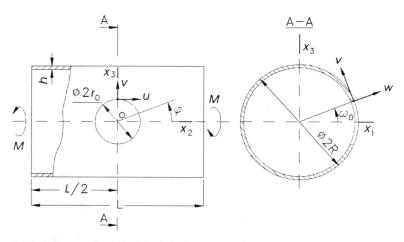

Figure 2.27. Scheme of a cylindrical shell subjected to a torsion

in Figure 2.24 but with thickness $h = 9$ mm was used as object for such comparison. The corresponding numerical study of displacement components in the stress concentration region was also carried out.

The experimental and numerical distributions of in-plane displacement components $\bar{d}_2$ and $\bar{d}_3$, obtained for a plane specimen of thickness $h = 9$ mm, almost coincide with those obtained for a specimen of thickness $h = 6$ mm, plotted in Figure 2.26. The increase of the specimen thickness from 6 to 9 mm influences the component $\bar{d}_1$ distribution only, and is shown in Figure 2.26 by curve 7 for net stress increment $\sigma_0 = 40$ MPa.

Comparison of the curves plotted in Figure 2.26 shows that the corresponding dependencies for in-plane displacement components d_2 and d_3 almost coincide with one another within the error resulting from measurement errors made during the experimental procedure, definition of the material mechanical properties and difference between the real specimen stiffness and that of the finite element mesh. The experimental dependencies of the normal displacement component d_1 obtained for specimens with different thicknesses differ from the corresponding numerical distributions by constant values caused by uniform thickness decreasing, resulting from Poisson's effect:

$$d_1^\infty = -\sigma_0 \mu \frac{h}{2E} \tag{2.77}$$

where σ_0 is the increment of net stresses applied to the specimen edges; E is the elasticity modulus; μ is the Poisson's ratio; and h is the thickness of a plane specimen. This fraction of a normal to the object surface displacement

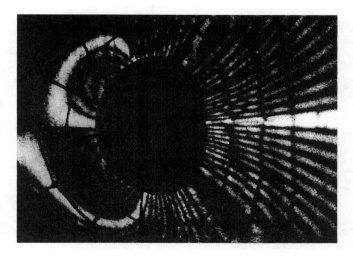

Figure 2.28. Interference fringe pattern obtained for shell torsion

component d_1 is not recorded on interferograms due to the use of a contact fixing of a photoplate on the object surface.

The values of d_1^∞ calculated from equation (2.77) for elasticity modulus $E = 71\,000\,\text{MPa}$, $\mu = 0.32$, $\sigma_0 = 40\,\text{MPa}$, $h = 6\,\text{mm}$ and $h = 9\,\text{mm}$ are equal to $0.58\,\mu\text{m}$ and $0.86\,\mu\text{m}$ respectively, and almost coincide with the difference of d_1 values obtained by subtraction of corresponding experimental and numerical data. It should also be noted that data obtained, concerning the d_1 component values, are in a good agreement with analogous experimental and theoretical investigation results presented in reference [56].

A deformation study of a hole boundary in a curved circular cylindrical shell represents the following characteristic example which can be used as a clear illustration of the situation when a normal and one from two tangential to the object surface displacement components have comparable magnitudes. The object investigated was a shell made of aluminium alloy and subjected to a torsional stress. The outer radius of the shell surface is $R = 30\,\text{mm}$, the hole radius $r_0 = 9\,\text{mm}$, the length of the thin-walled cylindrical part $L = 100\,\text{mm}$, the wall thickness $h = 1.5\,\text{mm}$ (see Figure 2.27). The origin of the Cartesian coordinate system is placed at the hole centre on the external surface. The direction of the x_2-axis coincides with the generatrix of the shell and the direction of the d_1-axis with the normal surface direction.

The reflection holograms are recorded in an opposite-beam arrangement with the plane-wave illumination directed along the x_1-axis. The device used for shell loading by torque moment allows us to record holographic interferograms with bright zero-motion fringes (see Section 2.2).

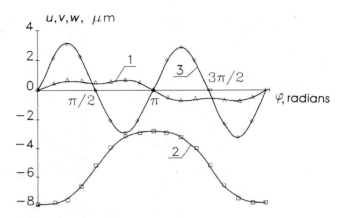

Figure 2.29. Distributions of the displacement components along the circular hole edge in cylindrical shell under torsion

A photoplate is mounted near the area investigated by means of a special mechanical holder without contact between the emulsion and shell surface. The corresponding air clearance should be adjusted as small as possible. Because the distance between the photoplate and curved shell surface is not constant over the area of interest, the observation angle ψ value should be chosen so as to ensure a view of the hole vicinity from all directions. The procedure of the fringe pattern recording and its subsequent quantitative interpretation completely corresponds to that described above for a plane specimen with a circular hole.

Figure 2.28 presents a three-exposure interferogram obtained for viewing directions specified by the parameters $\psi = 50°$ and $\alpha = 0$, with the torque moment increment during exposure $M = 1567$ N·m.

The solution of four-equation systems (2.32) enabled the determination of the Cartesian displacement vector components $\bar{d}_1$, $\bar{d}_2$ and $\bar{d}_3$ averaged according to equation (2.65) for $m = 4$. Then, the displacement components u, v and w in a cylindrical system (see Figure 2.27) were derived from

$$u = \bar{d}_2; \qquad v = \bar{d}_3 \cos\omega_0 - \bar{d}_1 \sin\omega_0$$
$$w = \bar{d}_1 \sin\omega_0 + \bar{d}_3 \sin\omega \tag{2.78}$$

where $\omega_0 = \sin^{-1}[r_0 \sin(\varphi/R)]$. The distributions of the components u, v and w along the hole edge are shown in Figure 2.29 by curves 1, 2 and 3 respectively. The errors made in the displacement component measurement in this case can be expressed as

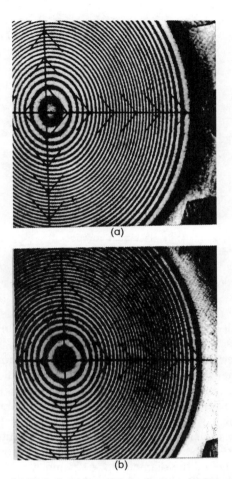

(a)

(b)

Figure 2.30. Holographic interferograms of the fragment of circular plate subjected to an internal pressure obtained for two viewing directions

$$\Delta u = \gamma_2; \qquad \Delta v = \gamma_3 \cos\omega_0 + \gamma_1 \sin\omega_0$$
$$\Delta w = \gamma_1 \cos\omega_0 + \gamma_3 \sin\omega_0 \qquad (2.79)$$

where γ_i ($i=1, 2, 3$) is derived from equation (2.66). In the case concerned, for $\psi = 50°$, $\theta = 60°$, $m = 4$ and $\Delta n = 0.2$ parts of fringe, the error values Δu, Δv and Δw are not more than 0.15, 0.15 and 0.05 μm respectively.

As will be shown in Chapters 4 and 5, the accuracy achieved in the displacement components determination allows determination of the strain and

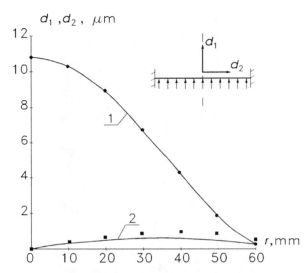

Figure 2.31. Distributions of the displacement components along a plate radius

stress distributions along the hole boundary to be comparable with the data of the analytical approach to the same problem solution.

The third characteristic metrological situation, when a normal to an object surface displacement component exceeds the tangential components by approximately ten times, always occurs if a thin plate is subjected to bending. To illustrate this problem, we shall consider the application of holographic interferometry to the displacement components determination in a circular plate of the constant thickness $h = 5$ mm loaded with a uniformly distributed pressure.

The outer plate edge of the diameter $D = 120$ mm was clamped. The plate was made of aluminium alloy. The reflection holographic interferograms were recorded in an opposite-beam arrangement with plane-wave illumination directed along the surface normal (x_1-axis). During the hologram reconstruction, 12 fringe patterns were recorded with the photoplate rotated through an angle $\alpha = 30°$ for the viewing directions specified by parameters $\psi = 60°$ and $\theta = 60°$. This permitted construction of four-equation systems (2.67) satisfying the C-optimality criterion for each point of interest. Two-exposure interferograms obtained for $\alpha = 0°$ and $\alpha = 180°$, with the internal pressure increment $q = 3$ MPa are shown in Figures 2.30(a) and 2.30(b) respectively. The origin of the absolute fringe orders counting can be at any point of the unmoved area of the plate which is clearly seen in the interferograms presented in Figure 2.30.

The normal $\bar{d}_1$ and radial $\bar{d}_2$ displacement component along the plate radius against the distance from the plate centre are plotted in Figure 2.31 by curves 1

Figure 2.32. Holographic interferogram of a deflection field of a circular plate under internal pressure

and 2 respectively. The errors of the displacement component determination according to the above analysis are $\Delta \bar{d}_1 = 0.05\,\mu\text{m}$ and $\Delta \bar{d}_2 = 0.10\,\mu\text{m}$. The technical theory of the thin bending plate deformation proves that the normal and circumferential displacement components must be connected through the following relation:

$$d_2 = \frac{h}{2}\frac{\partial d_1}{\partial r}.$$

The validity of this relation for the experimental data obtained demonstrates the high accuracy of the measurement procedure used.

The interferogram of the same plate, which is used in the above-presented example, shown in Figure 2.32 illustrates a conventional approach to the normal to the object surface displacement component by means of the optical arrangement in Figure 2.19.

In conclusion, it is interesting to consider a practical application of an optical interferometer system shown in Figure 2.19 for out-of-plane displacement component visualization in the presence of in-plane component exceeding the former by an order of magnitude. The tension of a plate with an open hole, as shown in Figure 2.26, is a very characteristic example of such a situation. The fringe patterns reconstructed with the optical system shown in Figure 2.19(b) near the holes in plane specimens of thickness $h = 6\,\text{mm}$ and $h = 9\,\text{mm}$

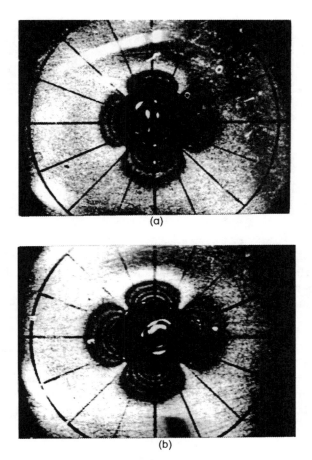

(a)

(b)

Figure 2.33. Holographic interferograms of relative thickness change of two plane specimens under tension: (a) thickness is 6 mm, (b) thickness is 9 mm

are presented in Figures 2.32(a) and 2.32(b) respectively. These interferograms correspond to net stress increment $\sigma_0 = 40$ MPa. In this case the interference fringes are the lines of the equal levels of the normal displacement component d_1.

Two special features of these interferograms must be noted. First, the symmetrical character of the fringe observed is in excellent agreement with the prediction of the elasticity theory and experimental results presented in reference [59]. Second, it is evident that the fringes, forming the right-hand petal of the rosette near the hole on the horizontal diameter in each photograph, have a low contrast in comparison with other fringes. This effect is more clearly expressed in Figure 2.32(b) since the value of the normal displacement component d_1 for the specimen with thickness $h = 9$ mm exceeds that of the specimen with thickness $h = 6$ mm by 1.5 times. The horizontal axis on the interferograms (x_2-axis) is the tension direction, and the left-hand side of the photographs corresponds to the unmoved parts of the specimens. Therefore, the difference of the axial displacement component d_2 at the left-hand point of the intersection of the hole edge and x_2-axis and the right-hand one is, as follows from the corresponding data of Figure 2.26, about $10\,\mu m$. The maximum value of the normal displacement component d_1 at the above-mentioned points are $1\,\mu m$ for the specimen of thickness $h = 6$ mm and $1.5\,\mu m$ for that of $h = 9$ mm.

It is important to point out that the technical implementation of the above-described approach to the normal displacement component determination based on a reflection hologram recording and a single fringe pattern reconstruction using an optical system (Figure 2.19) is too complex. Moreover, the fringe patterns thus obtained are capable of providing an estimation of the maximum value of the normal displacement component d_1 only. The approach based on the equation system (2.32) solution should be used for a more accurate and detailed description of the d_1 component distribution (see curves 3 and 7 in Figure 2.26).

The examples presented in this section confirm the high reliability and accuracy of the displacement components measurement procedure based on the use of optimal interferometer systems. Good results are achievable in a wide range of variations of the normal component value to the tangential one ratio. It should be noted that this range includes practically all possible measurement situations.

References

1. Erf, R. F. (ed.) (1974) *Holographic Nondestructive Testing*. Academic Press, New York.
2. Vest, C. M. (1979) *Holographic Interferometry*. Wiley, New York.
3. Schuman, W., Dubas, M. (1979) *Holographic Interferometry*. Springer Ser. Opt. Sci., 16. Springer-Verlag, Berlin.
4. Ostrovsky, Y. I., Butusov, M. M. and Ostrovskaya, G. V. (1980) *Interferometry by Holography*. Springer Ser. Opt. Sci., 20. Springer-Verlag, Berlin.
5. Abramson, N. (1981) *The Making and Evaluation of Holograms*. Academic Press, London.
6. Jones, R. J. and Wykes, C. (1983) *Holographic and Speckle Interferometry*. Cambridge University Press, Cambridge.
7. Schumann, W., Zurcher, J. P. and Cuche, D. (1985) *Holography and Deformation Analysis*. Springer Ser. Opt. Sci., 46. Springer-Verlag, Berlin.
8. Ostrovsky, Y. I., Shchepinov, V. P. and Yakovlev, V. V. (1991) *Holographic Interferometry in Experimental Mechanics*. Springer Ser. Opt. Sci., 60. Springer-Verlag, Berlin.
9. Rastogi, P. K. (ed.) (1994) *Holographic Interferometry*. Springer Ser. Opt. Sci., 68. Springer-Verlag, Berlin.

10. Iwata, K. and Nagata, R. (1979) Fringe formation in multiple-exposure holographic interferometry. *Opt. Acta* **26**, 995–1007.
11. Abramson, N. (1974) Sandwich hologram interferometry: a new dimension in holographic comparison. *Appl. Opt.* **13**, 2019–2025.
12. Odintsev, I. N., Shchepinov, V. P. and Yakovlev, V. V. (1988) Interferometric comparison of light waves, recorded on different holograms. *Zhur. Tekh. Fiz.* **58**, 990–991.
13. Stetson, K. A. and Powell, R. L. (1966) Hologram interferometry. *J. Opt. Soc. Amer.* **56**, 1161–1166.
14. Maclead, N. and Kapur, D. N. (1973) A kinematically designed mount for the precise location of specimens for holographic interferometry. *J. Phys. E: Sci. Instrum.* **6**, 423–424.
15. Bjelkhagen, H. I. (1993) *Silver-halide Recording Materials for Holography.* Springer Ser. Opt. Sci., 66. Springer-Verlag, Berlin.
16. Petrov, M. P., Stepanov, S. I. and Khomenko, A. V. (1991) *Photorefractive Crystals in Coherent Optical Systems.* Springer Ser. Opt. Sci., vol.59. Springer, Berlin, Heidelberg.
17. Powell, R. L. and Stetson, K. A. (1965) Interferometric analysis by wavefront reconstruction. *J. Opt. Soc. Amer.* **55**, 1593–1598.
18. Stetson, K. A. (1974) Fringe interpretation for hologram interferometry of rigid body motion and homogeneous deformation. *J. Opt. Soc. Amer.* **64**, 1–10.
19. Aleksandrov, E. B. and Bonch-Bruevich, A. M. (1967) Investigation of bodies surface deformations by hologram technique. *Zhur. Tekh. Fiz.* **37**, 360–369.
20. Ennos, A. E. (1968) Measurement of in-plane surface strain by hologram interferometry. *J. Phys. E: Sci. Instrum.* **1**, 731–734.
21. Sollid, J. E. (1969) Holographic interferometry applied to measurement of small static displacement of diffusely reflecting surfaces. *Appl. Opt.* **8**, 1587–1595.
22. Kopf, U. (1973) Fringe order determination and zero motion fringe identification in holographic displacement measurement. *Opt. Laser Technol.* **5**, 111–113.
23. Pryputnievicz, R. J. (1994) Quantitative determination of displacements and strains. In Rastogi, P. K. (ed.), *Holographic Interferometry*, pp. 33–74, Springer Ser. Opt. Sci., 68. Springer-Verlag, Berlin.
24. Abramson, N. (1972) The holo-diagram V: A device for practical interpreting of hologram interference fringes. *Appl. Opt.* **11**, 1143–1147.
25. Dhir, S. K. and Sikora, J. P. (1972) An improved method for obtaining general displacement field from a holographic interferogram. *Exp. Mech.* **12**, 323–327.
26. Borynyak, L. A., Gerasimov, S. I. and Zhilkin, V. A. (1982) Experimental techniques of interferogram recording and interpretation providing the required accuracy in strain tensor component determination. *Avtometriya* **N1**, 17–24.
27. Ennos, A. E. and Virdee, M. S. (1982) Application of reflection holography to deformation measurement problems. *Exp. Mech.* **22**, 202–209.
28. Wesolowski, P. (1985) Some aspects of numerical in-plane strain analysis by reflection holography. *Optik* **71**, 113–118.
29. Schibayama, K. and Uchiyama, H. (1971) Measurement of three-dimensional displacements by hologram interferometry. *Appl. Opt.* **10**, 2150–2154.
30. Voevodin, V. V. (1977) *Computational Basics of Linear Algebra.* Nauka, Moscow.
31. Faddeev, D. K. and Faddeeva, V. N. (1963) *Computational Methods of Linear Algebra.* Fizmatgiz, Moscow.
32. Nalimov, V. V. and Golicova, T. I. (1980) *Logical Foundations for Experiment Design.* Metallurgiya, Moscow.

33. Nobis, D. and Vest, C. M. (1978) Statistical analysis of error in holographic interferometry. *Appl. Opt.* **17**, 2198–2204.
34. Ek, L. and Biedermann, K. (1977) Analysis of systems for hologram interferometry with a continuosly scanning reconstruction beam. *Appl. Opt.* **16**, 2235–2545.
35. Lisin, O. G. (1981) About three-dimensional displacements measurements accuracy for diffuse objects by holographic interferogram data. *Opt. Spektrosk.* **50**, 521–531.
36. Osten, W. (1985) Some consideration on the statistical error analysis in holographic interferometry with application to an optimazed interferometer. *Opt. Acta* **32**, 827–838.
37. Matsumoto, T., Iwata K. and Nagata, R. (1973) Measuring accuracy of threedimensional displacements in holographic interferometry. *Appl. Opt.* **12**, 961–967.
38. Erler, K., Wenke, L. and Shreiber, W. (1980) Berechnung von Hologramminterferometern bestmoglicher Konditionerung. *Feingeratetechnik* **29**, 510–514.
39. Kohler, H. (1982) Holografische interferometrie: VI. Aspekte der Auswertungs-Optimierung. *Optik* **62**, 413–423.
40. Wernike, G. and Osten, W. (1982) *Holografische Interferometrie.* VEB Fachbuehverlag, Leipzig.
41. Indisov, V. O., Pisarev, V. S., Shchepinov, V. P. and Yakovlev, V. V. (1983) A comparison of holographic interferogram interpretation techniques with absolute and reference orders for deformation measurement. In *Deformation and Fracture of Materials and Structures in Nuclear Equipment*, pp. 45–54, Energoatomizdat, Moscow.
42. Indisov, V. O., Pisarev, V. S., Shchepinov, V. P. and Yakovlev, V. V. Application of interferometer based on reflection hologram for local deformation study. *Zhur. Tekh. Fiz.* **56**, 701–707.
43. Pisarev, V. S., Shchepinov, V. P. and Yakovlev, V. V. (1987) Optimized holographic interferometers for strain measurement by fringe pattern interpretation with reference orders. *Izmerit. Tekhn.* **3**, 13–15.
44. Pisarev, V. S., Shchepinov, V. P. and Yakovlev, V. V. (1987) Optimized holographic interferometers for strain measurement by fringe pattern interpretation with absolute orders. *Izmerit. Tekhn.* **10**, 23–25.
45. Balalov, V. V., Pisarev, V. S., Shchepinov, V. P. and Yakovlev, V. V. (1990) Holographic interference measurements of three-dimensional displacement fields and their use for stress determination. *Opt. Spektrosk.* **68**, 134–139.
46. Zhilkin, V. A., Gerasimov, S. I. and Sarnadsky, V. N. (1987) Accurracy of displacement measurements by overlay interferometer estimation. *Opt. Spektrosk.* **62**, 1385–1389.
47. Forsythe, G. and Moler, C. B. (1967) *Computer Solution of Linear Algebraic Systems.* Prentice-Hall, Englewood Cliffs, NJ.
48. Burmin, V. Y. (1976) The problem of experiment planning and conditioning of linear algebraic equations systems. *Izvest. AN SSSR, Tekhn. Kibern.* **N2**, 195–200.
49. Day, J. D. (1978) Errors in the computation of linear algebraic systems. *Int. J. Mathem. Educ. Sci. Technol.* **9**, 89–95.
50. Pisarev, V. S., Yakovlev, V. V., Indisov, V. O. and Shchepinov V. P. (1983) Design of experiment for strain determination by holographic interferometry. *Zhur. Tekh. Fiz.* **53**, 292–300.
51. Matsumoto, T., Iwata, K., Nagata, R. (1973) Simplified explanation of fringe formation in deformation measurements by holographic interferometry. *Bull. Univ. Osaka Prefect* **A22**, 101–109.

52. Zhilkin, V. A. (1981) Optical interference methods in strain studies. *Zavod. Labor.* **10**, 57–63.
53. Zhilkin, V. A. and Gerasimov, S. I. (1982) On a possibility of deformed state studying by overlay interferometer. *Zhur. Tekhn. Fiz.* **52**, 2072–2085.
54. Gorodnichenko, V. I. and Pisarev, V. S. (1990) Some lows of the initial stage of local cyclic deformation obtained by holographic interferometry. *Fiz.-Khim. Mekh. Mater.* (Soviet Material Science) **N5**, 106–113.
55. Boone, P. M. (1975) Use of reflection holograms in holographic interferometry and speckle correlation for measurement of surface displacement. *Opt. Acta* **22**, 579–589.
56. Newmann, D. B. and Penn, R. C. (1975) Off-table holography. *Exp. Mech.* **15**, 241–244.
57. Gorodnichenko, V. I., Grishin, V. I. and Pisarev, V. S. (1989) Three-dimensional strain state near the hole in strip subjected to tension by data of experimental, numerical and combined techniques. *Proc. of TsAGI* **20**, 67–75.
58. Peterson, R. E. (1974) *Stress Concentration Factors. Charts and relations useful in making strength calculations for machine parts.* John Wiley, New York.
59. De, Laminat P. M. and Wei, R. P. (1978) Normal surface displacement around a circular hole by reflection holographic interferometry. *Exp. Mech.* **18**, 74–80.

3 DISPLACEMENT MEASUREMENTS BY SPECKLE INTERFEROMETRY

An implementation of the speckle effect for determination of different parameters characterizing the deformation of an object under study is based on the fact that the resultant speckle pattern existing in the scattered light field or observed at some image plane of a recording system is capable of being a sum of two or more speckle fields. Such a superposition of speckle fields can be accomplished in two ways.

The first consists of a coherent superposition in the recording plane of a complex amplitude of simultaneously existing light waves. In this case, as a rule, an optical arrangement corresponding to some classical optical interferometer systems, for instance a Michelson interferometer, is used to form a resultant speckle pattern. The methods based on this principle can be classified as correlation speckle interferometry and speckle shearing interferometry (see Section 4.4).

The second way is established on the superposition of intensities of scattered light waves when a resultant speckle pattern is formed by speckle fields recorded at different time instants. In this case a recording procedure represents a conventional photographic recording of an optically rough surface illuminated with laser light. That is why this technique is called the speckle photography method.

A brief description of a fringe formation in the method of correlation speckle interferometry is presented in the first section of this chapter. The method of laser speckle photography most widely used in the field of strain analysis of various structures is considered in detail later. Much attention is given to establishing the range of practical implementation of this method. In many practical cases different types of rigid body motion or deformation of the structure to be investigated may result in decorrelation of speckle patterns forming a resultant speckle pattern. This fact, due to a degradation of fringe pattern contrast leads, in turn, to considerable errors made in displacement or slope component determination. An influence of different factors on the contrast of fringe patterns obtained by speckle photography technique is fully illustrated.

The unconventional technique of simultaneous recording of two speckle fields on a single hologram and the combined application of the image plane holography with speckle patterns recording, having a powerful capability of practical implementation to strain analysis, are also described in this chapter. The white light speckle photography technique is discussed at the end of this chapter.

In conclusion, it should be noted that the contents of this chapter are the results of our experience in the field of experimental strain analysis by means of the speckle photography method. The detailed description of various approaches to displacement and slope measurement based on correlation speckle interferometry, speckle photography, and speckle shear interferometry techniques is contained in the following monographs: Erf [1], Jones and Wykes [2], Dainty [3] and Sirohi [4].

3.1   Correlation speckle interferometry

The so-called correlation speckle interferometry methods are founded upon a coherent superposition of speckle fields formed by light waves scattered from an optically rough surface of the object under study, and the reference wave. A plane or spherical wave as well as other speckle fields can serve as a reference wave. Presence of the reference wave during recording of the resultant speckle pattern causes the intensity distribution in the image plane of the optical system to be dependent on the relative phase shift of the superimposing speckle fields. A deformation of the object surface in this case will result in a variation in the intensity distribution of the resultant speckle pattern.

Consider the speckle interferometer shown in Figure 3.1. The plane wave is split into two components by the beam splitter BS. The wave reflected by the beam splitter BS front illuminates the optically rough surface of the object O. The wave passing through the beam splitter is reflected by the mirror M onto the photographic plate FP and serves as the reference wave. The lens L constructs the focused image of the object surface in the plane of the photoplate FP. Therefore the wave scattered from the rough surface O can be considered as the object wave.

The complex amplitude A_0 of the object wave in the plane of the photographic plate FP can be written as

$$A_0 = a_0 \exp(-i\phi_1)$$

where a_0 and ϕ_1 are the amplitude and phase of the wave scattered by the object in its original state, which represent the random functions of the spatial

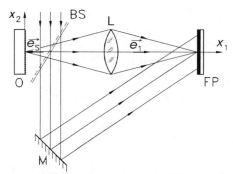

Figure 3.1. Optical arrangement of a speckle interferometer for measuring of the normal to the object surface displacement component

coordinates. The analogous complex amplitude A_r of the reference wave has the form

$$A_r = a_r \exp(-i\phi_2)$$

where a_r and ϕ_2 are the amplitude and phase of the reference wave.

Now we can represent the intensity distribution in the image plane where the photoplate is placed:

$$I_1 = |A_0 + A_r|^2 = a_0^2 + a_r^2 + 2a_0 a_r \cos\varphi \qquad (3.1)$$

where $\varphi = \phi_1 - \phi_2$ is the random value. Expression (3.1) describes the initial speckle structure which corresponds to the unstrained or undisplaced state of the object surface.

When the object surface undergoes stressing or moving, the corresponding intensity distribution, which is identified by I_2, is given by

$$I_2 = a_2^2 + a_r^2 + 2a_0 a_r \cos(\varphi + \delta) \qquad (3.2)$$

where δ is the strain-induced or displacement induced phase change of the object wave in the photoplate (image) plane.

Now let's introduce the correlation function $\rho(\delta)$ of two random functions I_1 (3.1) and I_2 (3.2) [3]:

$$\rho(\delta) = \frac{(<I_1 I_2> - <I_1><I_2>)}{(<I_1^2> - <I_1>^2)^{\frac{1}{2}}(<I_2^2> - <I>^2)^{\frac{1}{2}}} \qquad (3.3)$$

where $<\ >$ denotes an averaging over many points in the scattered field. Substituting equations (3.1) and (3.2) into equation (3.3), introducing an assumption that a_0^2, a_r^2 and φ are independent variables and can thus be averaged separately and that $<I_1> = <I_2> = <I>$, and taking into account that $<\cos\varphi> = <\cos(\varphi+\delta)> = 0$ and $<I^2> = 2<I>^2$, yields

$$\rho(\delta) = \frac{1+\cos\delta}{2}. \qquad (3.4)$$

Thus the correlation between the intensities I_1 and I_2 is unity when

$$\delta = 2\pi n_c \qquad (3.5)$$

and zero when

$$\delta = (2n_c + 1)\pi \qquad (3.6)$$

where n_c is the integer number.

As follows from equations (3.5) and (3.6) variations in the correlation functions of speckle patterns can be made to appear as a fringe pattern. These fringe patterns are usually called speckle correlation fringes or correlation fringes. The method of the displacement measurement based on the use of these fringes is called correlation speckle interferometry.

There are two methods for observing correlation fringes. The first of these techniques was proposed by Leendertz [5] and consists of a photographic recording of the initial speckle pattern of the object surface image on a photoplate. Then the developed negative must be accurately replaced in its original position. The deformed object should be observed through the photoplate. Over the areas of maximum correlation of speckle patterns I_1 (3.1) and I_2 (3.2) a minimum transmission of light scattered by the deformed object surface will occur. Therefore the variation in correlation is shown as a real-time correlation fringe pattern caused by a variation in the transmission of the light through the negative. The visibility of the fringes thus obtained is, however, rather low and does not exceed the value of 1/3 [2].

The second way of viewing speckle correlation fringe patterns is based on two-exposure recording and spatial filtering of the resultant speckle field in the Fourier plane [6]. The speckle patterns of two images of the object surface in its original and deformed state are recorded on the same photoplate FP (Figure 3.1), but a special lateral image shift has to be made between exposures. This procedure can be performed by the calibrated rigid translation of the photoplate FP in its own plane. Thus, two slightly different speckle patterns are recorded on the same photoplate by the two-exposure technique.

If now the double-exposure photoplate FP is placed on the optical arrangement which is capable of carrying out the Fourier spatial filtering procedure of the transmitted light wave, as shown in Figure 3.2, then the fringe pattern representing a system of the uniformly spaced straight lines will be observed in the focal Fourier plane of the lens. This fringe pattern is caused by

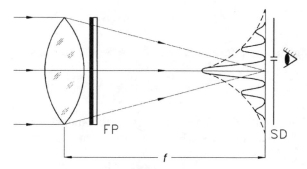

Figure 3.2. Schematic diagram of a Fourier-filtering set up for observing a speckle correlation fringe pattern

a uniform image shift that must be sufficient to form at least two fringes in the diffraction halo. Evidently, the photoplates fragments where the correlation of the speckle patterns achieves its maximum values will form the above-described fringes and the regions of minimum correlation will not form a fringe pattern in the diffraction halo. The latter areas will result in a diffraction halo appearance. A dashed line in Figure 3.2 shows the intensity distribution in this halo. If now a slit aperture is located at the centre of any dark fringes, no light is transmitted through the aperture from areas of maximum correlation. In this case the image formed by the lens in the Fourier plane will be covered with a fringe pattern, in which the intensity maxima (or bright fringes) correspond to the lines of minimum correlation, i.e. where condition (3.6) is valid.

Using the principal relation of the phase difference determination (2.24) and taking into account the orientations of the unit vectors $\vec{e}_1$ and $\vec{e}_s$, it is possible to show that the optical system of correlation interferometer presented in Figure 3.1 is capable of measuring the out-of-plane displacement component d_1 only:

$$d_1 = \frac{\lambda(2n_c + 1)}{4}. \tag{3.7}$$

Figure 3.3 represents the optical arrangement of the speckle correlation interferometer which is sensitive with respect to the in-plane displacement component d_2 only [5]. In this optical system the object surface, which lies in the (x_1, x_2) plane, is illuminated by two plane waves inclined at equal and opposite angles θ with respect to the surface normal which coincides with the x_1-axis. A speckle pattern of the object surface image, which is recorded on the photoplate, is formed by the coherent superposition of two speckle fields resulting from two illuminating wavefronts scattering. The phase difference for each illuminating beam $\delta_i(i = 1,2)$ can be derived from equation (2.24) using the notation of Figure 3.3:

$$\delta_1 = \frac{2\pi}{\lambda}(\vec{e}_1 - \vec{e}_{S1})\vec{d}, \quad \delta_2 = \frac{2\pi}{\lambda}(\vec{e}_1 - \vec{e}_{S2})\vec{d}$$

Thus the change of the phase difference $\delta = \delta_1 - \delta_2$ can be written as

$$\delta = \frac{4\pi}{\lambda} d_2 \sin \theta. \tag{3.8}$$

As follows from equation (3.8), the fringes of the speckle correlation in this case represent the lines of the levels of the displacement components d_2 equal values:

$$d_2 = \frac{n_c \lambda}{2 \sin \theta}. \tag{3.9}$$

Sirohi [4] developed an optical arrangement of correlation speckle interferometer with a special five-hole mask placed before the image-forming lens, that allows us to measure all three displacement components d_1, d_2 and d_3.

Apart from the method of holographic interferometry, the method of correlation speckle interferometry is capable of recording fringe patterns which characterize the in-plane displacement components only without being influenced by the out-of-plane displacement component. The sensitivity threshold of the correlation speckle interferometry techniques with respect to the in-plane displacement components measurement can vary from $0.3\,\mu m$ to $10\,\mu m$. The analogous characteristic for the out-of-plane displacement component lies in the range 0.3–$30\,\mu m$.

Moreover, the recording medium which would be used to realize the correlation techniques does not need to have a high spatial resolution compared with that required for holographic interferometric techniques. This results from the fact that in the former case the speckle pattern of the object surface image, which is formed by individual speckles of average size approximately 5–$80\,\mu m$, must be determined. The above-mentioned range of the minimum speckle size to be determined shows that a standard TV camera

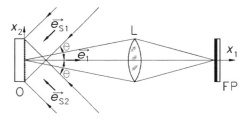

Figure 3.3. Optical arrangement of a speckle interferometer for measuring the in-plane displacement component

may be used to record the correlation fringe patterns. Thus, a video recording procedure can be applied to generate correlation fringes equivalent to those obtained photographically. The intensity correlation in this case is observed by means of the process of video signal substraction or addition. This method is known as electronic speckle pattern interferometry (ESPI)—see, for example, Jones and Wykes [2] and Wykes [7].

3.2 Recording of double-exposure specklegrams

The method of double-exposure speckle photography developed by Archbold *et al.* [8] and Archbold and Ennos [9] is today the most widely used technique for in-plane displacement component measurement based on the speckle effect. The principles of this method are based on a sequential recording of two speckle patterns of the object surface image on a single photographic plate. If one of these speckle patterns is related to the unstrained (or undisplaced) state of the object surface and the other corresponds to the strained object surface, the intensity distribution of the superimposed speckle pattern in the image plane can be considered as an intensity sum of two speckle patterns, one of which is displaced with respect to the other by an amount depending on the displacement value and the position of the recording plane. By using several methods of spatial filtering of a light wave (transmitted through the double-exposure specklegram) the intensity variation of the above-mentioned superimposed speckle pattern can be converted into an interference fringe pattern that allows the in-plane displacement component to be determined.

Let's consider the procedure of double-exposure specklegram recording in detail, beginning from the optical system shown in Figure 3.4. The lens L of the focal length f_0 and numerical aperture $F=f_0/D$ (where D is the diameter of the lens pupil) forms in the plane B the image of the object surface O, which is illuminated by a divergent coherent wavefront emerging from the point

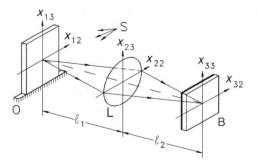

Figure 3.4. Schematic configuration of the speckle photography technique

source S. According to the principles of the geometrical optics, the distance l_1 between the object surface and the lens and the distance l_2 between the lens and the recording plane are connected with the focal length f_0 through the relationship

$$l_1^{-1} + l_2^{-1} = f_0^{-1}. \tag{3.10}$$

If we designate $A_0(x_{12}, x_{13})$ the complex amplitude of the illuminating wavefront near the object surface, the complex amplitude $A_B(x_{32}, x_{33})$ of the light field in the recording plane can be expressed in the following form [10]:

$$A_B(x_{32}, x_{33}) = \{[A_0(x_{12}, x_{13}) * h_1(x_{12}, x_{13})]$$
$$\cdot \exp\left[-i\frac{\pi}{f_0\lambda}(x_{22}^2 + x_{23}^2)\right] P_L(x_{22}, x_{23})\} * h_2(x_{22}, x_{23}) \tag{3.11}$$

where

$$h_j(x_{j2}, x_{j3}) = \frac{\exp\left(i\dfrac{2\pi}{\lambda} x_{j1}\right)}{i\lambda x_{j1}} \exp\left[i\pi\frac{x_{j2}^2 + x_{j3}^2}{h_{ji}\lambda}\right] \tag{3.12}$$

is the impulse response of the free space ($j = 1, 2$); $i^2 = -1$.

$$\exp\left[-\frac{i\pi}{f_0\lambda}(x_{22}^2 + x_{22}^2)\right]$$

is the transmittance function of a thin lens;

$$P_L(x_{22}, x_{23}) = \begin{cases} 1 \text{ when } (x_{22}^2 + x_{23}^2) \leqslant \dfrac{D}{2} \\ 0 \text{ otherwise} \end{cases}$$

represents the aperture function of the lens; * is the sign which denotes the convolution. Transforming equation (3.11) in the shown consequence yields

$$A_B(x_{32}, x_{33}) \propto \exp\left[\frac{i\pi}{l_1\lambda}(x_{32}^2 + x_{33}^2)\right]\left\{A_0\left(\frac{x_{32}}{M}, \frac{x_{33}}{M}\right)\right.$$
$$\left. \cdot \exp\left[\frac{i\pi}{Ml_2\lambda}(x_{32}^2 + x_{33}^2)\right]\right\} * F_p(x_{32}, x_{33}) \tag{3.13}$$

where

$$F_p(x_{32}, x_{33}) = \iint_{-\infty}^{\infty} P_L(x_{22}, x_{32}).\exp\left[-\frac{2\pi i}{l_2\lambda}(x_{22}x_{32} + x_{23}x_{33})\right]dx_{22}dx_{23} \quad (3.14)$$

is the Fourier transform of the lens aperture function $P_L(x_{22}, x_{23})$; and M is the magnification of the recording optical system.

The intensity distribution of the light field over the object surface image, which can be called the initial speckle pattern, can be defined as a square of the module of the complex amplitude:

$$I_B(x_{32}, x_{33}) = |A_B(x_{32}, x_{33})|^2 \propto \left|\left\{A_0\left(\frac{x_{32}}{M}, \frac{x_{33}}{M}\right)\right.\right.$$

$$\left.\left.\exp\left[\frac{i\pi}{Ml_2\lambda}(x_{32}^2 + x_{33}^2)\right]\right\} * F_p(x_{32}, x_{33})\right|^2.$$

Taking into account the random character of the complex amplitude $A_0(x_{12}, x_{13})$ distribution and averaging over many points in the scattered field, one can obtain [11]

$$I_B(x_{32}, x_{33}) \propto \left|A_0\left(\frac{x_{32}}{M}, \frac{x_{33}}{M}\right)\right|^2 * |F_p(x_{32}, x_{33})|^2. \quad (3.15)$$

When the object under investigation is deformed so that some points of its surface are displaced by a sufficient amount exceeding an average speckle size, the complex amplitude of the scattered light field changes. In this case the complex amplitude of the scattered light near the strained object surface can be described by the function $A_0(x_{12} + u, x_{13} + v)$, where u and v are the in-plane components of the displacement vector $\vec{d_\tau}$, which represents a function of two coordinates x_{12} and x_{13}. As the specklegram technique can be used for measurement of displacement components whose values exceed an average speckle size, it is evident that this method has a high sensitivity with respect to tangential to the object surface displacement vector components only. Therefore, further consideration will be restricted to the case when u, $v \gg w$, where w is the normal to the object surface displacement vector component. The expression of the complex amplitude $A_B(x_{32} + u, x_{33} + v)$ in the recording plane can obtained from equation (3.11) by analogy:

$$A_0(x_{32} + u, x_{33} + v) = \{[A_0(x_{12} + u, x_{13} + v) * h_1(x_{12}, x_{13})]$$

$$\cdot \exp\left[-\frac{i\pi}{f_0\lambda}(x_{22}^2 + x_{23}^2)\right]P_L(x_{22}, x_{23})\} * h_2(x_{22}, x_{23}). \quad (3.16)$$

The corresponding (3.16) intensity distribution in the recording plane $I_B(x_{32} + u, x_{33} + v)$ can be derived the same way as equation (3.15) and has the form

$$I_B(x_{32} + u, x_{33} + v) \propto \left| A_0\left(\frac{x_{32} + u}{M}, \frac{x_{32} + v}{M}\right) \right|^2 * |F_p(x_{32}, x_{33})|^2. \qquad (3.17)$$

A total exposure of the photoplate for equal duration of each from two parts of the speckle pattern recording procedure is given by

$$E(x_{32}, x_{33}) = \tau[I_B(x_{32}, x_{33}) + I_B(x_{32} + u, x_{33} + v)]. \qquad (3.18)$$

As follows from equation (1.3), after photographic development, the double-exposures specklegram will have the properties of a transparency whose amplitude transmittance function $T(x_{32}, x_{33})$ is proportional to exposure (3.18):

$$T(x_{32}, x_{33}) = b_0 + b_1\tau[I_B(x_{32}, x_{33}) + I_B(x_{32} + u, x_{33} + v)]. \qquad (3.19)$$

Expression (3.19) illustrates that when the resolution of the recording medium used is sufficient in order to separate an average size of an individual speckle, two almost identical but slightly displaced speckle patterns will be recorded on the photoplate together with the object surface image.

3.3 Observation of fringe patterns and displacement determination in the speckle photography method

Determination of displacement components of object surface points can be made by several optical techniques which require spatial filtering of the light wave passed through the double-exposure negative whose amplitude transmittance is characterized by function (3.19). These techniques enable the Fourier transform of the monochromatic wavefront to be displayed as a variation of the intensity of the speckle pattern in the observation plane.

Let's consider a classical set-up of an optical filtering of the spatial frequencies in the focal plane of a lens (see Figure 3.5). The plane light wave passing from the input plane (x_{02}, x_{03}) where the specklegram is placed is transformed by the lens L_1 into its own Fourier-transform (spatial spectrum) in the output plane (x_{F2}, x_{F2}) which is located on the focal distance f_1 from the principal plane of the lens L_1. This plane is known as the Fourier plane or plane of spatial frequencies. The lens L_2 of focal length f_2, the front focal plane of which coincides with the back focal plane (x_{F2}, x_{F3}) of the lens L_1, performs the inverse Fourier transform and forms in its back focal plane the inverse image of the illuminated object surface superimposed with a speckle pattern and scaled by the factor f_2/f_1.

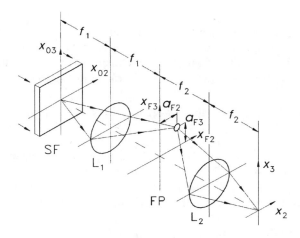

Figure 3.5. Schematic diagram of a whole-field Fourier-filtering arrangement

An optical filtering of the spatial frequencies can be carried out by using a non-transparent mask placed in the Fourier plane of the lens L_1 (see Figure 3.5) which contains a small aperture. This mask blocks the whole set of plane waves of different spatial frequencies (i.e. travelling at different angles) with the exception of a narrow beam passing through this aperture. If a filtering aperture is located at the point with coordinates a_{F2}, a_{F3}, the image formed in the output plane of the lens L_2 will consist of the components of the spatial spectrum with frequencies $a_{F2}/\lambda f_1$ and $a_{F3}/\lambda f_1$.

If the double-exposure specklegram whose amplitude transmittance is described by function (3.19) is placed in the input plane (x_{02}, x_{03}) of the lens L_1 and illuminated by a plane wavefront, the complex amplitude immediately behind the transparency can be written as

$$A(x_{02}, x_{03}) \propto \tau[I_B(x_{32}, x_{33}) + I_B(x_{32} + u, x_{33} + v)]. \tag{3.20}$$

The complex amplitude $A_F(x_{F2}, x_{F3})$ of the light wave in the back focal plane of the lens L_1 (Fourier plane) can be derived by a Fourier transformation of function (3.20):

$$A_F(x_{F2}, x_{F3}) \propto \tau \iint_{-\infty}^{\infty} [I_B(x_{02}, x_{03}) + I_B(x_{02} + u, x_{03} + v)]$$

$$.\exp\left[-\frac{i2\pi}{\lambda f_1}(x_{F2}x_{02} + x_{F3}x_{03})\right] dx_{02}\, dx_{03}. \tag{3.21}$$

Assuming that the displacement components $u(x_{02}, x_{03})$ and $v(x_{02}, x_{03})$ are the continuous functions of the Cartesian coordinates x_{02} and x_{03}, it is possible to use the main properties of the Fourier transform and represent the complex amplitude of the light wave in the Fourier plane in the following form:

$$A_F(x_{F2}, x_{F3}) \propto \tau \left\{ 1 + \exp\left[\frac{i2\pi}{\lambda f_1 M} (u x_{F2} + v x_{F3}) \right] \right\}.A_{F0}(x_{F2}, x_{F3}) \qquad (3.22)$$

where $M = l_2/l_1$ is the magnification of the optical arrangement which was used for recording the speckle patterns (see Figure 3.4) and $A_{F0}(x_{F2}, x_{F3})$ is the complex amplitude of the light wave in the Fourier plane diffracted by the initial speckle pattern which corresponds to the unstrained (or undisplaced) object surface.

The intensity distribution $I_F(x_{F2}, x_{F3})$ in the plane of the spatial frequencies (Fourier plane) can be obtained as a square of the module of the complex amplitude (3.22):

$$I_F(x_{F2}, x_{F3}) = |A_F(x_{F2}, x_{F3})|^2$$
$$\propto 4\tau^2 \cos\left[\frac{\pi}{f_1 \lambda} M(u x_{F2} + v x_{F3}) \right] I_{0F}(x_{F2}, x_{F3}) \qquad (3.23)$$

where

$$I_{0F}(x_{F2}, x_{F3})$$
$$\propto \left| \iint_{-\infty}^{\infty} |A_B(x_{02}, x_{03})|^2 \exp\left[\frac{i2\pi}{\lambda f_1} (x_{02} x_{F2} + x_{03} x_{F3}) \right] dx_{02}, dx_{03} \right|^2. \qquad (3.24)$$

The integral expression inside the square module of equation (3.24) represents the square of the autocorrelation of the aperture function. This function for a circular aperture can be expressed in the explicit form [12]

$$I_{0F}(x_{F2}, x_{F3}) \propto \left\{ \cos^{-1}\left(\frac{r_F l_1}{f_1 D} \right) - \frac{r_F l_1}{f_1 D} \sqrt{1 - \left(\frac{r_F l_1}{f_1 D} \right)^2} \right\}^2 \qquad (3.25)$$

where $r_F = \sqrt{x_{F2}^2 + x_{F3}^2}$ is the distance between the arbitrary point at the Fourier plane and an optical axis of the spatial filtering system and l_1 is a distance between the object surface and the objective when the speckle patterns are being recorded (see Figure 3.4).

As mentioned above, the lens L_2 (see Figure 3.5) make the inverse Fourier transform of the complex amplitude of the light wave (3.22). If the spatial frequencies are not blocked in the Fourier plane of the lens L_1, the interference fringes cannot be observed in the surface image which is formed in the back focal plane of the lens L_2, due to an averaging of the harmonic factor in expression (3.23).

The Fourier filtering procedure described above denotes the fact that the harmonic factor in expression (3.23) can be considered as a function of two independent variables u and v only, which represent the in-plane displacement components. These displacement components are not random functions and hence a form of the harmonic factor depends on the spatial coordinates of the centre of filtering aperture a_{F2} and a_{F3}. Evidently, the value and direction of the radius vector $\vec{r}_F = (a_{F2}, a_{F3})$ defines the sensitivity of the optical system used with respect to the displacement components. The second factor $I_{0F}(x_{F2}, x_{F3})$ in equation (3.23) is assumed to be constant within the small aperture area and describes an image brightness in the output plane (x_{02}, x_{03}). Therefore the object surface image in the output plane of the optical Fourier filtering arrangement shown in Figure 3.5 is modulated by a square cosine function that results in a fringe pattern formation.

The bright interference fringes occur when

$$\vec{d}_\tau \vec{r}_F = \frac{\lambda f_1}{M} n \qquad (3.26)$$

and for the dark fringes appearance the following condition has to be met:

$$\vec{d}_\tau \vec{r}_F = \frac{\lambda f_1}{M} \left(n + \frac{1}{2}\right) \qquad (3.27)$$

where $n = 0, \pm 1, \pm 2, \ldots$, is the number of the interference fringe and $\vec{d}_\tau = (u, v)$ is the total in-plane component of the displacement vector.

By sequentially locating the filtering aperture on two coordinate axes at the positions $\vec{r}_F = (a_{F2}, 0)$ and $\vec{r}_F = (0, a_{F3})$, we can obtain two independent fringe patterns which depict the in-plane displacement components u and v, respectively. The corresponding relations for the displacement components determination through the use of the bright fringes have the form

$$u = \frac{\lambda f_1}{M a_{F2}} n_{x2} \qquad v = \frac{\lambda f_1}{M a_{F3}} n_{x3} \qquad (3.28)$$

To make the same procedure using the dark fringes the following expressions should be applied:

$$u = \frac{\lambda f_1}{M a_{F2}} \left(n_{x2} + \frac{1}{2}\right), \qquad v = \frac{\lambda f_1}{M a_{F3}} \left(n_{x3} + \frac{1}{2}\right) \qquad (3.29)$$

where $n_{x2} = 0, \pm 1, \pm 2, \ldots$ and $n_{x3} = 0, \pm 1, \pm 2, \ldots$ are the fringe numbers along the coordinate axes x_2 and x_3 respectively.

Thus, in order to determine the in-plane displacement components of some plane and optically rough surface by means of the Fourier spatial filtering procedure, it is necessary to sequentially place a small filtering aperture in two orthogonal axes in the Fourier plane of the lens. In this case a quantitative interpretation of the fringe patterns obtained should be performed by using expressions (3.28) or (3.29).

The optical arrangement shown in Figure 3.2, which is a more simplified variant of the Fourier spatial filtering procedure, is widely used in research. In this optical system a double-exposure specklegram is located immediately behind the lens L, which is able to perform a spatial filtering of a plane wavefront to within an accuracy of some phase factor. However, this phase shift of a complex amplitude of a filtered wave does not influence the intensity distribution in the Fourier plane.

Another way of optical spatial filtering which results in the display of an interference fringe pattern is the pointwise filtering of a double-exposure specklegram by an unexpanded laser beam (see Figure 3.6). A narrow laser beam LB is brought to a double-exposure specklegram at the point of the object surface image under investigation. If the diffraction image of this point, which exists in the space behind the photoplate, is projected onto a non-transparent or ground-glass screen mounted in some plane parallel to the plane of the photoplate, a bright circular spot will be observed on the screen surface. This spot is known as the diffraction halo and its diameter is inversely proportional to an average speckle size. The presence on the photoplate of two practically identical speckle patterns, but slightly shifted with respect to the other, results in the appearance of a fringe pattern which can be observed on the screen as a set of straight and uniformly spaced lines. Such a fringe pattern contains information about direction and value of the total in-plane component $\vec{d}_\tau$ of the displacement vector.

The complex amplitude $A_S(x_{S2}, x_{S3})$ of the light wave, which resulted from diffraction of the narrow laser beam on the double-exposure specklegram, in the screen plane x_{S2}, x_{S3} located at the distance l from the photoplate is proportional to a convolution of the amplitude transmittance function (3.19) with an impulse response $h_S(x_{S2}, x_{S3})$ in the form (3.12):

$$A_S(x_{S2}, x_{S3}) \propto t(x_2, x_3) * h_S(x_{S2}, x_{S3}). \tag{3.30}$$

Substituting equation (3.19) into (3.30) and multiplying the result obtained on its complex conjugated analogue yields an intensity distribution in the screen plane $I_S(x_{S2}, x_{S3})$:

$$I_S(x_{S2}, x_{S3}) = |A_S(x_{S2}, x_{S3})|^2$$

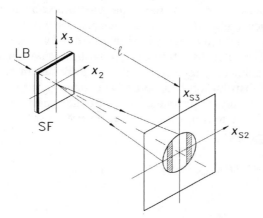

Figure 3.6. Schematic diagram of a pointwise filtering arrangement

$$\propto \tau^2 \left| \left\{ \left[\left| A_B\left(\frac{x_{32}}{M}, \frac{x_{33}}{M}\right) \right|^2 + \left| A_B\left(\frac{x_{32}+u}{M}, \frac{x_{33}+v}{M}\right) \right|^2 \right] \right. \right.$$
$$\left. \left. * |F_p(x_{32}, x_{33})|^2 \right\} * h_S(x_{S2}, x_{S3}) \right|^2 . \tag{3.31}$$

For the following analysis we must take advantage of a constancy over a small area illuminated with an unexpanded laser beam of the functions

$$\left| A_B\left(\frac{x_{32}}{M}, \frac{x_{33}}{M}\right) \right|^2 \text{ and } \left| A_B\left(\frac{x_{32}+u}{M}, \frac{x_{33}+v}{M}\right) \right|^2 .$$

Then, assuming that these functions are continuous and rather smoothed over the laser beam diameter, and taking into account the known properties of the Fourier transform, we have the following form of expression (3.31):

$$I_S(x_{S2}, x_{S3}) \propto \tau^2 [P_L(x_{32}, x_{33}) \otimes P_L(x_{32}, x_{33})]^2 \cos^2\left(\frac{\pi}{2l\lambda} M \vec{d}_\tau \vec{r}_s\right), \tag{3.32}$$

where $\otimes$ denotes autocorrelation; $r_S = (x_{S2}, x_{S3})$ is the radius vector of an arbitrary point in the plane of the screen; and l is the distance between the photoplate and screen.

The square root of the autocorrelation of the aperture function in square brackets in equation (3.32) describes the uniform diffraction image of the illuminated area of the double-exposure specklegram corresponding to the first exposure (the speckle pattern of the initial unstrained or undisplaced state of the object surface under study). The intensity distribution of this image is modulated by the harmonic factor having a form of a cosine square function. Expression (3.32) reveals that an intensity distribution within a diffraction halo range is mainly defined by the form of the aperture function. An explicit form of this intensity distribution can be derived for some geometrical form of a recording aperture only. In particular, for a circular aperture of the diameter D with uniform amplitude transmittance we can obtain [13]

$$I_{S0}(x_{S2}, x_{S3}) \propto [P_L(x_{S2}, x_{S3}) \otimes P_L(x_{S2}, x_{S3})]^2 = \left[\frac{2J_1\left(\frac{|\vec{r}_S| \bar{S}_\tau}{\lambda l} \right)}{\frac{|\vec{r}_S| \bar{S}_\tau}{\lambda l}} \right]^2 \tag{3.33}$$

where $\bar{S}_\tau$ is an average speckle size (1.9) and J_1 is the first-order Bessel function of the first kind.

Function (3.33) is depicted in Figure 3.7 by curve 1. If the specklegram contains two relatively displaced speckle patterns, the intensity oscillation described by the harmonic factor in (3.32) will be observed on the screen inside the diffraction halo boundary. The frequency of this oscillation, which can be represented as a fringe spacing, is given by the following equation:

$$\cos\left(\frac{\pi}{2l\lambda} M \vec{d}_\tau \vec{r}_S \right) = 0. \tag{3.34}$$

Distribution (3.34) is illustrated by curve 2 in Figure 3.7. The coordinates $\vec{r}_S$ of the dark fringes of an order n in the screen plane is related to the module of the total in-plane displacement component in the following way:

$$|\vec{d}_\tau| = \frac{\lambda l}{|\vec{r}_S| M \cos \theta} \left(n - \frac{1}{2} \right) \tag{3.35}$$

where $n = 1, 2, 3, \ldots$ is the order of the interference fringe and θ is the angle between vectors $\vec{d}_\tau$ and $\vec{r}_S$. Expression (3.34) depicts a set of uniformly spaced straight lines. The direction of the displacement vector $\vec{d}_\tau$ coincides with the normal to fringes direction to within sign.

For a more detailed understanding of fringe formation in the method of speckle photography, an image of the surface under study can be represented as a set of two small hole systems which are identical, but one of them is displaced with respect to the other by a small amount. The diameter of each hole is assumed to be equal to an average speckle size, which is assumed to be much less than the diameter of the unexpanded laser beam. If two super-imposed hole systems are now illuminated by a narrow laser beam a fringe pattern caused by the diffraction of the light of each couple of displaced holes

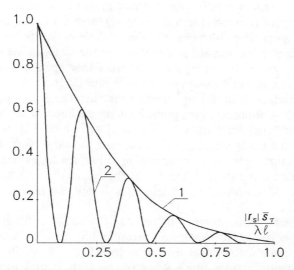

Figure 3.7. Intensity distributions of light resulting from pointwise filtering of a double-exposure specklegram

within a beam size will appear on a screen placed behind the plane containing conventional hole systems. This phenomenon is equivalent to that resulting from diffraction by two holes or slits, and for this reason the fringe patterns corresponding to condition (3.34) are often referred to as Young's fringes.

From a standpoint of the practical application of point-by-point scanning (pointwise) technique, it is convenient to transform expression (3.35) so that the measurement of the fringe spacing t_F (t_F is the distance between two neighbouring fringes in the direction of its normal) is used for the displacement components determination instead of using the fringe coordinates in the screen plane. The required expressions for the in-plane displacement components have the form

$$u = \frac{\lambda l \cos \theta_0}{M t_F}, \quad v = \frac{\lambda l \cos \theta_0}{M t_F} \tag{3.36}$$

where θ_0 is the angle between the x_2-axis and the normal to the fringes direction.

A rigid rotation of the plane specimen around the axis normal to its surface is, apparently, the most effective example to use as an illustration of the practical implementation of the two above-described techniques of spatial optical filtering of the light wave passing through a double-exposure specklegram. In this case the value of the total in-plane displacement component is proportional to the distance from the point under consideration and the centre of rotation. Figures 3.8(a)–(c) show Young's fringes obtained by means of pointwise filtering of the double-exposure specklegram and corresponding to the surface points lying on the straight line passing through the rotation centre. It can be seen that the fringe spacing decreases with the growth of the distance between the point of interest and rotation axis, while the fringe orientation is found to remain constant. The fringe pattern shown in Figure 3.8(d) is obtained in the point lying on the line whose direction does not coincide with the direction of the line along which Figures 3.8(a)–(c) are recorded, but its origin again is the centre of rotation. Note that the distance between the surface points corresponding to Figures 3.8(c) and (d) and that the rotation centre is the same. It can be clearly seen in Figure 3.8(c) that the fringe spacing coincides with that in Figure 3.8(d), but the inclination of fringes is different.

The fringe patterns obtained by means of the Fourier filtering procedure of the double-exposure specklegram corresponding to a shearing stress of the rectangular plate with central circular open hole are shown in Figure 3.9. Note that the drawing and loading scheme of this plate is presented in Figure 5.6. The filtering aperture is sequentially placed in the two positions lying on two mutually perpendicular lines at the same distance from the point of intersection of the optical axis of the filtering arrangement and Fourier plane. It can be seen that the fringe orientation in Figures 3.9(a) and 3.9(b) corresponds to the in-plane displacment component d_2 and d_3, respectively. These figures illustrate how the whole-field filtering technique can be implemented for determining the strain-induced displacement components.

The fringe patterns obtained by both filtering techniques described above are characterized by a noise caused by so-called secondary speckles having a relatively large size. The occurrence of these speckles on a fringe pattern results from the necessity to use either a small aperture or a narrow laser beam during implementation of the spatial filtering procedure. The influence of secondary speckles on the accuracy of a quantitative fringe pattern interpretation mainly has an effect upon fringes obtained by Fourier spatial filtering, since in this case the number and centre of the each fringe must be determined. The fringe patterns observed as a result of pointwise filtering of a double-exposure specklegram are usually interpreted by means of an averaging of fringe spacing over a diffraction halo. This leads to more accurate results.

Let's consider a general case of deformation of a curved body surface when the normal to the object surface displacement component w is comparable with the maximum value of one of the tangential displacement components u or v. In this case it may happen that a focused image of all object surface points cannot be obtained in a single (photoplate) plane. Moreover, in some cases a specially introduced defocusing is used for a measurement of various deformation parameters. When a photoplate is placed out of the plane of a

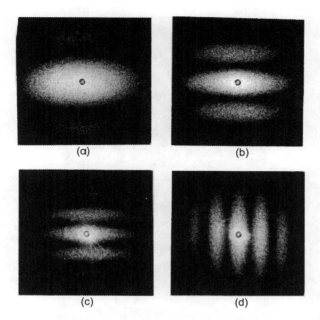

Figure 3.8. Young's fringes corresponding to rigid body rotation of the plane specimen around the axis normal to its surface obtained for the points lying on two lines passing through the rotation centre: (a, b, c) one line, (d) another line

focused image, the displacements of speckles in a recording plane are due to both the tangential u, v and normal w displacement components. The deflection gradient grad$w$ and the defocusing value also influence the kinematics of speckles. The rigorous consideration of speckle displacement in the aperture plane for the optical arrangement shown in Figure 3.10 leads to the following relation which describes a speckle kinematics in a recording plane [14]:

$$\vec{d}_s = -M\left(\vec{d}_\tau \frac{|\vec{r}_1|w}{l_1}\right) - \left(\frac{\Delta l_2}{l_2} - M\frac{w_0}{l_1}\right)$$

$$\cdot \left[\vec{d}_\tau - l_1\left(\vec{e}_2\frac{\partial \vec{m}_0\vec{d}}{\partial x_2} + \vec{e}_3\frac{\partial \vec{m}_0\vec{d}}{\partial x_3}\right)\right] \tag{3.37}$$

where $\vec{d}_S$ is the vector of speckle displacements in the recording plane; $M = (l_2 + \Delta l_2)/l_1$ is the transverse magnification of the recording system; $\vec{d}_\tau = (u, v)$ is the tangential projection of the displacement vector $\vec{d} = (v, u, v)$; $\vec{r}_1$ is the radius vector which connects the origin of the coordinate system with the surface point under investigation; l_1 and l_2 are the distances between the object surface, the lens and the image plane, respectively; Δl_2 is the defocusing depth;

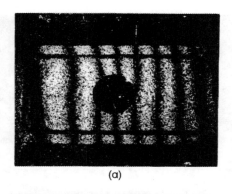

(a)

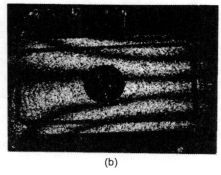

(b)

Figure 3.9. Strain-induced fringe patterns corresponding to the longitudinal (a) and transversal (b) in-plane displacement component obtained by means of whole-field filtering procedure

w_0 is the parameter depending on the surface curvature; $\vec{m}_0 = \vec{e}_s + \vec{e}_1$ is the sum of the illumination and observation unit vectors of the object surface point under study; and $\vec{e}_2$ and $\vec{e}_3$ are the unit vectors of the coordinate axes x_2 and x_3, respectively.

Expression (3.37) shows that when a focused image of the plane surface ($\Delta l_2 = 0$, $w_0 = 0$) is used for plane deformation study ($u, v \gg w$), the speckle displacements in the recording plane to within a sign and magnification factor M coincide with the speckle displacements in the object plane:

$$\vec{d}_S = -M\vec{d}_\tau. \tag{3.38}$$

The other particular case, which is of great importance and often used in practice for a deformation analysis of bending plates, occurs when a defocused image of the plane object surface is recorded. In this case, if condition $w \gg u$, and v is valid, the optical arrangement becomes sensitive to the deflection gradient only. This approach will be considered in detail in Section 4.5.

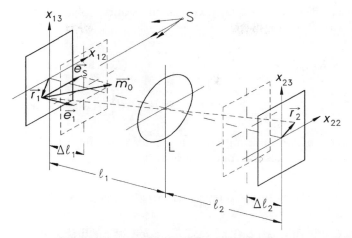

Figure 3.10. Schematic diagram of a speckle pattern formation in the case of arbitrary surface deformation and recording plane position

3.4   Time-average speckle photography

The method of speckle photography can be implemented in the study of moving objects during exposure as well as in holographic interferometry [15, 16]. Specklegrams thus obtained contain information on the character of an object motion and its deformation. Assume that the total tangential displacement vector component $\vec{d}_\tau(x_2, x_3, t)$ is a harmonic function with cyclic frequency ω:

$$|\vec{d}_\tau(x_2, x_3, t)| = [u_0(x_2, x_3) + v_0(x_2, x_3)] \sin \omega t \qquad (3.39)$$

where $u_0(x_2, x_3)$ and $v_0(x_2, x_3)$ are the vibration amplitudes of the tangential displacement components $u(x_2, x_3)$ and $v(x_2, x_3)$ respectively.

   In order to consider this important practical case in detail, Figure 3.4 should be referred to again. So far as the object surface position varies in time in accordance with the harmonic law (3.39), the expression for exposure $E(x_2, x_3)$ in accordance with relation (3.18) takes the form

$$E(x_2, x_3) \propto \int_0^\tau |A_B[x_{32} + u_0(x_{32}, x_{33}) \sin \omega t, x_{33} + v_0(x_{32}, x_{33}) \sin \omega t]|^2 dt \quad (3.40)$$

where τ is the exposure duration. The amplitude transmittance of such a specklegram will be inversely proportional to the exposure (3.40). When the specklegram thus obtained is placed in the Fourier filtering set-up (Figure 3.5), the intensity distribution $I_F(x_{F2}, x_{F3})$ in the Fourier plane can be written by analogy with equation (3.23):

$$I_F(x_{F2}, x_{F3}) \propto \left| \int_0^\tau \exp\left\{ -\frac{i\pi M}{\lambda f_1}[x_{F2} u_0(x_{32}, x_{33}) \sin \omega t \right.\right.$$
$$\left.\left. + x_{F3} v_0(x_{32}, x_{33}) \sin \omega t]\right\} dt \right|^2 I_{0F}(x_{F2}, x_{F3}). \tag{3.41}$$

When the exposure duration is much greater than the period of oscillations, expression (3.41) takes the following form:

$$I_F(x_{F2}, x_{F3}) \propto J_0^2\left[\frac{\pi M}{\lambda f_1}(x_{F2} u_0 + x_{F3} v_0)\right] \cdot I_0(x_{F2}, x_{F3}) \tag{3.42}$$

where $J_0(...)$ is the zero-order Bessel function of the first kind. As a result of Fourier filtering, a diffraction image of the object surface will be modulated with the set of bright and dark interference fringes. The brightness of bright fringes will be varied over the surface under study. The dark fringes correspond to the following equation:

$$\frac{\pi M}{\lambda f_1}(x_{F2} u_0 + x_{F3} v_0) = N_i \qquad i = 1, 2, 3 \tag{3.43}$$

where N_i is the ith root of the Bessel's equation $J_0(N_i) = 0$.
Expression (3.43) gives the following result:

$$\vec{r}_F \vec{d}_\tau = \frac{N_i \lambda f_1}{\pi M} \qquad i = 1, 2, 3. \tag{3.44}$$

By placing the filtering aperture substantially on the coordinate axis x_{F2} and x_{F3}, it is possible to select the corresponding tangential vibration amplitudes $u_0(x_2, x_3)$ and $v_0(x_2, x_3)$.

When the pointwise procedure is used for specklegram filtering, the intensity distribution within the diffraction halo, apart from the two-exposure technique (see equation (3.33)), is described with the zero-order Bessel function of the first kind. In this case the vibration amplitude at the surface point under consideration can be found through the relation

$$|\vec{d}_\tau| = \frac{\lambda l N_i}{2\pi |\vec{r}_S| \cos \theta} \tag{3.45}$$

where l is the distance between the specklegram and the screen where the interference fringe is observed; N_i is the ith root of the Bessel equation; $\vec{r}_S$ is the radius vector of an arbitrary point located at the centre of the ith interference fringe and θ is the angle between the vectors $\vec{d}_\tau$ and $\vec{r}_S$.

Figure 3.11 illustrates the interference fringe pattern resulting from pointwise interpretation of the specklegram recorded in accordance with a time-average procedure. It is clearly seen that the set of straight interference fringes is not uniformly spaced as it follows from the solution of Bessel's equation. The vibration amplitude $|\vec{d}_\tau|$ of the surface point of interest can be connected with the distance between the interference fringes in the following way:

$$|\vec{d}_\tau| = \frac{\lambda l N_i}{M t_i} \tag{3.46}$$

where t_i is the distance between the two fringes of the ith order which corresponds to the root of Bessel's equation with the number N_i.

3.5   Decorrelation of speckle patterns in the speckle photography method

The parameters of a speckle pattern which are formed in the recording plane depend on the surface roughness, the polarization of the illuminating light wave and the aperture function of the recording system. A variation of one or several of these parameters between exposures, as well as rigid-body translations and rotations and/or strain-induced displacement of the object surface, leads to the partial or complete decorrelation of speckle patterns recorded on the double-exposure specklegram. This circumstance gives rise to the degradation of a contrast of fringe patterns obtained both by the Fourier spatial filtering technique and pointwise filtering of a doubly exposed specklegram. This contrast degradation process may be finished with the

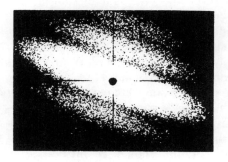

Figure 3.11. Typical fringe pattern resulting from a pointwise filtering of time-average specklegram

complete disappearance of the interference fringes. Note that in this section it is assumed that the surface microrelief (or roughness) remains constant between exposures.

An analysis of the influence of the above-mentioned parameters on a decorrelation of speckle patterns should be performed to establish the applicability range of speckle photography method in different fields of experimental mechanics. Moreover, a change in direction of polarization of the illuminating wave, as well as a form of the aperture transmittance function of the recording system, can serve as an implement of active control of the degree of speckle patterns decorrelation. This approach, as described below, can be used in some measurement techniques.

Each specific implementation of a speckle pattern corresponds to the definite direction of the illuminating wavefront. A change of this direction results in a random change of the initial speckle ensemble in the recording plane. Physically, this phenomenon can be responsible, among other things, for the fact that any slight disturbance of the light wave illuminating the object surface gives rise to a large change of the spatial spectrum of the scattered wave. As a speckle pattern represents by itself the sum of the set of light waves of different spatial frequencies scattered by the optically rough surface, the disturbance of the illuminated wave results in a formation of a new speckle pattern, which is not identical to the initial one.

For a metallic surface having a high roughness (the values of the CLA roughness are not less than $\bar{h} = 1\,\mu m$), the fringe contrast γ depends on the variation of the illuminating wave direction in the following way [17]:

$$\gamma = \exp\left[-\left(\frac{2\pi\bar{h}}{\lambda}\right)\Delta\beta\sin\beta\right] \tag{3.47}$$

where $\bar{h}$ is the r.m.s. deviation of the surface high peaks and β is the inclination angle of the illuminating wave with respect to the optical axis of the recording arrangement. When the value of γ becomes equal to 0.1 the interference fringes practically cannot be observed.

The influence of the value of the illumination beam tilting angle $\Delta\beta$ on the parameters of the speckle patterns correlation was investigated by using the optical arrangement shown in Figure 3.12(a). The change of the illumination direction was made by rotating the mirror M by the angle φ, so that $\Delta\beta = 2\varphi$.

Double-exposure specklegrams were recorded in the following way. The first exposure produces the speckle pattern of the initial undisplaced object surface state. The second exposure is made after the specimen's rotation around the axis normal to its surface and introducing a small tilt of the mirror M. Note that the specimen's rotation should be introduced between exposures to provide the so-called reference fringe pattern, which serves as a primary standard for estimation of the degree or the decorrelation of the speckle patterns.

Figures 3.12(b)–(d) represent Young's fringes obtained by means of pointwise filtering of three doubly exposured specklegrams corresponding to the illumination angle increment $\Delta\beta = 0$; 1.25×10^{-2} and 5×10^{-2} rad, respectively. The absence of fringes in Figure 3.12(d) denotes a complete decorrelation of two speckle patterns of the object surface image. This result is in good agreement with the data of formula (3.47). Indeed, for $\bar{h} = 20\,\mu m$; $\lambda = 0.633\,\mu m$ and $\beta = 15°$ the value of $\Delta\beta = 4.5 \times 10^{-2}$ gives rise to a complete fringe disappearance.

The optical system shown in Figure 3.13(a) was implemented to study the influence of polarization of the light wave scattered by the object surface (object wave) on the correlation properties of speckle patterns. The first exposure, as in the previous case, corresponds to the initial object surface state. The second exposure is carried out after a rotation of the specimen around the axis normal to its surface and changing the polarizer position by angle φ. Figures 3.13(b) and (c) show Young's fringes which correspond to two different angular positions of the polarizer. The absence of fringes in Figure 3.13(c) denotes the complete decorrelation of speckle patterns corresponding to two transversely polarized light waves ($\varphi = 90°$). The same result was obtained in reference [17].

A change of the aperture function of the recording system also leads to the decorrelation of the speckle patterns. Consider the case when the circular apertures of different diameters D_1 and D_2 are used during the first and second exposures respectively. The reference fringe pattern is again introduced by means of rotation of the plane specimen under study around the axis normal to its surface. The speckle patterns formed with two circular apertures of different diameter differ one another both in average speckle size (see equation (1.10)) and in the configuration of individual speckles. The quantitative description of the speckle patterns correlation in this case can be performed by means of a simple geometrical analogy. The essence of this analogy consists of the fact that the normalized area of two-aperture overlap can serve as a characteristic of the degree of the correlation of two speckle patterns formed with two different aperture functions (cross-hatched area in Figure 3.14) [12, 18]. The intensity cross-correlation functions plotted against the distance from the pupil centre for two circular apertures of different diameter, which are presented in Shchepinov et al. [19], show that these graphs have a relatively smooth character and, in accordance with the form of the aperture function, reach the maximum meaning at different points of the aperture overlap region.

Figure 3.15 shows Young's fringes obtained by means of a pointwise filtering of the double-exposure specklegram using the above-described two-aperture technique. The first and second exposures correspond to the numerical aperture number $F_1 = 4.5$ and $F_2 = 8$, respectively. The interference fringes presented are of a low contrast and illustrate a partial decorrelation of the speckle patterns superimposed.

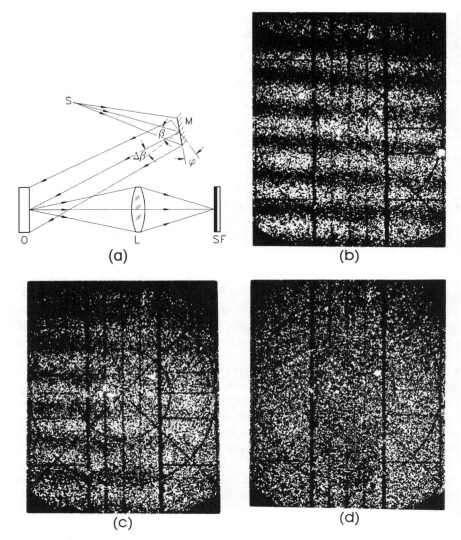

Figure 3.12. Influence of a variation of the illuminating wave direction between exposures on a speckle pattern correlation: recording set up (a), initial (b), partially (c) and completely decorrelated fringe pattern (d)

Consider now the other group of factors which exert its influence on the speckle patterns decorrelation. The ultimate meaning of the normal to the object surface displacement component w is directly related the transverse size of an individual speckle (1.10) and the complete decorrelation of speckle patterns will occur when the following conditions are valid:

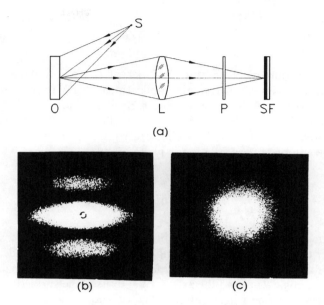

Figure 3.13. Influence of a variation of the illuminating wave polarization between exposures on a speckle pattern correlation: recording set up (a), initial Young's fringes (b) and decorrelated speckle patterns (c)

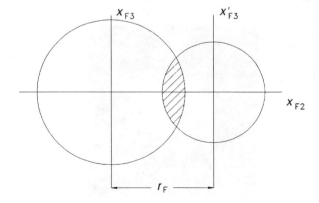

Figure 3.14. Geometrical analogue of a speckle patterns correlation

$$|w| \gtrsim 2F^2 \frac{\lambda}{M^2} \qquad (3.48)$$

where F is the numerical aperture number of the recording lens; λ is the wavelength; and M is the transverse magnification factor of the recording system.

A rotation of the object with a plane surface around the axis lying in this surface results in a considerable decorrelating influence on the resultant speckle pattern. The value of the angle φ characterizing this rotation which not yet leads to the complete decorrelation is given by [9]

$$\varphi = \frac{M}{2.4F(1 + M)}. \qquad (3.49)$$

The experimental illustration of the correlation properties of speckle patterns in the latter case was obtained by using a specimen of a special design having the form of a composed plate (see Figure 3.16). This plate is capable of being rotated by an angle θ around the axis normal to its plane surface. This rotation produces the reference fringe pattern. The central segment of the composed plate having the form of a thin rectangular plate is capable of being either rotated by an angle φ around the axis lying in its plane or strained as a cantilever beam whose bending can be caused with the deflection w of the plate's free edge.

The fringe patterns obtained by means of Fourier filtering of the light field diffracted by double-exposure specklegrams are presented in Figure 3.17. These interferograms illustrate the influence of the rotation angle φ on the degree of correlation of two speckle patterns.

Double-exposure specklegrams are recorded in the following way:

Figure 3.15. Influence of a change of an aperture function of the recording set up between exposures on a speckle patterns correlation

- The first exposure is made for the unstrained object.
- After the first exposure the composed plate is rotated by the angle $\theta = 1.16 \times 10^{-3}$ rad as a rigid body and then the central segment of the plate is tilted by the angle φ around the axis lying in its plane.
- The second exposure is made for the displaced object.

The contrast of the reference fringe pattern in the central part of the plate decreases with the growth of the φ-angle value—see Figure 3.17(a) ($\varphi = 0.44 \times 10^{-3}$ rad) and Figure 3.17(b) ($\varphi = 1.77 \times 10^{-3}$ rad). The value of the angle $\varphi = 2.65 \times 10^{-3}$ rad corresponds to the complete decorrelation of the resultant speckle pattern; in this case the interference fringes cannot be observed in Figure 3.17(c). It should be noted that the experimentally obtained value of the angle φ, which leads to the complete decorrelation, is less by approximately 10 times than the analogous value resulting from relation (3.49).

The displacement of points on the object surface in the direction normal to the surface w and the object rotation around the axis lying in the surface plane by the angle φ usually occurs in many cases of object deformation and often may restrict the range of possible practical applications of the method of laser speckle photography.

The sensitivity threshold of the method of speckle photography is defined by the minimum value of the mutual displacement of two speckle patterns at the point under consideration. This gives rise to the formation of two interference fringes within the diffraction halo as a result of pointwise filtering of double-exposure specklegrams. Moreover, the quality of these two fringes should be

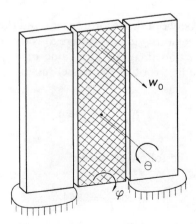

Figure 3.16. Diagram of the composed specimen and possible rotations and deformations of its parts

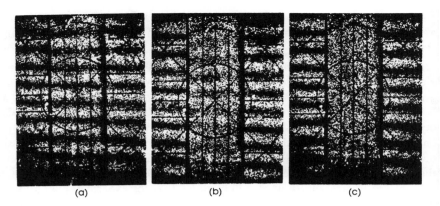

(a)   (b)   (c)

Figure 3.17. Influence of an object rotation around the axis lying in its own plane between exposures on a speckle pattern correlation: correlated (a), partially (b) and completely (c) decorrelated speckle patterns

sufficiently high to ensure a reliable and accurate measurement of the distance between them. It is evident that this distance must be less than the maximum dimension of the diffraction image of the surface area under investigation. For far-field recording conditions we can assume that an angular distance between two neighbouring interference fringes in the diffraction halo α_p can be expressed in the following form:

$$\alpha_p \simeq \frac{t_K}{l}$$

where l is the distance between the recording lens and the object surface and t_K is the fringe spacing. Obviously $\alpha_p \leqslant \alpha_l$ (where α_l is the angular distance of the diffraction halo (diffraction image)). The value of α_l according to equation (3.33) is defined by the first root of the Bessel function

$$J_1\left(\frac{|\vec{r}_S|\bar{S}_\tau}{\lambda l}\right) = 0.$$

This equation leads to the following result:

$$\alpha_l \simeq \frac{2r_S}{l} = 2.44\frac{\lambda}{\bar{S}_\tau}$$

which, taking into account equation (1.10), can be represented as

$$\alpha_l \simeq \frac{2M}{F(M+1)}.$$

Now, keeping in mind expression (3.36), we can obtain the minimum possible in-plane displacement value $d_{\tau\min}$ that can be measured by using the speckle photography method given by

$$d_{\tau\min} = \frac{\lambda F(1+M)}{1.8\, M^2}. \tag{3.50}$$

Expression (3.50) establishes the relation between the sensitivity threshold and the parameters of the recording optical arrangement used in speckle photography techniques.

Figure 3.18 shows the dependencies between the sensitivity threshold $d_{\tau\min}$ and transverse magnification factor of the recording system M for two values of the numerical aperture $F = 4.5$ (curve 1) and $F = 2.8$ (curve 2). In both cases $\lambda = 0.633\,\mu m$. The great influence of the magnification factor M on the sensitivity threshold can be seen when the value of M becomes less than two. Indeed, an increase of the value of M from 1 to 2 is accompanied by a double decrease of the sensitivity threshold, but further double decreasing of the latter value occurs when the magnification factor becomes more than five.

The maximum displacement value which is capable of being measured is defined by a relation between the fringe spacing within a diffraction halo t_K and an average size of secondary speckles $\bar{S}_\tau$ observed on the screen. If the angular distance between two neighbouring fringes becomes comparable to an analogous parameter of a secondary speckle of an average size, the interference fringe patterns cannot be observed. Combining expressions (1.10) and (3.36) and taking into account that a secondary fringe structure appears as a result of scanning a specklegram with an unexpended laser beam having a diameter d_0, we can obtain

$$|\vec{d}_{\tau\max}| \leqslant 0.15 d_0 \tag{3.51}$$

A rigid rotation of the plane specimen around the axis normal to the object surface can be characterized by the tangential displacement gradient along an arbitrary line passing through the centre of rotation. Young's fringes in this case cannot be observed when the displacements $\vec{d}_{\tau 1}$ and $\vec{d}_{\tau 2}$ of two points of a specklegram, which are maximally distant from one another within the area of the illuminated beam, correspond to the fringe patterns having a difference by one fringe order:

$$|\vec{d}_{\tau 1}| = R\theta = \frac{\lambda l N}{t_n M} \tag{3.52}$$

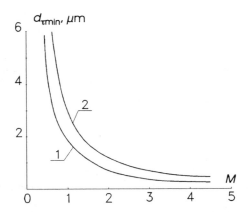

Figure 3.18. Dependencies between the threshold limit of the speckle photography technique $d_{\tau min}$ and transverse magnification factor of recording set up M obtained for the two values of numerical aperture F = 4.5 (curve 1) and F = 2.8 (curve 2)

$$|\vec{d}_{\tau2}| = (R + d_0)\theta = \frac{\lambda l(N + 1)}{t_N N} \qquad (3.53)$$

where d_0 is the diameter of the laser beam; R is the distance between the surface point under study and the axis of rotation; θ is the angle of the object rotation between exposures; λ is the wavelength; M is the magnification factor of the recording system; l is the distance between the photoplate and the screen; and t_N is the distance between N interference fringes. Expressions (3.52) and (3.53) give the ultimate value of the rotation angle θ_{max} which results in the complete decorrelation of speckle patterns:

$$\theta_{max} \simeq \frac{\lambda}{MFd_0} \qquad (3.54)$$

where F is the numerical aperture of the recording system. The obtained estimation (3.54) represents the upper possible error limit and is valid for surface points rather distant from the rotation axis.

A rotation of the object around the axis normal to its plane surface also leads to a non-uniform decorrelation of the speckle patterns forming the resultant speckle pattern in a double-exposure specklegram. The radius r_0 of the circle within which the speckle patterns remain with their correlation properties can be defined by the following relation [9]:

$$r_0 = \frac{\lambda f}{d_0 \theta} \qquad (3.55)$$

where f is the focal length of the lens which makes the Fourier transform of the light field diffracted by the specklegram and d_0 is the diameter of the filtering aperture.

Figures 3.19(a) and (b) illustrate the fringe patterns obtained for the plane specimen which is rotated by angle θ around the axis normal to its plane surface between exposures. The interference fringes presented in Figure 3.19(a) cover the whole object surface, while Figure 3.19(b) shows the fringes on the central part of the object surface only. The values of the radius r_0 obtained as a result of the experiment is found to differ from those resulting from expression (3.55) by 20 per cent.

An object deformation between exposures is also a reason for speckle pattern decorrelation caused by the surface displacement gradients. To analyse the influence of this factor on a correlation capability of two speckle patterns corresponding to an unstrained and strained object surface, we can use the approach introduced above for the definition of the rotation angle θ_{max} (3.54). The displacements of the most mutually distant points lying within the diameter of the scanning laser beam on a specklegram can be written by analogy with expressions (3.52) and (3.53):

$$|\vec{d}_{\tau 1}| = \frac{\lambda l N}{t_N M},$$

$$|\vec{d}_{\tau 2}| = |\vec{d}_{\tau 1}| + \varepsilon d_0 = \frac{\lambda l (N+1)}{t_N N}$$

where d_0 is the diameter of the laser beam and ε is an average strain value within the diameter of the laser beam. Combining these expressions gives the

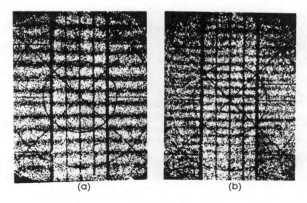

(a)   (b)

Figure 3.19. Influence of an object rotation around the axis normal to its surface on a speckle pattern correlation: partially (a) and completely (b) decorrelated speckle pattern

required relation which determines the ultimate strain value leading to a complete decorrelation of the resultant speckle pattern:

$$\varepsilon_{max} \simeq \frac{\lambda}{MFd_0} \tag{3.56}$$

where F is the numerical aperture of the recording system; M is the transverse magnification factor; and λ is the wavelength. For instance, if we assume that $F = 4.5$, $\lambda = 0.633\,\mu m$ and $d_0 = 1$ mm, the deformation ultimate value will be $\varepsilon_{max} = 1.5 \times 10^{-4}$.

Experimental investigation of the strain value influence on the correlation properties of two speckle patterns was carried out by using the composed plate shown in Figure 3.16. Remember that this plate has a central segment which is capable of being bent as a cantilever beam with the deflection w_0 of its free edge. Double-exposure specklegrams are recorded in this case in the following way:

• The first exposure is made for the unstrained plate surface.
• Then the plate is rotated around the axis normal to its surface in order to form a reference fringe pattern and bent with the deflection w_0 of its free edge.
• The second exposure is made for the rotated and strained object.

When the maximum deflection of the plate w_0 is equal to 50 μm (see Figure 3.20(a)) interference fringes of high contrast are observed over the whole plate surface. Figure 3.20(b), which corresponds to the maximum deflection value $w_0 = 250\,\mu m$, reveals the contrast fringe patterns in the region of the minimum plate deflection near the clamped plate edge only. It should be noted that in the latter case a loss of the fringe contrast near the free edge of the bending plate is mainly caused by the tilt φ (see expression (3.49)) of the corresponding cross-sections of the plate top part in the same way as a rigid body ($\varphi_{max} > 2.4 \times 10^{-3}$ rad). On the other hand, the decorrelation of the resultant fringe structure in the region of the plate fixing is due to bending strains of the plate surface ($\varepsilon_{max} \simeq 0.8 \times 10^{-3}$).

3.6 Recording of two displacement fields on a single specklegram

Two main approaches to speckle pattern recording which are capable of recording information about two or more displacement fields are known and used today. Note that the term 'displacement field' means an interferometric or correlation comparison of two different states of the object surface, for instance unstrained and strained.

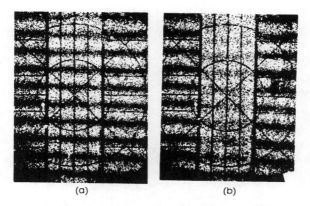

(a)   (b)

Figure 3.20. Influence of an object deflection on a speckle pattern correlation: correlated (a) and partially decorrelated (b) speckle pattern

One of these approaches consists of recording speckle patterns oriented along some direction or speckle patterns with a spatial carrier onto a single photoplate [17,20]. The other approach is based on using two photoplates and amplitude or spatial dividing of an object wave [21,22].

The simplest way to produce the speckle pattern oriented along a definite direction consists of locating a slit aperture (a × b) when a specklegram is recorded (see Figure 3.21(a)). The object O is illuminated by the divergent wavefront from the point source S. The light wave scattered by the optically rough object surface is transformed by the lens L into the focused object image in the plane of the photoplate SF. Due to the presence of the slit aperture immediately behind the lens this image is modulated by the spatially oriented speckle pattern. The magnified image of this speckle pattern is shown in Figure 3.21(b).

The average size of an individual speckle for the obtained speckle pattern can be determined by equation (1.10). The form of each speckle is geometrically similar to that of the recording slit aperture. The maximum dimension of most of the speckles coincides with the direction perpendicular to the slit axis. The geometrical anisotropy of the speckle pattern results in the anisotropy of sensitivity with respect to the in-plane displacement components. Indeed, the maximum sensitivity occurs in the direction of the slit axis in which an average speckle size is minimum. When the slit is rotated around the axis normal to its plane by the angle φ_i, a new speckle pattern, which is statistically independent with respect to the previous one, can be obtained as a result of the focused image recording procedure. Thus, a sequential rotation of the slit aperture located in the image recording arrangement allows us to record a set of speckle patterns in a single photoplate.

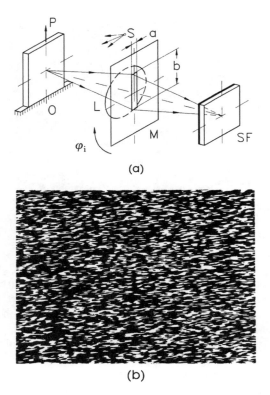

(a)

(b)

Figure 3.21. Schematic diagram of a speckle interferometer with a slit aperture (a) and recorded speckle pattern with individual speckles oriented along the slit direction (b)

The non-transparent mask M, which is placed in front of the lens, with two small circular apertures of diameter a, located symmetrically with respect to the optical axis, is one of the special features of the recording set-up shown in Figure 3.22. The distance b between apertures should be much more than the aperture diameter. In this case the recorded speckle pattern will be modulated with a carrier, the spacing of which is defined as

$$t_b = \frac{\lambda l_2}{b}$$

where l_2 is the distance from the lens to the recording plane. After such a recording procedure each individual speckle will be covered with a set of uniformly spaced straight fringes. The orientation of these fringes coincides with the direction normal to the line connecting the aperture centres. An average speckle size in this case is defined by the diameter of a small aperture a

$$S_\tau = \frac{\lambda l_2}{a}.$$

It is obvious that $S_\tau \gg t_b$.

The above-described measurement technique developed by Duffy [20] is sensitive to the in-plane displacement component directed in parallel with the common axis of two apertures. Due to the carrier fringes formed within each individual speckle, the resultant fringe pattern caused by the object surface displacement or deformation is usually of a high contrast, since the image of the object surface is formed by the narrow spatial frequency band.

A rotation of the double-aperture mask by the angle φ_i around the axis of the recording optical system allows us to form a new speckle pattern with its own characteristic carrier. The carrier fringes will have the same spacing but will be turned by angle φ_i with respect to the previous carrier fringe set. If the condition

$$\varphi_i > \frac{2a}{b}$$

is valid (see Figure 3.22), after the filtering of the light waves diffracted by the photoplates, two diffraction images corresponding to a different orientation of the aperture couple will be spatially selected which, in turn, allow us to observe two different fringe patterns. This provides us with the possibility, in principle, to record a set of displacement fields on a single photoplate [17].

An optical arrangement which is capable of recording two displacement fields onto two photoplates, and based on the amplitude division of the object wave, is shown in Figure 3.23(a) [22]. The object surface O is illuminated by a divergent wavefront emitted by the point source S. The lens L and beam splitter BS form two focused images of the surface O in the planes of two photoplates SP1 and SP2. By blocking in turns with a non-transparent mask two directions of the object wave propagation, one can seek subsequently for the required sequence of the exposures in order to record two double-exposure specklegrams corresponding to two displacement fields.

The optical system founded upon a spatial division of the object wave was proposed in reference [21]. In this arrangement (see Figure 3.23(b)) the prism with two mirror sides is located immediately behind the recording lens L. This prism allows us to form the focused image of the object surface illuminated by a divergent wave-front from the point source S in the planes of two photoplates SP1 and SP2 simultaneously. A non-transparent mask is also needed to ensure the required sequence of exposures.

The main disadvantage of the optical systems shown in Figure 3.23 is its inability to identify the surface point under investigation on two different photoplates. Moreover, each arrangement involved is quite bulky in

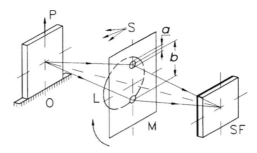

Figure 3.22. Schematic of the optical arrangement for recording of a specklegram with spatial carrier

comparison with the simpler optical system of specklegram recording shown in Figure 3.4.

Let's now consider the technique which is capable of ensuring the recording of two displacement fields on a single four-exposure specklegram. This technique is based on pair correlation of the speckle patterns. The analysis presented in Section 3.5 leads to the conclusion that a spatial separation of interference fringes corresponding to different displacement fields can be achieved by a variation of the aperture function of the recording optical system.

The fact that in a general case the direction of the total in-plane displacement vector is previously unknown over the surface under investigation restricts the class of aperture functions which can be used for spatial separation of fringe patterns to the class of axisymmetric functions. The most convenient class for a practical implementation are the circular and annular apertures.

The optical arrangement for specklegram recording with the variable-aperture function is shown in Figure 3.24. Four-multiple exposures of the photoplate SF are made in the following manner [19]:

- The initial state of the object surface is recorded on the photoplate with an exposure duration τ_1 and aperture function of the recording lens $P_{L1}(x_{23}, x_{33})$.
- To ensure the validity of the condition of the complete decorrelation of the two couples of the speckle patterns, some parameters (e.g. illumination direction) should be changed between the first and second exposures.
- The initial state of the object surface is again recorded on the photoplate with an exposure duration τ_2 and aperture function $P_{L2}(x_{23}, x_{33})$.
- The investigated object is displaced or strained due to the influence of some conventional factor Q_1 and for this reason the surface points are displaced by the value $\vec{d}_{\tau 2}(x_{12}, x_{13}) = [u_2(x_{12}, x_{13}), v_2(x_{12}, x_{13})]$, then in order to record the corresponding state of the object surface the third exposure is made with a duration τ_2 and aperture function $P_{L2}(x_{23}, x_{33})$.

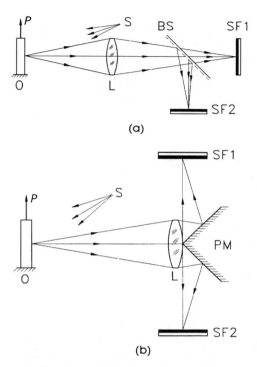

Figure 3.23. Schematic diagrams of the optical set up for recording of two displacement fields on a single specklegram based on amplitude (a) and spatial (b) division of the object wave

- The initial parameters of the recording arrangement, for instance the illumination direction, must be reconstructed and an adjustment of the initial aperture function $P_{L1}(x_{23}, x_{33})$ must be made.
- The object is displaced or stressed with another conventional factor Q_2 and for this reason the surface points are displaced by the value $\vec{d}_{\tau 1}(x_{12}, x_{13}) = [u_1(x_{12}, x_{13}), v_1(x_{12}, x_{13})]$, then the fourth exposure is made with a duration τ_1.

A specklegram thus obtained contains two pairs of mutually correlated speckle fields and, in addition, each correlated couple has its own average speckle size. As the speckle patterns recorded with different aperture functions are completely statistically independent, the total exposure of the photoplate can be expressed as sum of the relations of form (3.18) and the amplitude transmittance function takes the form

$$T(x_{32}, x_{33}) = b_0 - b_1 \sum_{i=1}^{2} \tau_i \cdot [I_{Bi}(x_{32}, x_{33}) + I_{Bi}(x_{32} + u_i, x_{33} + v_i)] \qquad (3.57)$$

where $I_{Bi}(x_{32}, x_{33})$ and $I_{Bi}(x_{32} + u_i, x_{33} + v_i)$ are the intensity distributions in the recording plane obtained for the different aperture functions.

Assume now that the specklegram characterized by the amplitude transmittance function (3.57) undergoes a pointwise filtering procedure by means of the optical arrangement shown in Figure 3.6. In this case two interference fringe patterns will be observed in the plane of the screen simultaneously, the intensity distribution of which is given by

$$I_S(x_{S2}, x_{S3}) \propto \tau_1^2 \{P_{L1}(x_{S2}, x_{S3}) \otimes P_{L1}(x_{S2}, x_{S3})\} \cdot \cos^2\left(\frac{\pi M}{2l\lambda} \vec{d}_{\tau 1} \vec{r}_s\right)$$

$$+ \tau_2^2 \{P_{L2}(x_{S2}, x_{S3}) \otimes P_{L2}(x_{S2}, x_{S3})\} \cdot \cos^2\left(\frac{\pi M}{2l\lambda} \vec{d}_{2\tau} \vec{r}_s\right). \qquad (3.58)$$

If the aperture functions $P_{L1}(x_{23}, x_{33})$ and $P_{L2}(x_{23}, x_{33})$ are axisymmetric, the corresponding diffraction images of each point of the specklegram are also axisymmetric and co-axial. Further considerations will be made for the circular and annual aperture functions, an implementation of which is the most convenient from a practical point of view.

Now we shall introduce the aperture function couple of a circular aperture of diameter D_i in the following way:

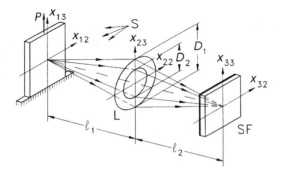

Figure 3.24. Schematic of the four-multiple exposure method for recording of two displacement field on a single specklegram

$$P_{Li}(x_{23}, x_{33}) = \begin{cases} 1 \text{ when } \sqrt{x_{23}^2 + x_{33}^2} \leqslant \dfrac{D_i}{2} \\ \\ 0 \text{ when } \sqrt{x_{23}^2 + x_{33}^2} > \dfrac{D_i}{2} \end{cases} \tag{3.59}$$

where $i = 1, 2$. Substituting into equation (3.58) the expressions for the autocorrelation of the aperture functions (3.33) gives the intensity distributions over the diffraction halo:

$$I_S(x_{S2}, x_{S3}) \propto \tau_1^2 \left[\frac{2J_1\left(\frac{|\vec{r}_S| \bar{S}_{\tau 1}}{\lambda l} \right)}{\frac{|\vec{r}_S| \bar{S}_{\tau 1}}{\lambda l}} \right]^2 \cos^2\left(\frac{\pi M}{2l\lambda} \vec{d}_{\tau 1} \vec{r}_S \right)$$

$$+ \tau_2^2 \left[\frac{2J_1\left(\frac{|\vec{r}_S| \bar{S}_{\tau 2}}{\lambda l} \right)}{\frac{|\vec{r}_S| \bar{S}_{\tau 2}}{\lambda l}} \right]^2 \cos^2\left(\frac{\pi M}{2l\lambda} \vec{d}_{\tau 2} \vec{r}_S \right). \tag{3.60}$$

In order to receive the quantitative estimation of the contrast of the interference fringes caused by the displacement fields $d_{\tau 1}$ and $d_{\tau 2}$ we can use expression (2.10) which in the case involved has the following form:

$$\gamma_S(x_{S2}, x_{S3}) = \frac{I_{max}(x_{S2}, x_{S3}) - I_{min}(x_{S2}, x_{S3})}{I_{max}(x_{S2}, x_{S3}) + I_{min}(x_{S2}, x_{S3})} \tag{3.61}$$

where I_{max} and I_{min} are maximum and minimum intensity values respectively. To derive the contrast function (3.61), it is necessary to eliminate any ambiguity in the duration of the exposures τ_1 and τ_2. The ratio of these durations, as follows from expressions (3.58) and (3.60), exerts its main influence on the contrast of the interference fringes (3.61).

As the aperture functions (3.59) are characterized by the uniform transmittance, the intensity distribution in the recording, if other conditions are equal, will be proportional to the actual pupil area. The fact that the total exposure is proportional to the product of the pupil area by the exposure duration should also be taken into account. Therefore the optimal ratio between exposure durations $\tau_i (i = 1, 2)$ can be established proceeding from the condition of the exposures equality. This means that the energy of the light waves coming to the recording medium during each individual exposure must be equal to one another:

$$D_i^2 \tau_i = \text{const}, \quad i = 1, 2 \tag{3.62}$$

For the circular aperture functions (3.59) the validity of condition (3.62) leads to the following relationship:

$$\tau_2 = \tau_1 \left(\frac{D_1}{D_2}\right)^2 = \tau_1 \left(\frac{F_2}{F_1}\right)^2 \tag{3.63}$$

where $F_i = f/D_i$ ($i = 1, 2$) are the numerical apertures of the recording lens.

Figure 3.25(a) displays the diffraction image of the small area of the four-exposure specklegram recorded according to condition (3.63) which corresponds to the intensity distribution (3.60). The internal diffraction halo ($|r_S| < D_2$) contains a number of the interference fringes which is less than that in the annular diffraction halo ($D_1 < |r_S| < D_2$). Substituting the maximum and minimum intensity values $I_S(x_{S2}, x_{S3})$ from expression (3.60), which correspond to the maximum and minimum values of the harmonic factors in expression (3.61), yields the magnitudes of the contrast of the interference fringes, which characterize two different displacement fields [23,24]. In the internal part of the diffraction halo the contrast of the interference fringes $\gamma_{S1}(x_{S2}, x_{S3})$ which result from the displacement field $\vec{d}_{1\tau}(x_{12}, x_{13})$ is rather low (see curve 1 in Figure 3.26), while in the annual diffraction halo region the meaning of this function is practically equal to unity. In the central area of the diffraction image $|\vec{r}_S| \leqslant D_1$ the interference fringes caused by the displacement field $\vec{d}_{\tau 2}(x_{12}, x_{13})$ can be observed for the most part. The contrast of these fringes is described by curve 2 in Figure 3.26. Note that in the range $\gamma_{S2}(x_{S2}, x_{S3})$ the fringes corresponding to the displacement field $\vec{d}_{\tau 2}(x_{12}, x_{13})$ almost cannot be seen.

Assume now that one exposure couple, which corresponds to the displacement field $\vec{d}_{1\tau}(x_{12}, x_{13})$, is made with the annular aperture $P_{L1}(x_{23}, x_{33})$, which is defined by the following aperture function:

$$P_{L1}(x_{23}, x_{33}) = \begin{cases} 1 \text{ when } \dfrac{D_2}{2} \leqslant \sqrt{x_{23}^2 + x_{33}^2} \leqslant \dfrac{D_1}{2} \\ 0 \text{ when } \dfrac{D_2}{2} > \sqrt{x_{23}^2 + x_{33}^2} > \dfrac{D_1}{2}. \end{cases} \tag{3.64}$$

The other exposure couple, which allows us to record the displacement field $d_{\tau 2}(x_{12}, x_{13})$, is carried out with the circular aperture (3.59). The external diameter of the circular aperture coincides with the internal diameter of the annular aperture. If these diameters are related through the relationship

$$D_2 = \sqrt{2} \, D_1 \tag{3.65}$$

the condition of the exposure equality (3.62) will be valid for equal exposure durations $\tau_1 = \tau_2$.

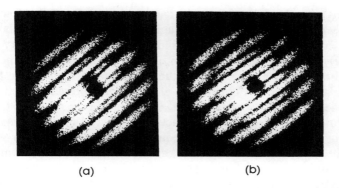

(a) (b)

Figure 3.25. Young's fringes obtained by means of pointwise filtering of the four-multiple exposure specklegram recorded with two circular apertures (a) and circular and annular apertures (b)

A typical fringe pattern which represents the result of a pointwise filtering of the obtained four-exposure specklegram is shown in Figure 3.25(b).

The autocorrelation function of the annular aperture which is contained in expression (3.58) has a very awkward form. That is why it is convenient to use the numerical result of its determination presented in reference [18] instead of its explicit analytical form. The visibility functions $\gamma_{S1}(x_{S2}, x_{S3})$ and

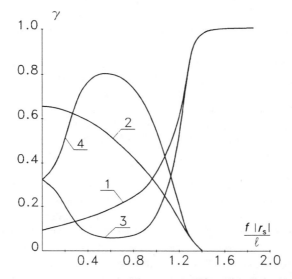

Figure 3.26. Distributions of the fringe contrast over diffraction halo in the case of the four-exposure recording of a specklegram with two circular apertures (1, 2) and circular and annular apertures (3, 4)

$\gamma_{S2}(x_{S2}, x_{S3})$ for Young's fringes obtained by means of the four-exposure specklegram and recorded with an annular aperture and circular aperture having a common circular edge are shown in Figure 3.26 by curves 3 and 4 respectively. Note that the geometrical dimensions of these apertures are similar to those which were used for the fringe contrast analysis in the case of two circular apertures. Comparison of the curves presented in Figure 3.26 reveals the shift of the peak of the function $\gamma_{S2}(x_{S2}, x_{S3})$ for the annular aperture out of the centre of the diffraction halo (see curve 4). This leads to an increase in the accuracy of the fringe interpretation procedure in the internal part of the diffraction halo when the combination of the annular and circular apertures are used during four-exposure recording in comparison with the case of two circular apertures. The intensity distribution in the range of the annular aperture is found to remain constant, and also results in the increase in the accuracy of the interference peaks identification [18]. Therefore the results presented clearly demonstrate that using the combination of the annular and circular apertures is preferred compared to the combination of two circular apertures in order to realize the four-exposure recording procedure.

Two-exposure speckle photography is capable of determining the in-plane displacement components to within a sign. The approach to physical displacement sign identification is known and based on the amplitude division of the object wave. In this case two double-exposure specklegrams are recorded on two individual photoplates. One of them must be displaced in the known direction between exposures by a definite amount [25]. A simultaneous analysis of two systems of Young's fringes obtained at the same surface point by a pointwise filtering of two specklegrams allows us to identify the displacement direction.

The problem of displacement sign identification can be effectively solved by means of the four-exposure speckle photography technique [26]. In accordance with this approach, the initial state of the object surface has to be recorded in the photoplate during the first and second exposures with the numerical apertures of the recording lens F_2 and F_1 respectively ($F_1 < F_2$). After object loading, two other exposures must be made. The third and fourth exposures are made with the numerical apertures F_1 and F_2 respectively, but between these exposures the object must be displaced in the known direction by the definite amount $|a|$. The pointwise filtering of such a recorded four-exposure specklegram results in the appearance of two diffraction images of the point under study. Young's fringes inside the internal circular diffraction halo are due to the superposition of the strain-induced object surface displacements and the uniform shift of the photoplate between exposures, while the fringe pattern in the annual region of the diffraction halo results from the photoplate uniform displacement. The vector diagram presented in Figure 3.27 gives rise to the determination of the displacement vector direction without ambiguity. Indeed, the vector sum of the vectors $\vec{a}$ and $\vec{u}$ leads to an incorrect result, since the vector $\vec{u}$ in this case lies in a direction which is not perpendicular to the straight fringes in the internal diffraction halo.

Figures 3.27(b) and (c) represent Young's fringes which are obtained by means of a pointwise filtering in two different points of the surface of the plane circular disk rotated around the axis normal to its surface. These points are located at an equal distance from the centre of rotation on the same disk diameter. The complete coincidence of the two fringe patterns obtained within the range of the annular region of the diffraction halo shows that the values of the displacement components in the two surface points under consideration are equal. An orthogonality of Young's fringes of equal spacing within the central circular part of the diffraction halo proves that the displacement components in the points considered have opposite directions.

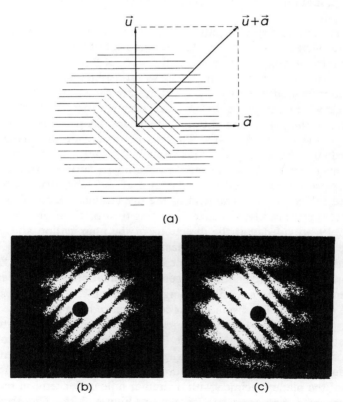

(a)

(b)   (c)

Figure 3.27. Vector diagram of a displacement sign determination by means of four-multiple exposure speckle photography technique (a) and Young's fringes obtained at the object surface points having opposite directions of displacement vectors (b) and (c)

3.7   Combined implementation of holographic interferometry and speckle photography for displacement measurement

The holographic method of recording light waves scattered by an optically rough surface allows simultaneous recording of the speckle patterns. This circumstance and similar ways for practical application of holographic interferometry and speckle photography two-exposure techniques sometimes allows simultaneous implementation of these techniques for separate recording of displacement vector components of the quasi-plane object surface displacement components induced by strain. As the various holographic interferometry and speckle photography techniques have a maximum sensitivity with respect to the different displacement components, the former method is recommended for measurement of the displacement component normal to the object surface, while the latter technique can be implemented for the determination of the displacement components tangential to the surface under study.

The first time that speckle patterns recorded onto double-exposure reflection holograms were implemented in the measurement of displacement components tangential to the object surface was in the work of Boone [27]. In this approach the emulsion is placed almost in contact with the object surface and, if coherent illumination is used, the forming structures are called objective speckles. Two exposures which are made before and after the deformation result in the recording on the photoplate of two objective speckle patterns, one shifted with respect to the other. A spatial filtering procedure of the plane wavefront diffracted by the hologram thus obtained allows us to determine the displacement components tangential to the object surface through the use of the same procedure as well as in the case of the subjective speckle patterns described in Section 3.3. The field of the displacement component normal to the object surface can be visualized by using the optical arrangement shown in Figure 2.19 provided that the directions of the illumination and observation coincide with the normal to the surface under study.

In order to illustrate the above-described technique we can use again the problem of the displacement component determination near a hole in the plane specimen subjected to a tensile loading (see Section 2.5). Figure 3.28(a) shows the field of the in-plane displacement component u lying in the direction of the tension. Figure 3.28(b) presents the fringe pattern corresponding to the other in-plane displacement component v. These interferograms were obtained by means of the Fourier spatial filtering procedure (see Section 3.3) of the plane wavefront transmitted through the double-exposure reflection hologram. The images of the plane specimen with a circular hole under tension reconstructed with the same hologram are shown in Figure 2.25. The fringe pattern characterizing the field of the displacement component normal to the plane specimen surface is presented in Figure 2.33(a). Note that the fringe patterns obtained by means of the above-described combined techniques have a direct mechanical interpretation in the same way as the lines of equal levels of the corresponding displacement components. But a practical implementation of this approach is rather complex in comparison to the usual procedure of reflection hologram reconstruction. Moreover, Figures 2.25, 2.33(a) and 3.28 clearly illustrate that the quality of interference fringes obtained as a result of the usual reflection hologram reconstruction greatly exceeds the quality of the fringes corresponding to the separate displacement component determination procedure.

It is evident that image plane holography is the most natural way to realize the combined application of holographic interferometry and speckle photography methods for displacement components measurement [28, 29]—see Figure 1.3(a). The interference fringe pattern describing the field of the normal to the surface displacement component can be revealed by means of the optical arrangement shown in Figure 2.19(b). This optical system represents in itself the system for optical filtering of low spatial frequencies.

Let the diameter of the central filtering aperture in the non-transparent mask located in the Fourier plane satisfy the condition

$$d_F < \frac{f_1}{M}\sqrt{\frac{\lambda}{2d}}$$

where M is the transverse magnification factor of the image plane hologram recording system (see Figure 1.3(a)); f_1 is the focus length of the recording system; d is the absolute value of the displacement vector; and λ is the wavelength. In this case the fringe pattern covering the holographically reconstructed image of the object under investigation will contain information about the normal to the object surface displacement component only. Two other in-plane displacement components can be determined through the use of the fringe patterns obtained by either Fourier filtering of the light wave diffracted by the hologram which provides filtering of the high spatial frequencies (see Figure 3.5) or pointwise filtering.

A combination of the technique for recording two displacement fields on a single photoplate (see Section 3.6) and the image plane hologram recording technique was implemented in reference [30] to study the deformation kinetics of a steel plate with a cut-out (see Figure 3.29). The geometrical dimensions of the specimen were: $a = 90$ mm, $b = 140$ mm and thickness $h = 30$ mm. The specimen was subjected to a step-by-step loading by the concentrated force Q (see Figure 3.29).

Simultaneous recording of the image plane holograms and specklegrams was made on the single photoplate by means of the single optical arrangement shown in Figure 3.30. The hologram of the initial state of the object surface (O_n), which was illuminated in the direction of the unit vector $\vec{e}_{S1}$, was recorded

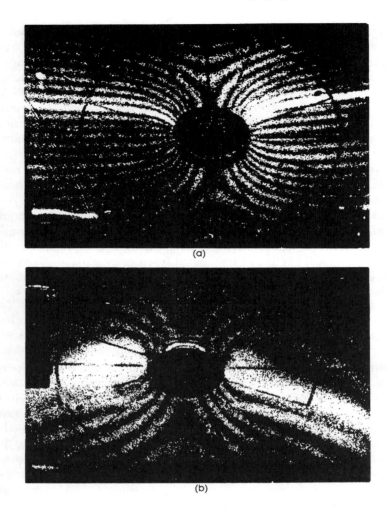

(a)

(b)

Figure 3.28. Fringe patterns corresponding to the in-plane displacement components in the direction of specimen tension (a) and in the perpendicular direction (b)

on the photoplate FP during the first exposure of duration τ. Then the illumination direction was changed into the new position $\vec{e}_{S2}$ through the use of the mirror M_1 mounted on the special precision kinematic device. The specimen was sequentially loaded by the forces Q_n and Q_{n+1} and the second and third exposures of duration 2τ were made without the reference wave. After the specimen unloading the initial illumination direction $\vec{e}_{S1}$ and reference wave were reconstructed by removing the mirror M_1 and mask

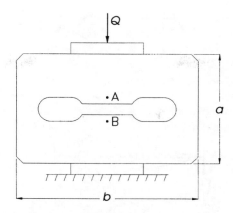

Figure 3.29. Sketch of the specimen and load applying

MA and the fourth exposure of duration τ was made to record the image plane hologram of the final object surface state (O_{n+1}).

The photoplate thus recorded contains information which is capable of being used for the image plane reconstruction and also four speckle patterns corresponding to different states of the object surface. Each pair of speckle patterns recorded under the same illumination conditions will be correlated. Pointwise filtering of this four-exposure specklegram results in the appearance of the diffraction halo which contains, in a general case, two systems of Young's fringes. These two systems have almost the same orientation but different fringe spacing, and describe the total and residual tangential displacement components corresponding to each loading step. However, the appropriate choice of exposure duration allows us to obtain the fringe pattern in the diffraction halo due to an increment of the total tangential displacement components only.

A reconstruction of the fringe pattern corresponding to the normal to the object surface displacement component can be carried out by means of illumination of the photoplate with the light wave conjugated with respect to the initial object wave. In the case involved the obtained displacement field characterizes the residual displacements. The process of the specimen deformation was investigated until the maximum load value $Q_{\mathrm{max}} = 200\,\mathrm{kN}$ by means of step-by-step loading with concentrated force increment $Q = Q_{n+1} - Q_n = 20\,\mathrm{kN}$. The dependence between the relative vertical displacement of surface points A and B (see Figure 3.29) and the load value Q obtained through the use of the speckle photography data is shown in Figure 3.31. This dependence has a linear character when the force value is less than $Q = 1.2 \times 10^5\,\mathrm{N}$. The non-linearity of the deformation process which is observed for the greater loading values is due to the plastic strains. The

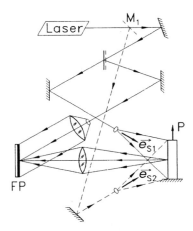

Figure 3.30. Schematic of the optical arrangement for simultaneous recording of image-plane hologram and specklegram

holographic fringe pattern corresponding to the unloading after the maximum load peak, which is presented in Figure 3.32, confirms the presence of plastic strains.

In conclusion, it should be noted that combination of the methods of holographic and speckle interferometry can be powerfully implemented for the solution of various problems dealing with strain and stress determination of various structures.

3.8   White-light speckle photography

A speckle pattern represents a random array of bright and dark spots resulting from a cross-interference of coherent light waves scattered by different points of an optically rough surface. Such a spot array can be considered as a reference grating attached to the object surface which serves as a basis for various speckle photography techniques used for the measurement of displacement components. On the other hand, the random variation of the reflection properties of the surface under investigation can be created by spraying this surface with a layer of special paint. In this case a fine pattern of random speckles can appear when the surface is illuminated by white light. The magnified images of the surface covered with a special fine powder which are shown in Figures 3.33(a)–(c) confirm the above-mentioned fact. The method which utilizes either artificially created or naturally present random patterns as a reference grating for displacement measurement is called white-light speckle photography.

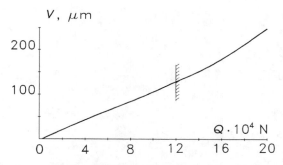

Figure 3.31. Dependence between the relative displacement of the specimen surface points A and B and applied load

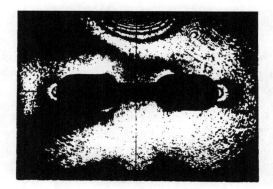

Figure 3.32. Holographic interferogram of the specimen residual strain

For the first time white-light speckle photography techniques were implemented through the use of a so-called retroreflective paint [31, 32]. This paint is composed of tiny glass spheres with diameters from 15 to 200 μm combined with an aluminium powder. These microspheres provide a scattering of light from the object surface within a narrow solid angle, the axis of which almost coincides with the illumination direction. The aluminium powder serves to enhance the reflection properties of the object surface. An interference of light waves reflected and scattered by the microspheres results in the creation of a random set of bright and dark spots. Chiang and Asundi [33] proposed a surface preparation technique whereas the object surface is sprayed with a layer of white and then a layer of black paint. The black coat is applied by means of an aerosol can. The spray created a fine mist which generated a random pattern of black dots.

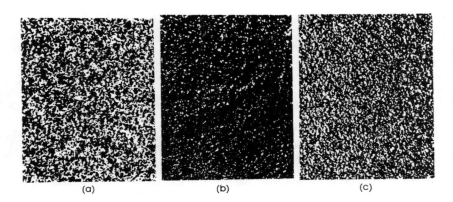

Figure 3.33. Magnified images of an aluminium object surface covered with a layer of black paint (a), special fine powder by brushing (b) and spraying (c)

The optical set-up for specklegram recording using white-light illumination is completely analogous to other set-ups using laser light (see Figure 3.4). When two artificial speckle patterns are superimposed by double exposure with relative in-plane displacement between them using white-light illumination, quantitative information about displacement components can be obtained by either the whole-field filtering technique (see Figure 3.5) or the pointwise technique (see Figure 3.6). In both cases the fringe patterns are observed by laser illumination. Figure 3.34 shows typical Young's fringes obtained by pointwise filtering of double-exposure specklegrams recorded with white-light illumination.

The set of special experiments was carried out to establish the sensitivity threshold and displacement measurement range of the method of white-light speckle photography. To reach these objectives it is necessary to estimate the influence of the transverse magnification factor of the recording optical system on the accuracy of the displacement component measurement procedure. This task was solved by determining the in-plane displacement components of the plane circular disk which was rotated around the axis normal to its surface. The main results of these test experiments are presented in reference [34].

When coherent laser illumination is used for recording speckle patterns, the average speckle size is mainly defined by the aperture number. In the case of a white-light recording procedure, a parameter that can be considered as an average speckle size mainly depends on the average dimensions of the spots applied to the object surface and the magnification factor of the recording lens. The value which is equivalent to an average speckle size (S_e) can be estimated through the angular diameter of the diffraction halo (α_0) by using the known relation $S_e = \lambda/M\alpha_0$.

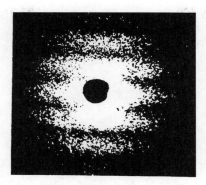

Figure 3.34. Typical Young's fringes obtained from a specklegram recorded under white light illumination

An analysis of the influence of the magnification factor on the metrological parameters of the white-light speckle photography technique showed that use of large size photoplates is one of the most important conditions for a powerful practical implementation of the technique.

The laser light wavelength is used in the white-light speckle photography method at the stage of fringe pattern observation and, hence, it can be considered a primary standard for displacement measurement. This allows us to compare the metrological characteristics of the holographic interferometry, laser, and white-light speckle photography methods. A diagram with an approximate estimation of the displacement measurement range of these methods is shown in Figure 3.35.

One of the most interesting applications of white-light speckle photography is its capability to measure the tangential displacement components in irregular zones of structures. In order to illustrate this, an investigation was carried out of the local displacements and strain near an open hole in the plane specimen subjected to shearing- stressing. A diagram of the specimen and the scheme of its loading is presented in Figure 5.6.

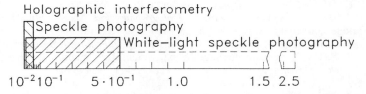

Figure 3.35. Schematic diagram of the measurement range of different optical techniques

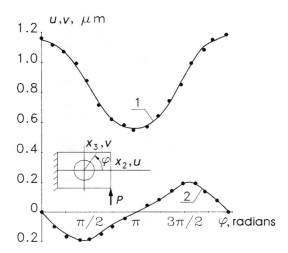

Figure 3.36. Distributions of the Cartesian in-plane displacement components along the hole edge in plate obtained by means of white-light speckle photography

The recording set-up used for double-exposure white-light specklegram recording had the following parameters: the numerical aperture of the lens $F = 2.8$; the focal length of the lens $f = 80$ mm; and the magnification factor $M = 0.5$. A square photographic film of dimensions 60×60 mm was used for speckle pattern recording. Young's fringe patterns were obtained by means of pointwise filtering at 16 points uniformly placed along the hole boundary. The distributions of the Cartesian in-plane displacement components u and v along the hole edge normalized to the concentrated force value $P = 10$ N are shown in Figure 3.36 by curves 1 and 2, respectively. The experimental results obtained are in excellent agreement with the data from the finite element method obtained for the same problem solution.

References

1. Erf, R. K. (1978) *Speckle Metrology.* Academic Press, New York.
2. Jones, R. J. and Wykes, C. (1983) *Holographic and Speckle Interferometry.* Cambridge University Press, Cambridge.
3. Dainty, J. C. (ed.) (1975) *Laser Speckle and Related Phenomena.* Springer, Berlin.
4. Sirohi, R. S. (1993) Speckle methods in experimental mechanics. In Sirohi, R. S. (ed.), *Speckle Metrology*, pp. 99–155, Marcel Dekker, New York.
5. Leenderz, J. A. (1970) Interferometric displacement measurement on surfaces utilizing speckle effect. *J. Phys. E: Sci. Instrum.* **4**, 277–279.
6. Butters, J. N. and Leendertz, J. A. (1971) A double exposure technique for speckle pattern interferometry. *J. Phys. E: Sci. Instrum.* **4**, 277–279.

7. Wykes, C. (1982) Use of electronic speckle pattern interferometry (ESPI) in the measurement of static and dynamic surface displacements. *Opt. Eng.* **21**, 400–406.
8. Archbold, E., Burch, J. M. and Ennos, A. E. (1970) Recording of in-plane surface displacement by double-exposure speckle photography. *Opt. Acta* **17**, 883–898.
9. Archbold, E. and Ennos, A. E. (1972) Displacement measurement from double exposure laser photographs. *Opt. Acta* **19**, 253–271.
10. Ablekov, V. I., Zubkov, P. I. and Frolov, A. V. (1976) *Optical and Electronic-optical Information Processing.* Mashinostroenie, Moscow.
11. Lowenthal, S. and Arsenault, H. (1970) Image formation for coherent diffuse objects: statistical properties. *J. Opt. Soc. Amer.* **60**, 1478–1483.
12. Khetan, R. P. and Chiang, F. P. (1976) Strain analysis by one-beam laser speckle interferometry. 1. Single aperture method. *Appl. Opt.* **15**, 2205–2215.
13. Gaskill, J. D. (1978) *Linear Systems, Fourier Transforms and Optics.* Wiley, New York.
14. Li, D. W., Chen, J. B. and Chiang, F. P. (1985) Statistical analysis of one-beam subjective laser speckle interferometry. *J. Opt. Soc. Amer.* **A2**, 657–666.
15. Chiang, F. P. and Juang, R. M. (1976) Vibration analysis of plate and shell by laser speckle interferometry. *Opt. Acta* **23**, 997–1009.
16. Chiang, F. P. and Lin, C. I. (1980) Stress analysis of in-plane vibration of 2-D structures by a laser speckle method. *Appl. Opt.* **19**, 2705–2708.
17. Francon, M. (1978) *La granularité laser (speckle) et ces applications en optique.* Masson, Paris.
18. O'Neill, E. L. (1956) Transfer function of annular aperture. *J. Opt. Soc. Amer.* **56**, 285–288.
19. Shchepinov, V. P., Vlasov, N. G. and Novikov S. A. (1990) Displacements and strains measurement by four-exposure speckle photography. *Zhur. Tekh. Fiz.* **60**, 43–50.
20. Duffy, D. E. (1974) Measurement of surface displacement normal to the line of sight. *Exp. Mech.* **14**, 378–384.
21. Hudson, R. R. and Setopoulos, D. P. (1975) A speckle interferometric method for the determination of time dependent displacements and strains. *Strain* **11**, 126–129.
22. Fagon, W. F. (1977) Novel speckle pattern camera and analyzer. In *Proc. VIth Int. Conf. on Experimental Stress Analysis*, pp. 456–461, Munich.
23. Shchepinov, V. P., Novikov, S. A. and Yakovlev, V. V. (1981) Elastic-plastic deformation determination by speckle photography. In *Physics and Mechanics of Deformation and Fracture*, pp. 82–86, Energoatomizdat, Moscow.
24. Novikov, S. A., Shchepinov, V. P. and Yakovlev, V. V. (1983) Displacement determination on the surface of deformed object by four-exposure speckle photography. In *Deformation and Fracture of Nuclear Machinery Materials*, pp. 54–61, Energoatomizdat, Moscow.
25. Marti, L., Patinio, A. and Ostrovsky, Y. I. (1981) Determination of the sign of displacement for points on the deformed surface in speckle interferometry. *Pizma Zhur. Tekh. Fiz.* **7**, 970–973.
26. Osintsev, A. V., Novikov, S. A. and Shchepinov, V. P. (1987) Displacement value and direction determination by four-exposure speckle photography. In *Strengh Analysis and Testing of Materials and Structures Elements of Nuclear Machinery*, pp. 48–51, Energoatomizdat, Moscow
27. Boone, P. M. (1975) Use of reflection holograms in holographic interferometry and speckle correlation for measurement of surface displacement. *Opt. Acta* **22**, 579–589.

28. Uozato, H., Iwata, K. and Nagata, R. (1977) Measurement of 3-dimensional displacements from a single image plane hologram using the combined holographic and speckle interferometry. *J. Appl. Phys.* **16**, 1689–1690.

29. Shaker, C. and Cirohi, R. S. (1978) Hologram interferometry and speckle photography combined for stress analysis. *Optik* **51**, 141–146.

30. Novikov, S. A., Shchepinov, V. P. and Yakovlev V. V. (1980) Combined application of holographic interferometry and speckle photography for deformation studing. In *Physics and Mechanics of Deformations and Fracture*, pp. 75–83, Atomizdat, Moscow.

31. Forno, C. (1975) White light speckle photography for measuring deformation, strain and shape. *Opt. Laser Technol.* **16**, 217–221.

32. Boone, P. and De Backer, L. C. (1976) Speckle methods using photography and reconstruction in incoherent light. *Optik* **44**, 343–355.

33. Chiang, F. P. and Asundi, A. (1981) A white light speckle method applied to the determination of stress intensity factor and displacement field around a crack tip. *Engineer. Fract. Mech.* **15**, 115–121.

34. Novikov, S. A., Pisarev, V. S. and Fursov, A. N. (1987) Stress determination in the contours of holes in flat structure elements by speckle photography in white light. *Probl. Prochn.* **N7**, 81–85.

4 DISPLACEMENT DERIVATIVES DETERMINATION

In most practical cases, the main objective of different holographic and speckle interferometric techniques implemented in strain analysis of structures is the quantitative values of strains and stresses on the object surface area of interest. As mentioned earlier, the methods of holographic and speckle interferometry are the most effective for measurement of displacement components fields on the surface under investigation. In order to determine strain or stress distributions both on the object surface and in the normal to the surface direction through a thin-walled structure, various combinations of both the displacement components and their partial derivatives with respect to curvilinear coordinates lying on the object surface must be used.

A discrete character of measurement procedure inherent in most optical interferometric methods leads to the fact that either a discrete set of experimental data obtained must be approximated with some analytical function or local numerical differentiation must be performed before strain and stress calculation. The known conventional difficulties related to the numerical differentiation procedure in the case when holographic and speckle interferometry is applied in a quantitative strain analysis may increase considerably due to a high sensitivity of these methods with respect to both strain-induced and rigid body displacement of real structures. That is why much attention was paid to establishing and justifying the most accurate and reliable ways of displacement component determination in Chapters 2 and 3.

However, it is well known that even the most accurate representation of a discrete data set including its correct approximation with some analytical dependencies can, in a general case, be incapable of ensuring an accurate determination of the derivatives of these values. Therefore, an illustration of some effective approaches of displacement derivatives determination is the main objective of this chapter. These approaches can be classified into three categories: traditional numerical differentiation techniques, geometrical methods that are based on the measurement of an orientation and curvature of the interference fringes and optical compensation techniques established upon real-time superposition of an actual and artificial displacement field. Much attention is given to displacement derivatives determination in irregular zones of structures which are of great interest from a strength analysis viewpoint.

An influence of both the form and power of the approximating power and trigonometrical series on the accuracy of strain derived from data obtained by holographic interferometric techniques is illustrated by using some classical problems of the elasticity theory as examples. A comparison of experimental and theoretical results allows us to introduce some criteria for a reliable choice of the power of the approximating series. The most powerful approaches to a local numerical differentiation of deflection distributions of bending plates obtained through the holographic data are also presented. The so-called semi-geometrical method for determining the first-order deflection derivatives which is most effective for stress concentration determination in bending plates is also discussed in this chapter.

A non-conventional approach for determination of slopes and curvatures of a deformed surface of regular bending plates by means of compensation holographic interferometry is proposed. The principle of this technique is that a quantitative interpretation of fringe patterns can be changed by measuring the geometrical parameters of the interferometer set up corresponding to the compensation instants.

In conclusion the methods of shear speckle interferometry and defocused speckle photography which are capable of directly determining the displacement derivatives with respect to the spatial coordinates are briefly considered.

4.1 Numerical differentiation of a discrete set of experimental data

The results of interpretation of holographic interference fringe patterns and speckle photographs, or speckle interferograms, in a general case represent the discrete functions of displacement components $u(\vec{r})$ $v(\vec{r})$ and $w(\vec{r})$ or derivatives $\partial u(\vec{r})/\partial \alpha_i$ and $\partial w(\vec{r})/\partial \alpha_i$ with respect to curvilinear orthogonal spatial coordinates α_i $(i = 1, 2)$. These functions are defined over a finite set of points $\{\vec{r}_i\}$ lying on the object surface. A different combination of displacement components and spatial derivatives are necessary for strain and stress determination for which the procedure will be described in Section 5.1.

If three displacement components are determined simultaneously from holographic fringe patterns, and when the pointwise procedure of speckle photograph interpretation is used, the displacement functions are usually defined at the nodes of a uniform mesh. When the direction of the displacement vector is known beforehand and, hence, its magnitude can be determined through the use of a single fringe pattern, as occurs, for instance, in the case of plate bending investigation, the points of set $\{\vec{r}_i\}$ coincide with the centres of interference fringes and the nodes where experimental data are known construct a non-uniform mesh.

The methods of discrete data approximation and their subsequent numerical differentiation when each individual measurement result can be characterized by a random error

$$h_i = y_i - g_i \qquad (4.1)$$

(where g_i is the true value of a function g at the point $\vec{r}_i$; y_i is its measured value at the same point) are described in numerous works both in the field of applied mathematics and in the field of the measurement data interpretation [1–7]. Therefore, in this section a brief description of the most effective and powerful approaches which use the results of holographic and speckle interferometric measurements as input data will be discussed.

The general concept of most numerical differentiation techniques consists of an approximation of discrete experimental data set with suitable continuous dependence and subsequent analytical differentiation of this dependence. The differences of possible approaches consist of a choice of approximating function and method of approximation.

The approximation procedure based on the least square method is more than adequate since this approach represents, as our practical experience has shown, a relatively simple, convenient and reliable way to construct a functional dependence describing a discrete data set. According to the least square method, the unknown function $g(\vec{r})$ must be substituted with r.m.s. regression $\bar{g}(\vec{r})$ which minimizes the residual square sum of the following form:

$$\sum_{i=1}^{n} [\bar{g}(\vec{r}_i) - y(\vec{r}_i)]^2 \Rightarrow \min. \qquad (4.2)$$

The linear regression is most frequently used:

$$\bar{g}(\vec{r}) = \sum_{p=1}^{m} C_p \varphi_p(\vec{r}) \qquad (4.3)$$

where φ_p are the basic functions and C_p the unknown coefficients to be determined. These coefficients can be found through the solution of a linear equation system which is formed by means of substitution of relations (4.3) into condition (4.2).

The methods of numerical differentiation of discrete functions with respect to a single spatial direction can be specified as global and local. In the former case a single approximating function $g(\vec{r})$ should be determined over the whole interval where an independent variable is defined. An analytical differentiation of this function immediately gives a derivative distribution. Moreover, the global approaches can be applied to solve two different problems, namely:

- Differentiation of functions defined over a finite interval
- The same procedure for functions defined over a closed line (contour)

The local techniques assume a pointwise determination of derivatives in the nodes of a mesh $\{\vec{r}_k^*\}$, which, generally, may not coincide with the nodes of the initial mesh $\{\vec{r}_k\}$ where the input data are defined. In the case concerned, in order to obtain the derivative value at some point $\vec{r}_k^*$, an individual (local) approximation should be constructed over a subset of mesh nodes contained in the nearest vicinity of the point $\vec{r}_k^*$. It should be noted that the values of functions $\bar{g}(\vec{r}_k)$ at the common point of two neighbouring intervals may differ.

Naturally, if some discrete function is defined over a two-dimensional domain, all possible combinations of global and local techniques can be used for determination of derivatives.

DIFFERENTIATION OF A DISCRETE FUNCTION DEFINED OVER A FINITE INTERVAL

The choice of the form and number of basic functions $\varphi_p(\vec{r})$ (4.3) for a global differentiation procedure demands, except for mathematical considerations, specific experience in solving particular problems. Difficulties may also be encountered following an apparently good approximation, i.e. meeting condition (4.2), which sometimes does not mean that an accurate determination of derivatives is automatically ensured. Therefore, particular recommendations concerning a reliable choice of the basic functions can be formulated in a general case only. A specific class of these functions should be adopted for solving each class of mechanical problems. For instance, the polynomial regression is widely used, in particular for strain and stress analysis of bending plates. In this case the one-dimensional function $\bar{g}(\vec{r})$ can be represented in the following form:

$$\bar{g}(x) = \sum_{p=0}^{m} C_p x^p. \tag{4.4}$$

Let's consider a particular example. Figure 2.32 shows a two-exposure interferogram of a thin circular plate with radius $R = 60\,\text{mm}$ and a thickness $h = 2\,\text{mm}$ which is subjected to a uniform internal pressure. This elastic problem has a reliable solution in terms of stress that can serve as a criterion of accuracy for the numerical differentiation procedure used. An axial-symmetrical character of deformation of the object provides us with the possibility to obtain the stress distribution in plate from the one-dimensional deflection distribution in the radial direction. In order to find the first $\partial w/\partial r$ and second $\partial^2 w/\partial r^2$ order derivatives, the discrete deflection function defined at the nodes, which are the points of intersection of the dark interference fringes and the plate radius, is approximated with a segment of power series of even powers:

$$w(r) = \sum_{p=0}^{m} C_p r^{2p}. \tag{4.5}$$

The accuracy of the differentiation procedure concerned greatly depends on the number m in expression (4.5). The residual square sum (RSS) can be used as a possible criterion to correctly choose the number of series (4.5) coefficients m:

$$\text{RSS} = \sum_{i=1}^{n} \bar{h}_i^2 \tag{4.6}$$

where $\bar{h}_i$ is defined by expression (4.1) substituting $g(\vec{r}_i)$ by $\bar{g}(\vec{r}_i)$.

A relative stabilization of the RSS value (4.6) can be considered as an adequacy criterion of the discrete data set approximation. In our example such a stabilization is observed when the RSS value becomes equal to—or more than—two. Thus, the deflection function in the radial direction can be represented as a bisquare function:

$$w = C_0 + C_1 r^2 + C_2 r^4. \tag{4.7}$$

An analysis of the RSS value should be completed with an analysis of the remainder function $\bar{h}_i$ (4.1). Figure 4.1 illustrates how the remainder values change along the plate radius. The distribution corresponding to $m = 1$ (curve 1) has a relatively regular character, whereas the analogous dependence for $m = 2$ (curve 2) can be considered as a random distribution denoting an adequacy of the approximation used. The dependencies approximating the experimentally obtained deflection values by means of regression (4.5) are shown in Figure 4.2 by curves 1 ($m = 1$) and 2 ($m = 2$) respectively. It is evident that the better approximation corresponds to $m = 2$. The first and second deflection derivatives with respect to the distance from the plate center are presented in Figure 4.3 by curves 1 and 2 respectively.

The distributions of radial σ_r and circumferential σ_φ stresses along the plate radius are shown in Figure 4.4 by curves 1 and 2 respectively. These data show good comparative results with corresponding data from the analytical solution of analogous elastic bending problem [8]. Consequently, this shows a confirmed reliability of the numerical differentiation procedure used. However, to complete the presented analysis, it should be noted that expression (4.7) for $m = 2$ coincides with the form of the analytical solution of the problem of a

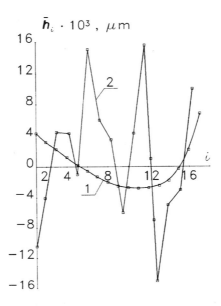

Figure 4.1. Distribution of the remainder values along a plate radius (curve 2 is multiplied by 10)

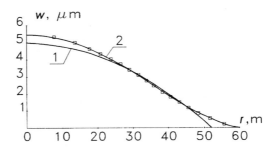

Figure 4.2. Discrete deflection distribution and its approximation with power series along a plate radius

circular plate bending caused by an internal pressure for arbitrary boundary conditions. This fact, beyond doubt, exerts its main influence on the correct approximation of the deflection distribution and shows that use of additional information concerning a character of deformation of the object under study leads to an increase in the accuracy of the results of numerical differentiation procedure.

For a more detailed illustration of the above-mentioned circumstance we shall consider a bending strain investigation of the structure shown in Figure

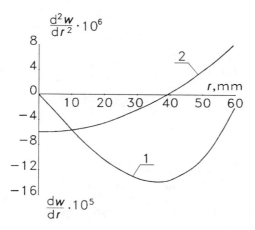

Figure 4.3. The first (1) and second (2) order radial deflection derivative distribution along a plate radius

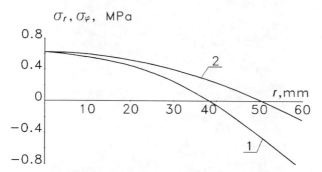

Figure 4.4. Distributions of the radial (1) and circumferential (2) stress along a plate radius

4.5. The annular segment of this structure represents a thin annular plate with, rigidly-elastically clamped edges of external radius $R = 32.5$ mm and internal radius $r_0 = 7.5$ mm edges. A typical fringe pattern corresponding to the plate loading with internal pressure $q = 0.006$ MPa obtained by using the optical set-up in Figure 2.19 is shown in Figure 4.6.

The best approximation of the experimental deflection distribution with series (4.5) corresponds to the power $m = 5$ (see curve 1 in Figure 4.7(a)). If we now reduce the experimental data set, namely exclude four experimental points near the outer plate fixing denoted by the crosses in Figure 4.7(a), the final result will be considerably changed. In this case the series (4.5) with the power

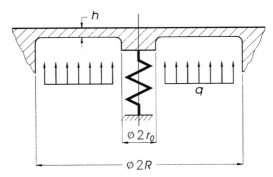

Figure 4.5. Sketch of the annular plate subjected to an internal pressure. (Reproduced by permission from Ostrovsky, Shchepinov and Yakovlev, Holographic Interferometry in Experimental Mechanics. Fig. 5.6a, vol. 60, copyright Springer Verlag GmbH & Co. KG)

Figure 4.6. Interferogram of the annular plate deflection. (Reproduced by permission from Ostrovsky, Shchepinov and Yakovlev, Holographic Interferometry in Experimental Mechanics. Fig. 5.5, vol. 60, copyright Springer Verlag GmbH & Co. KG)

$m = 5$ results as curve 2 in Figure 4.7(a), completely distorting the deformation character in the plate region considered.

Another situation can be observed when the form of theoretical solution for annular plate bending [8] is implemented for approximation of the discrete deflection distribution:

$$w(r) = C_0 + C_1 r^2 + C_3 \ln\left(\frac{r}{R}\right) + C_4 r^4 \ln\left(\frac{r}{r_0}\right). \tag{4.8}$$

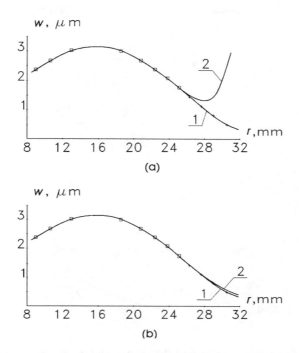

Figure 4.7. Approximation of the annular plate deflection with power series of even powers (a) and series having the form of theoretical solution (b) (Reproduced by permission from Ostrovsky, Shchepinov and Yakovlev, Holographic Interferometry in Experimental Mechanics. Fig. 5.6b, vol. 60, copyright Springer Verlag GmbH & Co. KG)

Unknown coefficients C_p, which can be determined by means of the least square method, are characterized by the actual geometrical parameters of plate, boundary and loading conditions. Figure 4.7(b) illustrates the regression of the form (4.8) for complete (curve 1) and reduced (curve 2) data sets. These two curves have almost the same form, so illustrating a high immunity of the regression based on theoretical considerations to perturbations of input data. This is of great importance when a determination of displacement derivatives must be performed at an end point of an interval where a discrete function is defined. Note that such boundary points are points of great interest, since in most cases some coincide with the region of maximum strain and stress concentration.

Distributions of the first- and second-order deflection derivatives along the plate radius obtained by using the complete data set are presented in Figures 4.8(a) and (b), respectively. Curves 1 and 2 correspond to the regression of form (4.5) and (4.8), respectively. It can be seen that the maximum distinction between two dependencies is about 50 per cent.

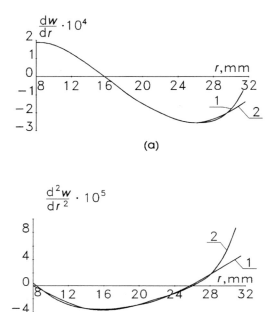

Figure 4.8. The first (a) and second (b) order radial deflection derivative distributions along a radius of the annular plate

NUMERICAL DIFFERENTIATION OF A DISCRETE FUNCTION DEFINED ON A CLOSED CONTOUR

The questions dealing with the accuracy of derivative determination near the end of an interval on which a discrete function is defined are automatically eliminated when strain and stress distributions have to be constructed along a smoothed closed contour. An investigation of strain and stress concentration is the most characteristic situation when the input data needed for the problem solution should be represented in the form of one-dimensional arrays defined on a closed line. For an obvious reason, the displacement component and its derivative distributions along a sufficiently smoothed closed line can be described by periodic functions. A Fourier or trigonometrical series are the most convenient and effective way of approximation of periodic functions of a general form. The coefficients of these series are usually determined by means of the least square technique.

It should be noted again that in most practical cases, by not approximating the function itself but its derivatives with respect to the spatial coordinates is of main interest. When the proper choice of approximating function has been made, the number of series discrete harmonics will exert its main influence on the accuracy of derivatives determination. This argument results from the fact that the term of a trigonometric series with the number p, for instance $A_p \cos px$, becomes $-pA_p \sin px$ after differentiation. If the coefficient A_p is small the following variants may be possible: (i) keeping a term of series with the number p is an unjustified assumption; (ii) rejecting the same term is an unjustified assumption. In both cases a mistake in the estimation of the validity of small coefficient A_p leads to a significant error of the first-order derivative determination. A calculation of the second-order derivative in such a case has no practical sense.

A brief analysis of the influence of measurement errors on the accuracy of a discrete periodical function approximation with trigonometrical series and subsequent derivatives determination, necessary from a practical point of view, will be presented below.

The general one-dimensional problem can be formulated in the following way. By using the trigonometrical polynomial of the order m of the form

$$\phi_m(x) = A_0 + \sum_{p=1}^{m} \{A_p \cos px + B_p \sin px\} \tag{4.9}$$

a continuous 2π-periodical function $f(x)$ has to be approximated. The least square approximation of the function $f(x)$ is ensured when the residual $\rho(f, \phi_m)$ reaches its possible minimum value:

$$\rho(f, \phi_m) = \sqrt{\frac{1}{2\pi} \int_0^{2\pi} [f(x) - \phi_m(x)]^2 \, dx} \Rightarrow \min. \tag{4.10}$$

The theory of Fourier series proves that the validity of condition (4.10) occurs when the coefficients of series (4.9) are expressed by

$$A_0 = \frac{1}{2\pi} \int_0^{2\pi} f(x) \, dx, \qquad A_p = \frac{1}{\pi} \int_0^{2\pi} f(x) \cos px \, dx,$$

$$B_p = \frac{1}{\pi} \int_0^{2\pi} f(x) \sin px \, dx, \qquad\qquad p > 0. \tag{4.11}$$

The coefficients (4.11) represent the coefficients of Fourier decomposition of the function $f(x)$ [9].

The above-mentioned fact is also valid if the function $f(x)$ is defined on a discrete set of points

$$x_i = \frac{2\pi i}{n+1} \qquad i = 1, 2, 3, \ldots .$$

In such a case coefficients (4.11) of the best approximating polynomial (4.9) take the following form:

$$A_0 = \frac{1}{n+1} \sum_{i=0}^{n} f\left(\frac{2\pi i}{n+1}\right),$$

$$\left\{\begin{matrix} A_p \\ B_p \end{matrix}\right\} = \frac{2}{n+1} \sum_{i=0}^{n} f\left(\frac{2\pi i}{n+1}\right) \left\{\begin{matrix} \cos\left(p\dfrac{2\pi i}{n+1}\right) \\ \sin\left(p\dfrac{2\pi i}{n+1}\right) \end{matrix}\right\}, \qquad p = 1, 2 \ldots m; \qquad m \leqslant \frac{n}{2}.$$

$$(4.12)$$

In a real physical experiment the values $f(x_i)$ can be determined to within a certain measurement error. These errors distort the polynomial obtained by the least square method. If we suppose that the construction of polynomial (4.9) with coefficients (4.11) or (4.12) is made by using measured values $f(x_i)$ characterized by random errors with zero average meaning and dispersion, which is independent of the point number n, the following relation is valid [10]:

$$D[\Gamma_m(x)] = \frac{2m+1}{n+1} \sigma^2 \tag{4.13}$$

where $\Gamma_m(x)$ is the random error of the approximating polynomial; $D[\Gamma_m(x)]$ is the dispersion of the $\Gamma_m(x)$; σ is the r.m.s. error of the $f(x_i)$ values; and $m \leqslant n/2$. The r.m.s. value of error $\Gamma_m(x)$ corresponding to dispersion (4.13) is given by

$$\sigma_m(x) = \sqrt{D[\Gamma_m(x)]} = \sigma\sqrt{\frac{2m+1}{n+1}}. \tag{4.14}$$

Apart from random errors of form (4.14), an accuracy of approximation by means of polynomial (4.9) is affected by distortion of Fourier coefficients (4.12) and, also, a mistake resulting from substituting the infinite Fourier series with its finite segment of the order m. An analysis of all three possible types of errors reveals the following conclusions. The growth of the number n, i.e. the number of points where measurement of the values $f(x_i)$ are carried out, decreases the dispersion of random error (4.13) and, thus, favours the decrease of errors resulting from Fourier coefficient distortion. For instance, for $n = 27$ and $m = 3$ according to estimation (4.14) the value of the r.m.s. error of the approximating polynomial will be less by two times the value of r.m.s.

measurement errors. This fact, in particular, illustrates the so-called smoothing effect of the least square method.

If the number n is fixed, the choice of a series power m, reduces to a trade-off of the random error increasing with a growth of m in accordance with equation (4.14) and, also, the error caused by a distortion of Fourier coefficients, the number of which is proportional to m, against the errors dealing with rejecting a Fourier series residual that decreases with an increase in the polynomial order m.

In a real holographic or speckle interferometric experiment the number of measurement points is usually the highest possible. However, this number depends on the error of coordinate determination on the object surface and, also, upon errors made in measurement of interference fringe pattern parameters (fringe spacing, fringe orders, etc.). In particular, when displacement or slope components measurement has to be made along an edge of a hole in a plate or shell, the number of measurement points is never below 16. Hence, the correct choice of the order of the approximating polynomial is the main problem that should be solved for an accurate determination of derivatives of periodical discrete functions.

Following this, we can say that an error analysis based on the approximation theory is capable of presenting some useful general recommendations, but cannot give a direct answer concerning the criteria of a proper choice of the order of the polynomial (4.9). Therefore, to estimate an accuracy of the numerical differentiation procedure some additional considerations should be made. These can be made, for instance, with an analysis of measurement errors in combination with a comparison of strain and stress values expressed in the experimentally obtained results with corresponding analytical strain and stress distributions resulting from known elastic problem solutions.

The value of residual (4.10) or analogous value of RSS, as mentioned above, can serve as a relatively reliable criterion to choose the order m of approximating polynomial (4.9). The value of residual (4.10) for the discrete function $f(x_i)$ takes the form

$$\rho(f, \phi_m) = \sqrt{\frac{1}{n+1} \sum_{i=1}^{n} [f(x_i) - \phi_m(x_i)]^2} = \sqrt{\| f \|^2 - A_0^2 - \frac{1}{2} \sum_{p=1}^{m} (A_p^2 + B_p^2)}$$

$$(4.15)$$

where

$$\| f \|^2 = \frac{1}{n+1} \sum_{i=0}^{n} f^2\left(\frac{2\pi i}{n+1}\right).$$

In practice, usually, a growth of power m from 1 to some value k leads to a notable decrease of residual value (4.15). In this case two variants are possible: (1) an addition of the term with the number $(k + 1)$ results in the residual increase; (2) an addition of the terms with numbers $(k + 1)$, $(k + 2)$, . . . decreases the residual value by not more than by 3–5 per cent. In both cases the proper choice of the m value requires additional considerations.

As mentioned above, a comparison of the experimental data obtained with the known data resulting from the solution of testing problems can be used to establish a criterion of choice of the order of the polynomial of form (4.9). Our practical experience shows that the best agreement is reached if in both the above-mentioned cases the last series term with the number $m = k$ is retained.

Further development of the approach, capable of increasing the reliability of the results of numerical differentiation, deals with, apart from a minimization of residual value (4.15), a comparative analysis of the value of coefficients A_p and B_p (4.12) and errors of its determination. These errors can be estimated from inequalities (2.58) and (2.59) taking (4.14) into account. Below we shall consider a similar procedure using some characteristic elastic problems.

We shall return to the classical elastic problem of a stress concentration determination near a circular open hole in plane strip subjected to tension—see Figure 2.24. Distributions of the Cartesian displacement components along the hole boundary corresponding to the tensile stress increment $\sigma_0 = 40\,\mathrm{MPa}$ are presented in Figure 2.26. Remember that the displacement components errors Δd_1, Δd_2 and Δd_3 can be obtained from estimations (2.58)–(2.59). Distribution of the circumferential strain along the hole edge is of main interest and can be expressed through the in-plane displacement components $d_2 \equiv u$ and $d_3 \equiv v$ according to relation (5.87).

Experimentally obtained values of the in-plane displacement components are approximated with trigonometrical series of form (4.9):

$$u = \sum_{p=1}^{m} A_p \cos p\varphi, \qquad v = -\sum_{p=1}^{m} B_p \sin p\varphi. \qquad (4.16)$$

The results of interpretation of three holographic interferograms in the elastic deformation range corresponding to the net stress increment $\sigma_0 = 40\,\mathrm{MPa}$ shows that the first and third harmonics are contained in an approximation series (4.16) only. These values are in the following ranges:

$$\begin{aligned} A_1 &= 9.8 \pm 0.1\,\mu\mathrm{m} & A_3 &\leqslant 0.2\,\mu\mathrm{m} \\ B &= 3.3 \pm 0.2\,\mu\mathrm{m} & B_3 &\leqslant 0.2\,\mu\mathrm{m}. \end{aligned} \qquad (4.17)$$

Note that addition of the third harmonics from (4.17) into the corresponding series (4.16) decreases the value of residual (4.15) by less than 2 per cent. Therefore, one can expect that rejecting the terms A_3 and B_3 when substituting expressions (4.17) and (4.16) into (5.87) will not influence the accuracy of the numerical differentiation procedure involved.

Naturally, the question arises whether it is possible to find the relation between threshold values of coefficients in equations (4.9) or (4.16) with the values of errors Δd_2 and Δd_3 from equation (2.59). It should be assumed that the coefficients exceeding the maximum possible range of the measurement errors must be kept for numerical differentiation. This range is equal to the double magnitude of the absolute value of the measurement error, Thus, the condition of the coefficients validity can be represented in the following form:

$$A_p, B_p \geqslant 2\Delta d_2 = 2\Delta d_3. \tag{4.18}$$

In the case concerned $2\Delta d_2 = 0.3\ \mu m$ that exceeds the values A_3 and B_3 from equation (4.17). Let's introduce two characteristic examples which can be considered as a justification of condition (4.18).

The distribution of relative circumferential strain $\varepsilon_\varphi/\varepsilon_0$ (where $\varepsilon_0 = \sigma_0/E$, σ_0 is the net stress increment) obtained experimentally for the elastic modulus value $E = 71000\ \mathrm{MPa}$ is shown in Figure 4.9 by curve 1. Note that this curve completely coincides with the dependence of relative normal strain $(-\varepsilon_0/\mu\varepsilon_0)$ [11]:

$$\varepsilon_z = \frac{2(d_1 - d_1^\infty)}{h} - \mu\varepsilon_0 \tag{4.19}$$

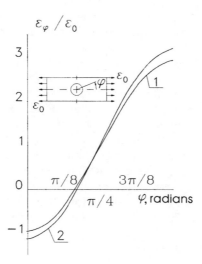

Figure 4.9. Experimental (1) and numerical (2) distribution of the relative circumferential strain along the hole edge in plane specimen under tension

where $(d_1 - d_1^\infty)$ is the experimentally obtained relative changing of the specimen thickness in the normal to the surface direction; ε_z is the usual strain in the same direction; h is the specimen thickness; and μ is Poisson's ratio of the specimen material.

The dependencies for the relative circumferential and normal strains which almost coincide with curve 1 in Figure 4.9 are obtained through the use of the finite element method (FEM) [12]. The calculations were carried out by using the FITING software package. The characteristics of the numerical model used are presented in Section 2.5. The distribution of relative circumferential strain in the specimen middle plane along the hole edge is shown in Figure 4.9 by curve 2. This distribution is in excellent agreement with the analogous distribution presented by Peterson [13]. This confirms the accuracy of the FEM solution used for comparison with experimental results.

Thus practical applicability of condition (4.18) in choosing the order of the polynomial (4.9) is justified by a comparison of the experimental data with the results of the solution of an analogous problem by FEM and also with the data of the elasticity theory. This results from the agreement of the experimental data obtained from expression (4.19) with curve 1 in Figure 4.9.

Apart from the comparison criterion discussed above, the equilibrium criterion having an evident mechanical sense can be effectively used to verify the correctness of the approach to the choice of the number of coefficients of series (4.9). The essence of this criterion consists of the fact that the total sum of forces acting in any cross-section of the specimen, for instance in the central cross-section of the specimen shown in Figure 2.24, have to be equal to the total applied load $P = \sigma_0 \cdot h \cdot b$ (where b is the specimen width).

We shall consider a deformation of the plane specimen analogous to that presented in Figure 2.24 but with a thickness $h = 3\,\text{mm}$ and hole radius $r_0 = 9\,\text{mm}$. The in-plane displacement components are determined at 16 points uniformly spaced along the hole radius and each circle line of radius $r = 11, 14, 16, 19, 21, 27$ and 30 mm. The discrete data obtained are approximated with series (4.16) the coefficients of which satisfy condition (4.18). The values of these coefficients corresponding to the tension in the elastic range with the net stress increment $\sigma_0 = 28\,\text{MPa}$ are listed in Table 4.1. Typical distributions of the in-plane cartesian components u (1,2) and v (3,4) along the hole edge (1,3) and circle line $r = 19\,\text{mm}$ (2,4) are shown in Figure 4.10.

In the points of the cross-section involved, when $\varphi = 90°$ (see Figure 2.24) the circumferential strain (5.87) is the deformation in the direction of tension which is denoted as ε_x. The maximum value of this deformation is reached on the hole edge $\varepsilon_x^A = 1.23 \times 10^{-3}$. This magnitude can be used to estimate the stress concentration factor value

$$K_\sigma = \frac{\varepsilon_x^A E}{\sigma_0}. \tag{4.20}$$

Table 4.1 The coefficients of decomposition (4.16) of the in-plane displacement component u(A_k) and v(B_k) around the hole in plane specimen under tension.

r, mm	A_1	A_3	B_1	B_3
9	11.10	—	4.33	—
11	10.60	-0.61	4.14	0.73
14	10.50	-0.83	4.08	0.96
16	10.70	-0.87	4.13	1.07
19	11.20	-0.84	4.38	1.07
21	11.70	-0.82	4.48	1.12
24	12.50	-0.76	4.69	1.20
27	13.20	-0.76	.98	1.22
30	14.10	-0.76	5.15	1.20

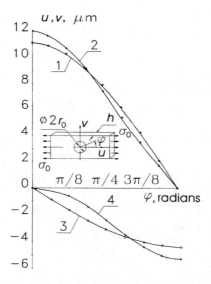

Figure 4.10. Typical distributions of the in-plane Cartesian displacement components around the hole edge in plane specimen under tension

For the elasticity modulus $E = 74\,000\,\text{MPa}$ and net stress increment $\sigma_0 = 28\,\text{MPa}$, calculation with expression (4.20) gives

$$K_\sigma = 3.26$$

that is in a good agreement with known handbook data $K_\sigma = 3.36$ [13].
The distribution of the strain ε_x along the cross-section $x_2 = 0$ is shown in

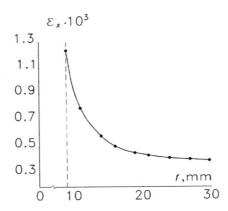

Figure 4.11. Distribution of tensile strains along the central cross-section of the plane specimen with circular hole

Figure 4.11. The load, acting normally to this cross-section, with a high degree of accuracy can be determined as follows:

$$P = 2 S_\varepsilon \cdot h \cdot E \qquad (4.21)$$

where S_ε is the area under the ε_x-curve in Figure 4.11. In the case concerned expression (4.21) yields

$$P = 2 \times 11.3 \cdot 10^{-3} [\text{mm}] \times 3 [\text{mm}] \times 74\,000 [\text{MPa}] = 5020\,\text{N}.$$

The load holding in the regular cross-sections of the specimen is $P_0 = \sigma_0 h \cdot b = 5040\,\text{N}$. These results almost coincide with one another.

The above example, which allows us to estimate the accuracy of the numerical differentiation procedure, is of a reasonably general character, but illustrates the strain determination on the plane object surface. However, the main advantage of holographic interferometry consists of the capability of studying deformation of objects with curved surfaces. The quantitative estimation of accuracy of numerical differentiation in such cases is illustrated through the local strain determination near the hole in a curved circular cylindrical shell made of aluminium alloy—see Figure 2.27.

Note that circumferential stress distribution along a circular hole edge is determined by means of relations (5.52) and (5.49). The former relation shows that the first-order differentiation of the tangential displacement components u and v have to be made to obtain the circumferential strain distribution needed for stress calculation.

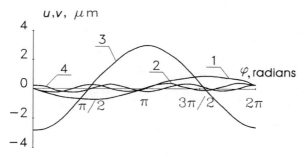

Figure 4.12. Distributions of harmonics of the tangential displacement components along the hole edge in cylindrical shell under torsion

Experimentally measured displacement component values u and v (2.78) shown in Figure 2.29 are approximated by means of a trigonometrical series of the form (4.9):

$$u = A_0 + \sum_{p=1}^{m} A_p \sin p\varphi, \qquad v = \sum_{p=0}^{m} B_p \cos p\varphi \qquad (4.22)$$

where coefficients A_p and B_p should be established by using the least square method. The testing for regression of the coefficients in equation (4.22) according to condition (4.18) for $2\Delta u = 2\Delta v = 0.2\,\mu m$ gives the following results:

$$A_1 = -0.75\,\mu m \qquad B_1 = -2.84\,\mu m$$
$$A_3 = -0.24\,\mu m \qquad B_3 = 0.24\,\mu m.$$

The distributions of harmonics of the tangential displacement components u and v with coefficients A_1, A_3, B_1 and B_3 illustrating their relative magnitudes are presented in Figure 4.12 by curves 1, 2, 3 and 4 respectively.

Let us analyse the influence of the harmonics of series (4.22) on the accuracy of the circumferential stress σ_φ determination. Figure 4.13 provides a basis for such a consideration. Curve 1 represents the relative circumferential stresses σ_φ/τ_0 on the external face along a hole edge constructed using the first and third harmonics of equation (4.22). The normalizing factor τ_0 is determined from equation (5.93). Curves 2 and 3 are obtained by substituting into equation (5.52) only the first and third harmonics respectively. Dependence 1 completely coincides with the data of the analytical solution of Van Dyke [14]. Dependencies 2 and 3, as follows from the same analytical data, can be

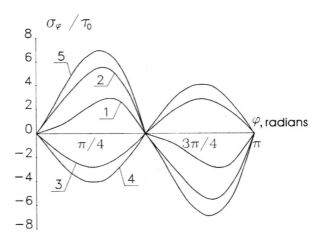

Figure 4.13. Distributions of the relative circumferential stress along the hole edge in cylindrical shell under torsion

considered as an approximate description of the relative membrane σ_φ^M/τ_0 (5.49) and bending σ_φ^B/τ_0 (5.50) stresses along the hole edge.

Therefore, we may affirm that an authentic determination of the coefficients of the first and third harmonics of decomposition (4.22) allows us to obtain the total circumferential stress σ_φ/τ_0 along the hole boundary on the external shell surface and, in addition, to estimate the values of the membrane and bending stresses. The latter fact enables us to construct an approximate distribution of circumferential stresses in the normal surface direction through the shell thickness (5.47).

The other, more accurate, approach, to the bending stress determination, as will be shown in Section 5.5, can be based upon using relations (5.53) and (5.50). Expression (5.53) as well as expression (5.52) contain the derivatives with respect to angular coordinate φ. The distinction of these two relations consists the necessity of double numerical differentiation of the normal to the object surface displacement component w (2.78) in equation (5.53). The fact that the error made in a determination of the normal displacement component w is less than the analogous error for tangential displacement components u and v by a factor of two (see estimations (2.58) and (2.59)) gives us reason to expect a tolerance of the second-order numerical differentiation procedure.

The distribution of the normal to the shell surface displacement component w along the hole edge (curve 3 in Figure 2.29) in accordance with equation (4.9) and taking into account condition (4.18), has the form

$$w = -C_2 \sin 2\varphi$$

where $C_2 = 3.05\,\mu$m. The corresponding distribution of the relative bending stress σ_φ^B/τ_0 is presented in Figure 4.13 by curve 4. The difference between curves 1 and 4 shows the refined version of the membrane stress σ_φ^M/τ_0 distribution depicted by curve 5. Curves 4 and 5 are found to compare excellently with the corresponding analytical data of Van Dyke [14].

LOCAL NUMERICAL DIFFERENTIATION

An approximation of the full set of discrete experimental data with a unified analytical relation defined on the whole interval to be considered is not always rewarding, especially when local intervals of a discrete function argument definition are of great interest from a practical standpoint. An absence of residual stabilization (4.15) with a growth of the order of polynomial (4.9) when a remainder distribution (4.1) has a non-random character is one of the most evident indicators of such a situation, which frequently occurs for complex-form experimental distributions.

When such a situation occurs, the growth of the order of polynomials (4.4) or (4.9) leads to an oscillation of the approximating function between its measured values. This fact is valid for problems where the actual deformation of the object has a different character on different intervals of the surface area to be investigated. In these cases a division of the whole interval into parts, where a deformation can be described with a definite type of approximating function, would be appropriate. These functions and their derivatives have to be linked on each border between two intervals. The spline functions embody the above-mentioned concept [15]. The cubic splines representing a set of partial cubic parabola with continuous first- and second-order derivatives are the most widely spread for approximation of experimental data. However, the appropriate choice of the end points of each interval is not such a simple problem and its correct solving should be based on practical experience. Some attempts to establish the principles of automatic choosing of dividing points of the investigated domain have been made (see, for instance, reference [16]).

The following step for localization of the numerical differentiation process consists usually of an implementation of the pointwise procedure. The principal distinction of this procedure from the global approach to numerical differentiation is that it results in a discrete set of function derivatives defined on some mesh instead of a single continuous derivative. Thus, the relatively simple approximations $\bar{g}(\vec{r})$ can be used to obtain the derivatives of complex form functions $g(\vec{r})$.

The degree of locality of each specific approach depends on the number of mesh nodes $\{\vec{r}_i^*\}$ which are used to calculate the derivative in the single node of interest $\{\vec{r}_j^*\}$. Usually this number is not more than five. Generally, handbooks contain the explicit forms of different relations used for a local approximation and corresponding derivative calculation. For instance, using a polynomial of the best least square approximation gives the following estimation of the first-order derivative with respect to the variable x [17]:

$$\bar{g}'(x_i) \simeq \frac{1}{\Delta x} \frac{3}{(n+1)(n+3) \cdot n} \sum_{j=-n}^{n} j \cdot y(x_{i+j}) \qquad (4.23)$$

where x_i is the uniform mesh with node spacing Δx.

When the integer value of absolute fringe orders are used for interference fringe pattern interpretation to obtain a one-dimensional displacement function, for instance deflection distribution $w(x)$ as it takes place for plate bending investigation, a transition to an inverse function $x(w)$ allows us to obtain a uniform mesh along the w-axis with spacing $\lambda/2$, and to use relation (4.23) for numerical differentiation.

A local differentiation procedure founded upon a polynomial approximation or interpolation, in particular upon interpolation with the second-order polynomials, is frequently used in research [4,5,7].

The simplest approach to the local estimation of derivatives is based on the linear interpolation

$$\bar{g}'(x_i^0) \simeq \frac{y_{i+1} - y_i}{x_{i+1}^* - x_i^*}, \qquad (4.24)$$

where $x_i^* \leqslant x_i^0 \leqslant x_{i+1}^*$. Sometimes the following expression is accepted:

$$x_i^0 = \frac{x_{i+1}^* + x_i^*}{2}$$

If the discrete function from equation (4.24) represents the values of plate deflection that are determined in the centres of interference fringes, the following condition is valid on intervals where the functions are smoothed:

$$y_{i+1} - y_i = \Delta w = \frac{\lambda}{2}$$

In such a case expression (4.24) takes the form

$$w'(x_i^0) \simeq \frac{\lambda}{2p_x} \qquad (4.25)$$

where p_x is the fringe spacing in the x-direction.

The accuracy of expression (4.24) depends greatly on the value of mesh spacing. The optimal magnitude of this spacing can be established in each case as a result of the counteraction of two trends. On the one hand, a decrease of mesh spacing leads to a growth in the influence of errors made in determination of discrete values on the final result. On the other hand, an increase of this spacing results in an increase of the influence of node locations. Moreover, when the mesh spacing value is comparable to the dimension of an interval to be investigated, the estimation is not capable of providing proper information about derivatives. The above-presented consideration is generally true for all local differentiation techniques. If the mesh spacing coincides with the spacing of interference fringes, as is accepted in expression (4.25), the optimal condition of local numerical differentiation of the deflection function w can be reached by varying the fringe spacing of the interference fringe pattern.

This approach can be implemented by superimposing the deflection field w with an additional artificial field $\tilde{w}$ of linear [18] or square [19] form. A description of the required technical aspects to realize this procedure will be presented in Section 4.3. When linear and square additional deflection fields are introduced simultaneously, relationship (4.25) takes the following form:

$$w'(x_i^0) \simeq \left[\frac{\lambda}{2p_x^*} - \frac{\partial \tilde{w}}{\partial x} \right]_{x=x_i^0} \tag{4.26}$$

where p_x^* is the fringe spacing of a superimposed deflection field w^*. If $\tilde{w} = \tilde{A} + \tilde{B}x + \tilde{C}y$ is the linear field (where $\tilde{A}$, $\tilde{B}$ and $\tilde{C}$ are the variables to be controlled), expression (4.25) may be rewritten as proposed by Wang, Hovanesian and Hung [20]:

$$w'(x_i^0) \simeq \frac{\lambda}{2} \left[\frac{1}{p_x^*} - \frac{1}{\tilde{p}_x} \right]_{x=x_i^0}$$

where $\tilde{p}_x$ is the uniform fringe spacing in the interferogram of the additional field $\tilde{w}$.

As an example of the practical implementation of the local numerical differentiation approach, consider the determination of the first-order deflection derivative for the circular plate clamped along its external edge and subjected to internal pressure (see Figure 2.32). Figure 4.14 presents an interferogram of the same plate for the same internal pressure as in Figure 2.32, but with superposition of the additional artificial linear deflection field $\tilde{w}$. The unstrained part of the specimen under study located outside the external plate boundary is covered by a set of uniform straight interference fringes, the spacing of which $\tilde{p}_x$ is defined by the linear term $\tilde{B}$.

The points in Figure 4.15 are obtained through use of expression (4.26) for local numerical differentiation in the mesh nodes. The continuous curve in the same figure reveals the result of global numerical differentiation (see curve 1 in Figure 4.3). Local differentiation data are well dispersed, but the character of the first-order deflection derivative is described well.

To decrease the derivative dispersion, it is necessary to approximate the discrete deflection distribution with local smoothing polynomials. By using the

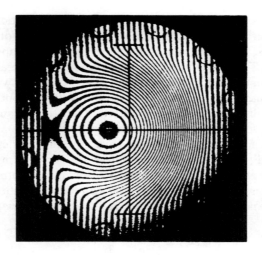

Figure 4.14. Interferogram of the superposition of the actual and artificial deflection fields

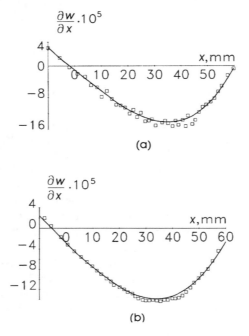

Figure 4.15. Results of local numerical differentiation of the deflection distribution along the plate radius: (a) initial, (b) after smoothing procedure

above-mentioned technique of transition to inverse functions $x(w)$ and expression (4.23) for $n = 2$ after subtraction of the constant value $\tilde{B}$ which corresponds to the derivative of the additional linear function, we can obtain the required distribution of the radial deflection derivative $\partial w/\partial r = \partial w/\partial x$ when $x \geqslant 0$ (see Figure 4.15(b)).

4.2 Determination of the deflection derivatives with respect to spatial coordinates by means of geometrical analysis of the holographic interferograms

As shown in Section 2.2, the principal relation of holographic interferometry is related to the absolute fringe order at the point of interest on the object surface with the linear combination of the unknown displacement components for a fixed observation direction. However, the function of the fringe order, which can be determined as a result of interferogram interpretation, represents only a part of the information contained in the holographic fringe pattern. In particular, in some cases such parameters of the fringe pattern as an orientation and curvature of interference fringes can be used for quantitative interpretations of interferograms, especially if the displacement derivatives with respect to spatial coordinates must be determined as a final result. Certainly, the above-mentioned parameters of a fringe pattern cannot be determined with as high an accuracy as the fringe orders, but sometimes its implementation may be very powerful for strain and stress analysis, especially in irregular zones of structures. The approaches involved were established on some geometrical relationships and will be called the geometrical methods of fringe pattern interpretation in the following.

Application of these geometrical methods shows great efficiency when the interference fringe pattern has a direct mechanical interpretation. Such a situation takes place when the bending of thin plates is holographically studied by means of the optical set-up shown in Figure 2.19. In this case the displacement vector direction coincides with the surface normal and the displacement vector value is defined by a single deflection component w. Therefore a form of deformed plate surface can be described by

$$z = w(x, y) \tag{4.27}$$

where the z-axis is directed along the surface normal.

If an interferometer with a constant sensitivity is used for recording holograms characterized by a deflection field, the interference fringes will represent the lines of equal meaning of deflection, which can be described by

$$w(x, y) = \text{const.} \qquad (4.28)$$

In this section the term 'an interference fringe' means the geometric place of the object surface points having the same—but may be non-integer—fringe order value.

Due to the assumption of small deflection and strain, which is the basis of various theories of thin plate bending, the first derivatives of deflection w with respect to Cartesian coordinates x and y in the plate plane surface represent the local angles of the surface normal rotation in the direction of the corresponding coordinate axes. These values are frequently called slopes. The second spatial derivatives of function (4.27) define the local curvatures and torsion of the deformed surface. Therefore, in this Chapter we will introduce the following notation:

$$\theta_x = \frac{\partial w}{\partial x} \qquad \theta_y = \frac{\partial w}{\partial y}$$

$$\ae_x = \frac{\partial^2 w}{\partial x^2} \qquad \ae_y = \frac{\partial^2 w}{\partial y^2} \qquad \ae_{xy} = \frac{\partial^2 w}{\partial x \partial y} \qquad (4.29)$$

Note that a more detailed description of the geometrical parameters of a deformed surface of a thin-walled structure and its connection with strains and stresses will be presented in Section 5.1.

The relationships between the local geometrical parameters of interference fringes and the geometrical parameters of the deformed surface (4.29), which define a strained and stressed state of a bending plate in an unique way, is presented in this section.

If the geometrical parameters of fringes are measured immediately from the interference pattern reconstructed by double-exposure, time-averaged, or combined holograms (see Section 2.1), these approaches can be specified as direct geometrical methods, some of which have been developed by Abramson [21]. A more accurate way of determining the geometrical parameters of fringes consists of an implementation of various compensation techniques. Some were introduced to exclude rigid body influence on strain-induced fringe pattern formation [22,23], visualization of local strains through noise of large surface form changing [24,25], and for selection of fringe patterns corresponding to individual displacement components.

The compensation technique which can be used for the measurement of derivatives (4.29) will be presented in this section. This technique is based on real-time introduction of an additional pseudo-displacement field during a stage of fringe formation. The parameters of this pseudo-displacement field can be varied in order to visualize the characteristic fringe pattern having the simplest geometrical forms in the surface region of interest. The parameters of the obtained resultant fringe patterns can be measured with a higher accuracy than those of the initial fringe pattern. The special optical arrangement needed to implement the compensation technique will be described below.

THE DIRECT GEOMETRICAL METHODS OF FRINGE PATTERN INTERPRETATION

Consider an arbitrary point $P(x, y)$ on the surface of a deformed object. We shall assume that this point is not a point of stationarity of the deflection field, i.e.

$$\vec{\text{grad}}\, w(x, y)|_P \neq 0 \tag{4.30}$$

The total derivative of deflection w with respect to the arc length S (in the case concerned arc S represents the interference fringe) is equal to zero in accordance with equation (4.28). This condition leads to the following relation between slopes:

$$\frac{dw}{ds} = \vec{\text{grad}}\, w \cdot \vec{t} = \theta_x t_x + \theta_y t_y = 0 \tag{4.31}$$

where $\vec{t}(t_x, t_y)$ is the unit vector tangential to the interference fringe. Condition (4.30) asserts that the unit vector $\vec{t}$ is defined at point P which lies in the interference fringe in a unique way. Therefore expression (4.31) relates to the components of local rotation angles θ_x and θ_y of an infinitesimally small surface element with the direction tangential to the interference fringe at this point, which is defined with direction cosines

$$\begin{aligned} t_x &= \cos \alpha_1 \\ t_y &= \cos \alpha_2 = \sin \alpha_1 \end{aligned} \tag{4.32}$$

where α_1, and α_2 are the angles between the tangential line to the interference fringe at the point $P(x, y)$ and the Cartesian coordinates x and y respectively. These angles are measured on the photograph of a fringe pattern.

In order to determine two independent unknowns θ_x and θ_y in the general case, it is necessary to use two equations containing these unknowns. The second relation can be obtained in the following way. The deflection field $w(x, y)$ should be superimposed with an artificially introduced linear field

$$\tilde{w} = \tilde{A} + \tilde{B}x + \tilde{C}y \tag{4.33}$$

where $\tilde{A}$, $\tilde{B}$ and $\tilde{C}$ are the known parameters. The tangential line to the new interference fringe at the point $P(x,y)$ corresponding to the total displacement field $w^* = w + \tilde{w}$ in the general case will be characterized with new direction

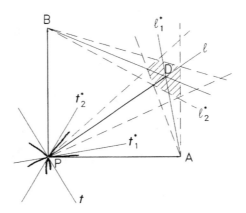

Figure 4.16. Schematic diagram of an implementation of the direct geometrical technique of the first order deflection derivatives determination

cosines t_x^* and t_y^* of the form (4.32). The relation analogous to (4.31) in this case can be represented in the following form:

$$\theta_x t_x^* + \theta_y t_y^* = -(\tilde{B} t_x^* + \tilde{C} t_y^*) \qquad (4.34)$$

Situations where at the some surface points equations (4.31) and (4.34) are in linear dependence to one another may occur. Therefore, to achieve a reliable quantitative interpretation of fringe patterns by means of the geometrical technique involved, we should use two equations of form (4.34) for different additional fields $\tilde{w}_1$ and $\tilde{w}_2$. These fields must meet the condition

$$\vec{\mathrm{grad}}\,\tilde{w}_1 \cdot \vec{\mathrm{grad}}\,\tilde{w}_2 = 0 \qquad (4.35)$$

If all three equations of forms (4.31) or (4.34) become linearly independent, the corresponding overdetermined system should be solved through the use of the least-square-method.

Figure 4.16 illustrates implementation of the direct geometrical approach to a determination of slopes θ_x and θ_y. In this figure the interference fringes corresponding to a different meaning of displacement gradients $\vec{\mathrm{grad}}\,\tilde{w}$ at point $P(x, y)$, and the tangent lines to these fringes, are shown. In particular, line t satisfies condition (4.30), and t_1^* and t_2^* correspond to fringes resulting from superposition of the initial deflection field and additional fields of form (4.33) which meet condition (4.35).

Let's now draw the vectors $\overrightarrow{PA} = -\vec{\mathrm{grad}}\,\tilde{w}_1$ and $\overrightarrow{PB} = -\vec{\mathrm{grad}}\,\tilde{w}_2$ from point P to points A and B respectively, and then draw the line l_1^* through point A and line l_2^* through point B. The two latter lines are perpendicular with respect to lines t_1^* and t_2^* which lie tangential to the corresponding interference fringes at point P. The line l passing through point P lies perpendicular with respect to the line t which is tangential to the initial interference fringe at point P formed before introducing an artificial field $\tilde{w}$. The vector $\vec{PD} = \theta_x \vec{i} + \theta_y \vec{j}$ represents the solution of the corresponding overdetermined equation system, consisting of one equation of form (4.31) and two equations of form (4.34), which defines two unknown values θ_x and θ_y. Note that point D is the point of intersection of three lines, l, l_1^* and l_2^*. The directions of these straight lines can be determined within certain errors only. The upper limits of these errors are shown in Figure 4.16 by dashed lines, and the cross-hatched polygon near point D reveals the values of maximum possible errors of the derivatives $\partial w / \partial x$ and $\partial w / \partial y$ determination.

We shall now consider a particular case when one of the components of rotation angle θ is *a priori* known to be equal to zero. This situation may be possible, for instance, when the x-axis is the symmetry axis of the deflection field $w(x, y)$; in this case $\theta_y(x, 0) = 0$. As the interference fringes are orthogonal with respect to the vector $\vec{grad} \, w$, equation (4.31) becomes identical. In order to determine the slope θ_x, one must use one equation of form (4.34) only, resulting from interpretation of the hologram obtained as superposition of the actual deflection field and one additional linear field $\tilde{w} = \tilde{A} + \tilde{C}y$. A solution of such an equation has the form

$$\theta_x = -\tilde{C} \text{tg} \alpha_1^* \tag{4.36}$$

where α_1^* is the angle between the tangential to the interference fringe at the point of interest and the x-axis.

This variant of the direct geometrical method of deflection derivatives determination was developed by Abramson [21]. The technique involved can be effectively implemented for the solution of bending problems having an axial symmetry. Let us illustrate this approach by using the circular thin plate clamped along an external edge and subjected to internal pressure which has been considered in Section 4.1. A typical interferogram of this plate, corresponding to superposition of the plate deflection and artificial linear field $\tilde{w} = \tilde{A} + \tilde{B}x$, is presented in Figure 4.14.

The values of the angle of local rotation of the surface normal in polar coordinate system θ_r are determined at the points of intersection of the vertical diameter in Fig. 4.14 (this axis is specified as the r-axis) with dark interference fringes. The distribution of tangential inclinations α_r^* along the r-axis is shown in Figure 4.17 by small squares. The value of parameter $\tilde{B}$, contained in equation (4.36), is defined through the uniformly spaced straight fringes which can be seen on the undeformed parts of the specimen under investigation. The distribution of the deflection derivative $\theta_r = -\partial w / \partial r$ obtained through the experimental data presented in Figure 4.17 is shown in Figure 4.18 by small

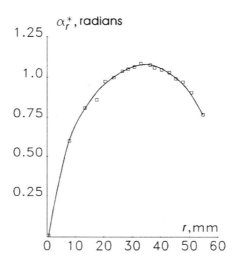

Figure 4.17. Distribution of the inclination of tangential to the interference fringes along the plate radius

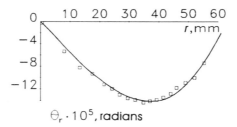

Figure 4.18. Example of the geometrical approach implementation for the first order differentiation

squares. The continuous line reveals the result of the global numerical differentiation was obtained in Section 4.1—see Figure 4.3.

The method combining the above-described geometrical relations and various techniques of numerical differentiation is of practical interest. This approach can be called the semi-geometrical method of fringe pattern interpretation.

The problem of deflection derivatives determination along a smoothed line L on the surface under study is the most classical problem which can be solved through the use of the above-mentioned technique. These problems frequently arise when stress concentration is the main subject of the investigation. In the case concerned one of the equations needed for slopes θ_x and θ_y determination is the geometrical relation of forms (4.31) or (4.34). The second equation can be derived after determination of the derivative dw/dq distribution along the line L by means of numerical differentiation of a discrete set of deflection values:

$$\theta_x q_x + \theta_y q_y = \frac{dw}{dq} \qquad (4.37)$$

where q_x and q_y are the Cartesian components of the unit vector tangential to the line L at the point under consideration. If equation (4.31) is used as a geometrical relation, two deflection derivatives can be determined by using a single conventional fringe pattern.

The main advantage of the semi-geometrical approach in comparison with the numerical differentiation procedure in two orthogonal directions is the lack of necessity for local calculation of deflection derivatives in the direction normal to line L. A local numerical differentiation in the latter case may lead to significant errors, especially when the tangential direction to the line L is very close to the direction of grad w.

In order to illustrate the above-described semi-geometrical method of fringe pattern interpretation, we shall consider determination of stress concentration in a bending plate. The object investigated was a circular aluminium plate of diameter $2R = 120$ mm, thickness $h = 4$ mm and with an eccentric circular cutout of diameter $2r_0 = 32$ mm. The distance between the plate centre and the hole centre was $a = 16$ mm. The elasticity modulus of the plate material was $E = 72\,000$ MPa, and Poisson's ratio $\mu = 0.32$. The plate is rigidly clamped along the external edge and subjected to internal pressure so that the loaded hole edge is capable of being displaced and rotated without limitations. The drawing of the plate and the coordinate system used are shown in Figure 4.19.

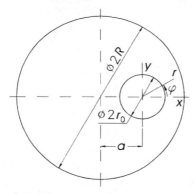

Figure 4.19. Drawing of the plate with eccentric hole

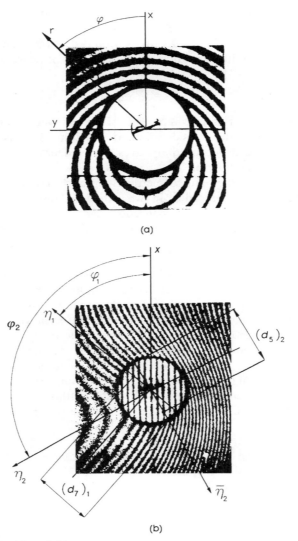

Figure 4.20. (a) and (b)

A typical fringe pattern which represents the system of infinite width fringes corresponding to the lines of equal deflection level obtained for internal pressure increment $p = 0.006\,\text{MPa}$ is shown in Figure 4.20(a). The term 'infinite width fringes' means a fringe pattern recorded without an additional linear field described by an equation of form (4.33). The Cartesian and polar

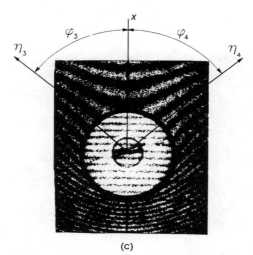

(c)

Figure 4.20. Interferograms of the deflection of the plate with eccentric hole subjected to an internal pressure: conventional (a) and superimposed with additional linear field in two directions (b),(c)

coordinate systems whose origins coincide with the centre of the hole are also shown in Figure 4.20(a).

The deflection distribution along the hole boundary is shown in Figure 4.21. An approximation of the discrete set of experimental data is performed through the use of a trigonometrical series of form (4.9). The power of polynomial which met the condition of the best approximation, and which was established in accordance with the procedure described in Section 4.1, is found to be equal to 2 ($m = 2$).

The values of angles α_r between the r-axis and the tangential line to interference fringes at the points of intersections of these fringes and the direction of the r-axis along the hole edge (below this angle α_r will be called the angle of orientation of interference fringes along the r-direction) are shown in Figure 4.22.

In principle, the following variant of semi-geometrical method can be realized. According to this approach, the measured magnitudes of α_r angles should be approximated with a trigonometrical series, and then an equation of form (4.31) should be solved to determine the radial derivative values $\theta_r = \partial w/\partial r$ by using $\partial w/\partial t = \partial w/r_0\partial\varphi$, t_r and t_φ as known coefficients. However, we should be aware that the errors made in the measurement of α_r-angle values are usually not so small as to ensure an accurate and reliable solution of the above-mentioned equation, especially in the ranges where $\alpha_r \simeq 90°$. The fringe orientation near the polar angle magnitudes $\varphi = 0$ and $\varphi = \pi$ is a clear illustration of this statement (see Figure 4.20(a)).

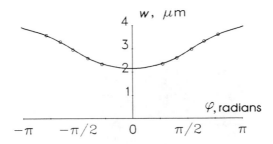

Figure 4.21. Deflection distribution along the eccentric hole edge

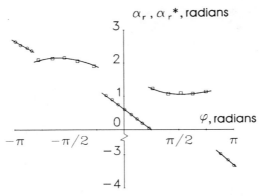

Figure 4.22. Distributions of the inclination of tangential to the interference fringes along the eccentric hole edge

The use of finite width fringe patterns provides a more accurate implementation of the above-described semi-geometrical procedure. The principle of this approach enables us to remove the points of uncertainty of the radial derivative value calculation out of the hole boundary which is the region of interest. One of the possible finite width fringe patterns, which corresponds to the additional artificial displacement field (4.33) with parameters $\tilde{B} = 0$ and $\tilde{C} = 12.4 \times 10^{-5}$, is presented in Figure 4.20(b). The parameter $\tilde{C}$, representing the value which is in inverse proportionality to the uniform fringe spacing caused by a slight tilt of the illumination beam, is determined through the use of fringe patterns on the unstrained and undisplaced body inside the hole. The corresponding distribution of fringe orientation angles near points $\varphi = 0$ and $\varphi = \pi$ is shown in Figure 4.22. Consequently, we have the complete set of experimental data needed in order to obtain the rotation angle θ_r distribution.

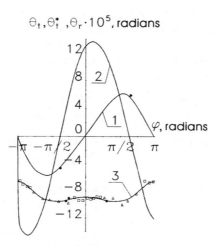

Figure 4.23. Distributions of the tangential (1,2) and radial (3) first order deflection derivatives along the eccentric hole edge

The distribution of derivative $\partial w / \partial t$ along the hole edge obtained by means of numerical differentiation of experimental deflection distribution is shown in Figure 4.23 by curve 1. Curve 2 depicts the analogous distribution of the derivative $\partial w^* / \partial t$ resulting from the solution of

$$\frac{\partial w^*}{\partial t} = \frac{\partial w}{\partial t} + \tilde{C} \cos \varphi$$

In order to find the distribution of the radial derivative $\partial w / \partial r$, it is necessary to use an equation of form (4.34). The corresponding values obtained from an infinite width fringe pattern (see Figure 4.20(a)) and a finite width fringe pattern (see Figure 4.20(b)) are given in Figure 4.23 by triangles and squares respectively. Curve 3 represents the corresponding dependence resulting from trigonometric series approximation.

In conclusion, we have considered another possible way of first-order derivatives determination in this subsection. In Figure 4.20(b) two radial directions η_1 and η_2 are indicated. These directions correspond to the points $\alpha_r^* = 0$ and $\alpha_r^* = \pi/2$ lying at the hole boundary. At these points the following relations are valid:

$$\theta_r(\varphi_1) = -\frac{d\tilde{w}}{d\eta_1}$$

$$\theta_t(\varphi_2) = -\frac{d\tilde{w}}{d\bar{\eta}_2}$$

where φ_1 and φ_2 are the angles between the x-axis and directions η_1 and η_2 respectively; $\bar{\eta}_2 \perp \eta_2$. The magnitudes of derivatives of function $\tilde{w}$ with respect to these directions η_1 and η_2 can be calculated through the corresponding projections of uniform fringe spacing onto the axes η_1 and $\bar{\eta}_2$ as shown in Figure 4.20 in accordance with the relation

$$\left|\frac{\partial \tilde{w}}{\partial \eta_1}\right| = \frac{\lambda}{2}\frac{n}{(d_n)_1}, \quad \left|\frac{\partial \tilde{w}}{\partial \bar{\eta}_2}\right| = \frac{\lambda}{2}\frac{n}{(d_n)_2}$$

where n is the number of fringes used for calculation and $(d_n)_1$ and $(d_n)_2$ are the distance between the jth and $(j+n+1)$th fringes along the axes η_1 and $\bar{\eta}_2$, respectively.

An analogous effect for θ_r determination can be achieved by introducing a linear field in the direction orthogonal to that used to produce Figure 4.20(b). This fringe pattern is shown in Figure 4.20(c). The corresponding directions are indicated as η_3 and η_4. The derivative values thus obtained are represented by circles in curves 1 and 3 in Figure 4.23. The above-described procedure can be applied at any point of the hole boundary by varying the parameters of additional field (4.33) and, thus, to construct derivative distributions along a hole boundary. Note that this technique can be classified as a variant of holographic compensation methods which will be considered in the following section.

The results of the circumferential strain determination based on the displacement derivatives shown in Figure 4.23 will be represented in Section 5.3.

Let's now consider the direct geometrical method of fringe patterns interpretation in terms of the second-order deflection field derivatives which represent the curvatures and torsion of the deformed surface. This approach is based upon relations between the values to be determined and parameters describing the curvature of the interference fringe considered as the lines of equal magnitudes of deflection field $w(x, y)$ which are the plane cross-sections of the deformed surface.

The curvature vector of the interference fringe can be introduced in the following way:

$$\vec{C}(C_x, C_y) = \frac{d\vec{t}}{ds}. \tag{4.38}$$

Representing the components of the vector $\vec{t}$ in accordance with equation (4.31) in the form

$$t_x = -\frac{\theta_y}{|\vec{\text{grad}}\, w|},$$
$$t_y = \frac{\theta_x}{|\vec{\text{grad}}\, w|}, \tag{4.39}$$

taking corresponding derivatives and substituting equation (4.39) into (4.38) yields

$$C_x = -\frac{\theta_x}{(\text{grad } w)^2}\, æ_t$$

$$C_y = -\frac{\theta_y}{(\text{grad } w)^2}\, æ_t$$

where

$$æ_t = \frac{d^2 w}{dt^2} = \frac{æ_x \theta_y^2 - 2æ_{xy}\theta_x\theta_y + æ_y\theta_x^2}{(\text{grad } w)^2} \tag{4.40}$$

is the second-order derivative of deflection $w(x, y)$ with respect to the t-direction. Note that the curvature vector (4.38) is inward-directed to the centre of the curvature at any point of the fringe involved.

We shall take into consideration the local vector basis $\{\vec{n},\ \vec{t}\}$, where

$$\vec{n} = \frac{\text{grad } w}{|\text{grad } w|}$$

is the unit vector directed normally to the interference fringe. Decomposing the vector $\vec{C}$ (4.38) on this basis and taking into account that $\theta_t = dw/dt \equiv 0$, we can obtain

$$\vec{C} = C_n \vec{n} = -\frac{æ_t}{|\text{grad } w|}\, \vec{n} \tag{4.41}$$

where $æ_t$ is derived from equation (4.40).

The radius of curvature of the interference fringe can be defined as

$$\rho = \frac{1}{C_n} \tag{4.42}$$

In the case involved the curvature radius (4.42) is an algebraic value; $\rho > 0$ when vector grad w is inward-directed and vice versa $\rho < 0$.

We shall denote ξ_1 and ξ_2 as the directions of the axes of principal curvature of the deformed surface in the point under consideration, $æ_1$ and $æ_2$ as the corresponding values of principal curvature, and β_1 as the angle between the ξ_1-axis and vector $\vec{t}$. Then the following relation is valid proceeding from the Euler theorem and expression (4.41):

$$æ_t = æ_1 \cos^2 \beta_1 + æ_2 \sin^2 \beta_1 = -\frac{|\text{grad } w|}{\rho} \tag{4.43}$$

Another relation can be written

$$|\vec{\text{grad }} w| = \frac{\theta_1}{\sin \beta_1} = \frac{\theta_2}{\cos \beta_1} \tag{4.44}$$

where θ_1 and θ_2 are the angles of rotation of the normal to the infinitesimal surface element in directions ξ_1 and ξ_2 respectively.

Expressions (4.43)–(4.44) can be directly implemented, in principle, for calculation of the second-order deflection derivatives through the use of measured values of the curvature radius of the interference fringe (if the first-order derivatives have been preliminary determined). It should be noted, however, that the above-presented direct geometrical approach is capable of providing only rough estimations of second-order deflection derivatives due to large errors related to the measurement of the curvature radius of the interference fringe. Moreover, if the values of both principal curvatures are independent variables and also the directions of both axes of the principal curvature are unknown, the direct geometrical method leads to a non-linear equation system.

In some cases another situation may become more simple. For instance, the direct geometrical technique is used in reference [21] for determination of second-order derivatives of the deflection of a rectangular plate subjected to cylindrical bending. In such a case it is *a priori* known that the longitudinal axis of the plate (x-axis) coincides with one of the axes of the principal curvature ξ_1. This means that for all points of the plate surface two relations are valid: $\beta_1 = \alpha_1$ and $\text{æ}_2 = 0$. When the infinite width fringe holographic interferogram is recorded, the interference fringes obtained are straight lines orthogonal to the ξ_1-direction and equation (4.43) becomes identical. In order to determine the value of principal curvature of the deformed surface $\text{æ}_1 = \partial^2 w / \partial \xi_1^2$, the deflection field has to be superimposed with an additional artificial field $\tilde{w} = \tilde{A} + \tilde{C}y$. In this case equation (4.43) can be represented as

$$\text{æ}_1 = \frac{|\tilde{C}|}{\rho |\cos^3 \alpha_1^*|}$$

where α_1^* is defined by analogy with equation (4.36). Obviously, the problem of accurate measurement of the curvature radius value still remains in this relatively simple case.

Typical interferograms of a thin rectangular plate subjected to a pure bending, and recorded by means of infinite and finite width techniques are shown in Figures 4.24(a) and (b) respectively. In contrast to the above-considered problem from Abramson's work, in this case the deformed plate surface takes an anticlastic form. This means that the following relation between the values of principal curvatures are valid:

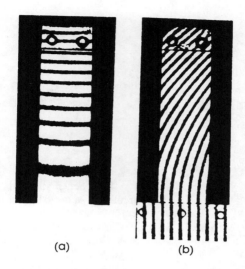

(a)   (b)

Figure 4.24. Finite (a) and infinite (b) fringe patterns resulting from pure bending of rectangular plate

$$æ_2 = -\mu æ_1 \neq 0 \qquad (4.45)$$

where μ is the material Poisson's ratio. For the line of longitudinal symmetry of the plate (x-axis) combination of expressions (4.43) and (4.45) gives

$$æ_1 = -\frac{|\tilde{C}|}{\rho(\cos^2 \alpha_1^* - \mu \sin^2 \alpha_1^*)|\cos \alpha_1^*|} \qquad (4.46)$$

Figure 4.25 illustrates the measurement procedure of the curvature radius of the interference fringe at a point P located on the x-axis. Three lines, n, n_1 and n_2 directed normal to the fringe should be constructed starting from point P and two points P_1 and P_2 located in the vicinity of point P. The length of the segment PD (where point D is the centre of the segment $D_1 D_2$) is considered to be the value of the curvature radius and, hence, $\rho = (P_1 D_1 + P_2 D_2)/2$. The calculated results of the magnitude of the principal curvature $æ_1$ in accordance with equation (4.46) are shown in Figure 4.26 by a dot located near the horizontal line. These data were obtained for $\mu = 0.32$. The distribution of the slope values θ_x (inclined line in Figure 4.26) obtained by means of Abramson's technique are also presented in this figure.

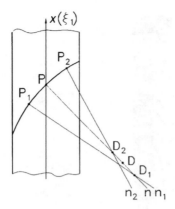

Figure 4.25. Schematic diagram of the measurement of a curvature of interference fringe

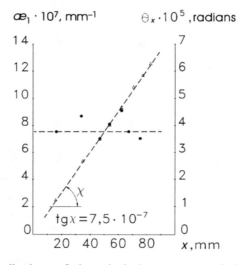

Figure 4.26. Distributions of the principal curvature and slope along the axis of cantilever beam under pure bending

4.3 The methods of compensation holographic interferometry

The previously described procedure of direct geometrical methods for differentiation (fringe patterns interpretation in terms of the first- and second-order derivatives) is based on the assumption that condition (4.30) is valid. We shall now consider the case when the slope values become zero at some object surface point simultaneously. As known, such points are the points of stationarity of the deflection field $w(x, y)$. These points of stationarity are the singular points on interference fringes. Let's assume that at some surface point not only are θ_x and θ_y equal to zero, but all derivatives from the first to the $(n-1)$th order become zero. The line described by equation (4.28) has at this point n tangentials, some of which may coincide with one another and some of which may be imaginary. When $n = 2$ and one of the principal curvatures of the deformed surface $\ae_1$ or $\ae_2$ differs from zero, there are two tangential lines to the interference fringe passing through the point under consideration. The inclinations of these tangential lines $\mathrm{tg}\alpha_1 = dy/dx$ are defined by

$$\ae_y \cdot \mathrm{tg}^2\alpha_1 + 2\ae_{xy} \cdot \mathrm{tg}\alpha_1 + \ae_x = 0$$

that can be rewritten, if the coordinate axes used coincide with the axes of principal curvature, in the following form:

$$\ae_2\mathrm{tg}^2\beta_1 + \ae_1 = 0 \tag{4.47}$$

Without loss of generality we can assume that $\ae_2 \neq 0$. In this case equation (4.47) gives

$$\mathrm{tg}^2\beta_1 = -\frac{\ae}{\ae_2^2} \tag{4.48}$$

where $\ae = \ae_1 \cdot \ae_2$ is the Gaussian curvature of the deformed surface in the point of interest.

The points of singularity are usually classified depending on the sign of $\ae$. The case when $\ae < 0$ describes a so-called nodal point that represents an intersection of two interference fringes. The inclinations of two real tangentials in the nodal point are of equal value but opposite sign. The fringes converging in such a point are quasi-symmetrical with respect to the axes of principal curvature ξ_1 and ξ_2.

When $\ae > 0$, both tangentials are imaginary and conjugated with one another. Such a singular point is called an isolated point. The axes of the principal curvature can again be considered as the axes of quasi-symmetry of the interference fringe pattern which represents a set of closed fringes. Also, if $\ae = 0$, both tangentials are real and coincide with one another. Such a singular point is called the point of reset or self-conjugation. The direction lying orthogonal to the common tangential is the direction of principal curvature axis ξ_2 that corresponds to the non-zero principal curvature value $\ae_2$.

The points of stationarity of the deflection field $w(x, y)$ when $\ae < 0$ (hyperbolic), $\ae > 0$ (elliptical) and $\ae = 0$ (parabolic) can be easily identified in the interference fringe pattern (see Figure 4.27). At each point the condition $\mathrm{grad}\, w = 0$ is satisfied. When the fringe pattern concerned corresponds to a

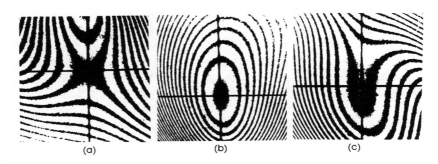

Figure 4.27. Interference fringe patterns obtained near the stationarity points of a deflection field: hyperbolic (a), elliptical (b) and parabolic (c)

superimposed deflection field obtained in accordance with condition (4.33), the following relations take place at the point of the stationarity:

$$\theta_x = -\tilde{B} \qquad \theta_y = -\tilde{C} \qquad\qquad (4.49)$$

Equations (4.49) are the basis for determination of the first-order deflection derivatives by means of holographic compensation techniques. Therefore, if it is possible to visualize in the interference fringe pattern a point of stationarity of any of the three types by varying parameters of the additional artificial displacement field, the values of deflection derivatives θ_x and θ_y can be determined by equation (4.49). A presence of a singular point in the fringe pattern means that the gradient of the (initial) deflection field is compensated by introducing an additional field at this surface point.

Note that in addition, the axes of principal curvature of the deformed object surface can be identified as the symmetry axes of the fringe pattern in the vicinity of a singular point.

Figure 4.14 can be considered as an illustration of this situation. This figure demonstrates two types of singular points in the interference fringes: the isolated point and the nodal point. When the gradient $\vec{\text{grad}}\ \tilde{w}$ of the additional field increases, these points move to one another and then coincide forming the reset point—see Figure 4.28. The distribution of radial deflection derivative $\theta_r(r)$ along the plate radius obtained through the use of this compensation technique is presented in Figure 4.29 by circles. The continuous curve in this figure shows the result of the global numerical differentiation procedure.

The compensation method of 'optical' differentiation is found to be very powerful. Note that the compensation approach, in comparison with numerical and direct geometrical approaches to derivative determination, is capable of extending the measurement range to give greater values of the deflection gradient. Indeed, the interference fringes near a singular point

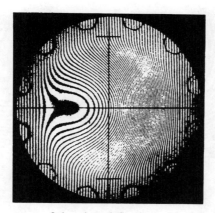

Figure 4.28. Interferogram of the plate deflection with the reset singular point

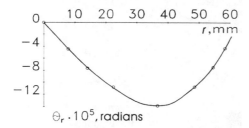

Figure 4.29. Distribution of the first order radial deflection derivative along the plate radius obtained by means of optical compensation technique

having a relatively low density are clearly observed whereas the fringes far from this point cannot be resolved.

By way of example of determination of the principal curvature axes of the deformed surface, consider a bending of a rectangular plate rigidly fixed along its edge and under uniform internal pressure. Due to the absence of axial symmetry of the structural element, the question of the principal curvature directions is not so clear as in the case of a circular plate subjected to uniform internal pressure. The interferograms of the plate deflection obtained in the infinite width fringes (a) and finite width fringes (b, c) are presented in Figure 4.30. By taking advantage of plate symmetry, holograms of only one quarter of the plate are recorded. The line arrow in these fringe patterns indicates the corner point and diagonals of the plate. The axes of quasi-symmetry of local fringe patterns are also indicated in the singular points in Figures 4.30(b) and (c). Remember that these axes represent the axes of principal curvature of the deformed plate surface. The results from interpretation of a set of similar fringe

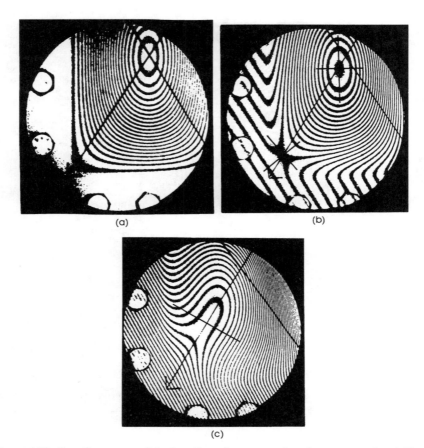

Figure 4.30. Interferograms of the bending of a rectangular plate: conventional (1) and visualizing the elliptical (b) and parabolic (c) points of deflection field stationarity

patterns obtained for different compensation conditions are shown in Figure 4.31. The values of angle β_1 (4.43) between one of the axes of principal curvature and the x-axis lying in the direction of the greater plate side (small triangles) is plotted against the distance d from the plate centre in the diagonal direction (curve 1). The distributions of the slopes θ_x (2) and θ_y (3) obtained from the same interferograms along the plate diagonal are also presented in Figure 4.31.

The principle of the holographic compensation technique to determine the curvatures $æ_x$ and $æ_y$ of a deformed object surface consists of (as in the case of slope measurement) recording fringe patterns formed by the deflection field and an additional artificial field with a known form. The parameters of the

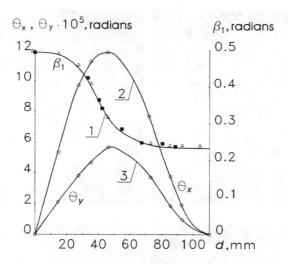

Figure 4.31. Orientation of the principal curvature direction and slope distributions along the diagonal of a rectangular plate under bending

latter field should be chosen taking into account some *a priori* infromation on the problem to be solved. A wide class of problems dealing with strain and stress analysis, material mechanical properties determination, etc., can be effectively solved through the use of a compensation technique based on an additional field in the form of a two-dimensional square function

$$\tilde{w} = \tilde{A} + \tilde{B}x + \tilde{C}y + \frac{\tilde{D}}{2}(x^2 + y^2) \qquad (4.50)$$

To determine the principal curvatures of the deformed surface, it is necessary to visualize a parabolic stationarity point at the surface point under study by the variation of the parameters of the additional field (4.50). As mentioned above, a common tangential to the interference fringe at the point concerned characterizing singularity of reset type defines one of two axes of the principal curvature of the total surface $z = w^*(x, y)$. If this axis is ξ_1^*, the corresponding value of principal curvature is equal to zero, $æ_1^* = 0$. Taking into account the form of the additional artificial displacement field (4.50), note that the direction ξ_1^* coincides with the direction ξ_1, i.e. introducing the field $\tilde{w}$ (4.50) into an initial fringe pattern does not change the directions of principal curvatures. If the compensation in the direction ξ_1 occurs when $\tilde{D} = \tilde{D}_1$, then equation (4.50) gives

$$æ_1 = -\tilde{D}_1 \qquad (4.51)$$

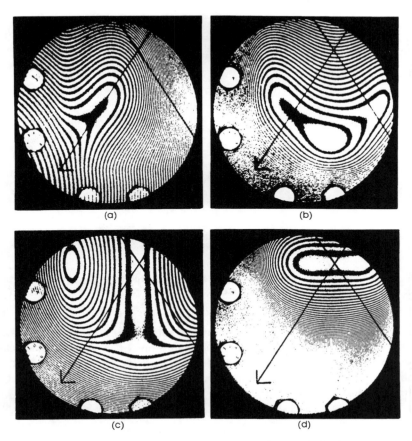

Figure 4.32. The instants of an optical compensation of two principal curvatures of the deformed rectangular plate surface in intermediate points on a diagonal (a,b) and in the plate center (c,d)

By analogy with equation (4.51) the corresponding relation in the case of compensation in the ξ_2 direction can be written

$$\ae_2 = -\tilde{D}_2$$

As an example of the above-described technique, Figures 4.32(a) and (b) illustrate two cases of compensation of two principal curvatures $\ae_1$ and $\ae_2$ in one of the diagonal points of the rectangular bending plate. Figures 4.32(c) and (d) demonstrate analogous compensation in the central point of this plate. In the latter two photographs the parabolic stationarity points in the fringe

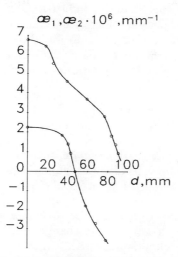

Figure 4.33. Distributions of the principal curvatures of the deformed surface along diagonals of a rectangular plate under bending

patterns correspond to the singularity points of self-conjugation in the interference fringes. The experimental distributions of the values of principal curvatures $æ_1$ and $æ_2$ along the plate diagonal are presented in Figure 4.33.

The values of angle β_1 between the principal curvature axis ξ_1 and the x-axis determined in the points of stationarity of the field w^* are denoted in Figure 4.31 by squares.

A determination of the second-order derivatives of the deflection field $w(x, y)$ along a defined direction has to be performed by visualization of the points of inflection of the interference fringes on a fringe pattern. Let's consider the conditions of occurrence of these fringe inflections.

Let the following condition be valid in some point of a bending plate surface

$$æ_\nu^* = 0 \tag{4.52}$$

where ν is an arbitrary direction that does not coincide with the tangential to the direction of principal curvature in the point under consideration. Assume, also, that the condition

$$\frac{dw^*}{d\nu} = \theta_\nu^* = 0 \tag{4.53}$$

is satisfied by varying the linear term of additional field (4.50). The validity of equation (4.53) means that the direction of the tangential in the point concerned coincides with the ν-direction.

We shall analyse how this situation influences the character of a fringe pattern in the vicinity of the point under consideration. For this purpose, let's consider a change of curvature vector $\vec{C}$ in the above-mentioned point. The corresponding expression can be obtained by means of differentiation of equation (4.38) taking into account equations (4.52) and (4.53):

$$\frac{d\vec{C}}{ds} = -\frac{d\text{æ}_t^*}{ds}\frac{\vec{n}}{|\vec{grad}\, w^*|} \tag{4.54}$$

If $d\text{æ}_t^*/ds \neq 0$, then æ_t^* changes its sign in such a point and vector $\vec{C}$ in accordance with (4.54) changes its direction on the opposite one. This point on the interference fringe pattern represents the point of inflection of the interference fringe, and the condition $\vec{C} = 0$ is valid in this point. In a general case when all derivatives $d^k\text{æ}_t^*/ds^k$ to $k = m$ are equal to zero, but $d^{m+1}\text{æ}_t^*/ds^{m+1}$ is not equal to zero, a sign change of æ_t^* occurs for the uneven number of m and an inflection is observed in the interference fringe. For an even number of m, the interference fringe has no inflection and the curvature value takes its extremal meaning in this point.

Therefore, to find non-extremal values of the second-order deflection derivative at the defined point with respect to direction ν, the deflection field should be superimposed with the special additional field. The parameter of this field should be varied to visualize the inflection of the interference fringe, the tangential to which in the point under consideration coincides with the ν-direction. It should be noted, however, that solution of the inverse problem of searching points with definite values of æ_ν is more convenient in practice. In this case for each different discrete value of parameter $\tilde{D}$ the points where the above-mentioned conditions are met should be found by varying of values $\tilde{B}$ and $\tilde{C}$ only.

One salient feature of the approach involved should be pointed out. Writing equation (4.43) for the total displacement field and substituting the obtained result into the expression for the module of curvature of the interference fringe yields the following relation in the inflection point:

$$\text{tg}^2\beta^* = -\frac{\text{æ}_1^*}{\text{æ}_2^*} \geqslant 0 \tag{4.55}$$

Relation (4.55) means that the inflection of the interference fringe cannot be observed in any elliptical point on the deformed surface. As, according to equation (4.44),

$$\text{tg}\beta^* = \frac{\theta_1^*}{\theta_2^*}$$

a condition the appearance of an inflection of the interference fringe can be written in the following form:

$$\left(\frac{\theta_1^*}{\theta_2^*}\right)^2 = -\frac{\ae_1^*}{\ae_2^*}$$

Bending of the circular plate rigidly fixed along its edge and subjected to internal pressure is again used to illustrate the technique of second-order optical differentiation. Figure 4.34 shows the interferogram of the total deflection field recorded in such a way that the condition $\vec{\text{grad}}\, w^* \neq 0$ is valid over the whole plate surface. The set of inflection points that may be called the l-line is clearly displayed in this fringe pattern. In accordance with the above consideration, at each point of the l-line the following conditions are satisfied:

$$\frac{\mathrm{d}^2 w}{\mathrm{d}(t^*)^2} = -\tilde{D}$$

Taking advantage of the axial symmetry of the problem, the value of the radial coordinate, for which the absolute value of curvature in the radial direction $\ae_r$ is equal to the value of parameter $\tilde{D}$, can be easily determined. It is evident that this value is equal to the radial coordinate value of the point belonging to the l-line for which the tangential to the interference fringe is directed along the plate radius. The configuration of the l-line and the location of the above-mentioned point are illustrated in Figure 4.34.

Let's return to the technique of determination of the principal curvature values $\ae_i$ by visualizing the reset points in the interference fringes. Note that by

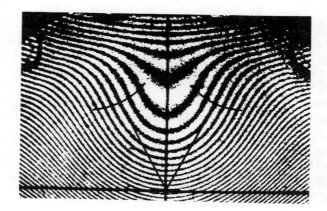

Figure 4.34. Interferogram of the superposition of actual and linear artificial deflection fields

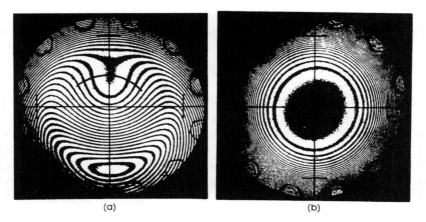

(a)   (b)

Figure 4.35. The instants of an optical compensation of intermediate (a) and extremal (b) values of the principal curvature of the deformed surface of a plate under bending

using the l-line, a more accurate determination of these points can be ensured, simultaneously representing the singularity points on the l-line. Figure 4.35(a) shows the case of compensation of $æ_r$ at an intermediate point lying on the plate radius with the creation of the parabolic point of stationarity. The corresponding l-line is drawn in this photograph. For a more detailed description, Figure 4.35(b) illustrates the case of compensation of the extremal radial curvature value $æ_r$ in the centre of the plate. The broad circular fringe at the plate centre vicinity shows a complete character of curvature compensation in the surface area, in contrast to the pointwise compensation shown in Figure 4.35(a).

In Figure 4.36 the radial curvature $æ_r$ is plotted against the distance from the plate centre r. This distribution is constructed by using the data of two types of compensation techniques for different values of the additional field parameter $\tilde{D}$. The line marked by -○- represents the approach based upon reset singular points visualization, and the line marked by -□- corresponds to the inflections of interference fringes. These graphs reveal a good agreement between the values of radial curvature obtained. The continuous curve in Figure 4.36 denotes the results of global numerical differentiation which were presented above.

Let's now consider the particular case when the deformed plate surface represents the second-order surface. This means that in the coordinate system ξ_1 and ξ_2 the deflection field has the form

$$w = A + B\xi_1 + C\xi_2 + \frac{æ_1\xi_1^2}{2} + \frac{æ_2\xi_2^2}{2} \tag{4.56}$$

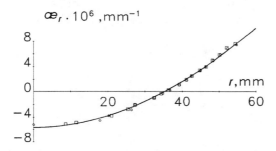

Figure 4.36. Distribution of the principal curvature along a radius of a plate under bending obtained by means of two optical compensation techniques

where A, B, C, $æ_1$ and $æ_2$ are the constants. This situation occurs, for instance, when a thin plate is subjected to pure bending. For an arbitrary value of the additional field parameter $\tilde{D}$ all the points of the total deformed surface corresponding to the total displacement field $z = w^*(\xi_1, \xi_2)$ will be either elliptical ($æ^* > 0$) or hyperbolic ($æ^* < 0$). When the parameters of the additional field meet the compensation conditions

$$\tilde{B} = -B \qquad \tilde{D} = -æ_1 \qquad (4.57)$$

the field w^* will be the function of the single variable ξ_2 and all surface points will be the points of parabolic type ($æ^* = 0$). Therefore, the corresponding fringe pattern will consist of straight fringes parallel to the ξ_1-axis. Moreover, each point belonging to the straight line $\xi_2 = (C - \tilde{C})/(æ_2 - æ_1)$ will be the point of stationarity of the total deflection field w^* and the singular point of fringe self-conjugation. When the parameter $\tilde{A}$ of the additional field (4.50) varies continuously, the interference fringes will either originate or disappear along this line.

Thus, the measurement of the principal curvature value $æ_1$ consists of the determination of the current meaning of the adjustable parameter $\tilde{D}$ which corresponds to the situation when the interference fringes being transformed from elliptical into hyperbolic form (or vice versa) become straight lines parallel to the ξ_1-axis. This type of transformation of interference fringe pattern means that the sign of the curvature of the total surface $z = w^*(\xi_1, \xi_2)$ has changed. By visualizing these fringes patterns it can be seen that the first condition (4.57) is met and accompanied by the appearance of the point of stationarity of the fringe pattern in the line $\xi_1 = 0$. The methods of compensation of square deflection fields having the form (4.56) and their application to the determination of material mechanical properties will be considered in detail in Section 7.1.

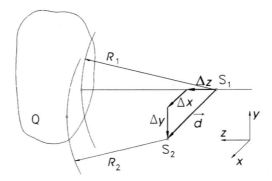

Figure 4.37. Schematic diagram of an artificial deflection field generation

In conclusion, we have described the main principles of construction of optical systems which are capable of implementing holographic compensation techniques. To derive the basic relations, consider the plane object surface Q which is illuminated with a divergent laser light wave emerging from the point source S_1 (see Figure 4.37). The wavefront surface at the distance R_1 from the source S_1 is given by

$$z_1 = R_1 - \sqrt{R_1^2 - (x_1^2 - y^2)} \qquad (4.58)$$

Taking into account the conditions of a real holographic interferometric experiment, we can assume that $R_1 > R$ (where R is the characteristic linear dimension on the surface Q investigated). Let's now move the point light source into the new position S_2. The value and direction of this displacement is described by the vector

$$\vec{S_1 S_2} = \vec{i} \cdot \Delta x + \vec{j} \cdot \Delta y + \vec{k} \cdot \Delta z \qquad (4.59)$$

The form of the corresponding wavefront surface coming to the object surface taking into account equations (4.58) and (4.59) can be written as

$$z_2 = R_2 - \sqrt{R_2^2 - (x - \Delta x)^2 - (y - \Delta y)^2} \qquad (4.60)$$

where $R_2 = R_1 + \Delta z$. Assuming that Δx, Δy, $\Delta z < R_1$, we can obtain from equations (4.58) and (4.60) the following relation for the optical path difference of two optical beams:

$$\Delta L^\circ \simeq A^\circ + B^\circ x + C^\circ y + D^\circ (x^2 + y^2) \qquad (4.61)$$

where

$$D° = -\frac{\Delta z}{2R_1 R_2}; \qquad B° = -\frac{\Delta x}{R_2}; \qquad C° = -\frac{\Delta y}{R_2}$$

(the term $A°$ in this case is not of interest).

The character of expression (4.61) allows us to interpret the fringe pattern corresponding to the path difference value $\Delta L°$ as a superposition of artificial deformation of plate surface Q caused by its pure bending and artificial rotation of the plate around two orthogonal axes lying in the plate surface. This superposition generates the additional deflection field of form (4.56):

$$\tilde{w} = -\frac{1}{R_2(1 + \cos\psi)}\left[\frac{\Delta z(x^2 + y^2)}{R_1} + \Delta x \cdot x + \Delta y \cdot y\right] + \tilde{A} \qquad (4.62)$$

where ψ is the observation angle.

The optical system that is capable of producing the artificial deflection field of form (4.62) will be called the three-parameter holographic interferometer. One possible set-up of this interferometer system is shown in Figure 4.38. The point source of spherical light waves is the focal point of a thin lens L_2 mounted on a special device. This device is capable of providing a precision translation of this lens along three orthogonal directions. Note that an analogous optical system was used in reference [19] to produce the carrier fringes on the object surface under study.

The function of the lens L_3 is to linearize the measurement procedure. The following illustration will make this clear. Consider the parameter $\chi = |\Delta\tilde{D}/\Delta z|$, representing the coefficient of interferometer sensitivity. It is obvious that without lens L_3 this coefficient is not constant—see equation (4.62). It can be shown that the presence of the lens L_3 in the optical system presented in Figure 4.38 results in the following expressions for sensitivity coefficient:

$$\chi = \left[\frac{f_3}{a(f_3 - b) + b^2}\right]^2 \qquad (4.63)$$

where f_3 is the focal length of the lens L_3. The equation (4.63) shows that $\chi = 1/f_3^2 \equiv const$ when $b = f_3$.

A minimal influence of errors made during interferometer adjustment, as follows from equation (4.63), is ensured when $a = 2f_3$ (where a defines the initial position of the lens L_3).

The interferograms of the additional artificial deflection field caused by displacement of the illuminating point source are shown in Figure 4.39. Figure 4.39(a) corresponds to the translation in the transverse direction with respect to the interferometer optical axis when $\Delta x = \Delta y$. Figure 4.39(b) reveals the interference fringes obtained as a results of point source translation along the

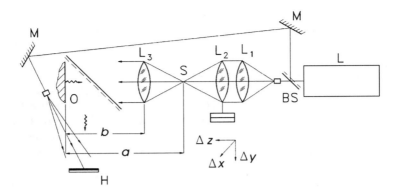

Figure 4.38. Optical arrangement of the 3-parameter holographic interferometer

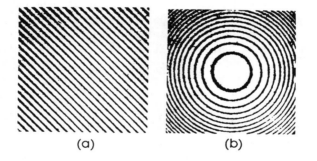

(a)   (b)

Figure 4.39. Interferograms of an artificial deflection field obtained by means of transverse (a) and longitudinal (b) translation of the point source in the 3-parameter interferometer

optical axis. As expected, in the first case the interference fringes are a set of uniformly spaced straight lines which are orthogonal with respect to the translation direction defined by the vector $(\Delta x, \Delta y)$. In the second case the fringe pattern represents a set of co-axial circular lines, the numbers and diameters of which are related by a square dependence.

4.4 Shearing speckle interferometry

The basic principles of shearing speckle interferometry were established in the initial work of Hung and Taylor [26] and were further developed in references [27] and [28]. The principles of this method consist of the double recording of two object images, one of which is slightly shifted with respect to the other.

One pair of these shifted images corresponds to the unstrained object surface and another pair are recorded after the object's deformation. It was established by Goodmen [29] that a coherent adding of two speckle fields having identical statistical parameters results in the formation of a new speckle field with the same statistical parameters. Observation of the interference fringes, as a result of a Fourier filtering procedure of the light wave passing through the specklegram recorded in the above-mentioned way, is related to the first-order derivatives of the displacement components with respect to spatial coordinates lying in the plane of the object surface.

A rich variety of methods for recording double images have been implemented. Among them are the following techniques:

- A two-aperture mask with two glass plane plates mounted near each aperture [26]. Image shearing is introduced by a tilting of these plates.
- Defocusing of an image formed with a two-aperture mask [30,31].
- The use of grating forming two images in the ±1 st diffraction orders [32].
- A two-aperture mask with a two Dauvais prisms and divided lens [27,33] for shearing of images having a radial symmetry.
- A wedge assembly being capable of image shearing in three directions [34].
- An annular wedge [35].

An optical system of a speckle shearing interferometer where an annular wedge is used as a shearing element is shown in Figure 4.40. The presence of wedge W immediately behind lens L allows two different points on the object surface O, for instance points B_1 and B_2 to be imaged as a single point in the image plane. It is evident that the light beams forming the images of these points, and passing through the wedge, acquire an additional optical path difference in comparison with the light beam passing outside the wedge.

Let us consider the complex amplitudes of light passing through the wedge and passing outside the wedge, respectively:

$$A_1 = a_0 \exp(-i\phi_1) \qquad (4.64)$$

$$A_2 = a_0 \exp(-i\phi_2) \qquad (4.65)$$

where ϕ_1 and ϕ_2 are the phase of the light waves emerging from points B_1 and B_2 respectively and a_0 is the amplitude of these waves.

The complex amplitude A of light at each point of the image is the sum of equations (4.64) and (4.65). The intensity distribution in the image plane can be determined conventionally:

$$I_1 = AA^* = (A_1 + A_2)(A_1^* + A_2^*) = 2a_0^2(1 + \cos \varphi) \qquad (4.66)$$

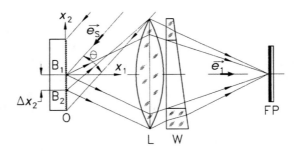

Figure 4.40. Schematic diagram of the optical shearing interferometer with annular wedge

where $\varphi = \phi_1 - \phi_2$ is the relative phase shift. After the object's deformation the relative phase shift will be changed and relation (4.66) takes the form

$$I_2 = 2a_0^2[1 + \cos(\varphi + \delta)] \tag{4.67}$$

where $\delta = \delta_2 - \delta_1$, δ_1 and δ_2 are the phases corresponding to displacement vectors of points B_1 and B_2 caused by the object's deformation.

The total exposure E of the photoplate is proportional to the sum of intensity (4.66) and (4.67):

$$E = \tau(I_1 + I_2) = 4a_0\left[1 + \cos\left(\varphi + \frac{\delta}{2}\right)\cos\frac{\delta}{2}\right] \tag{4.68}$$

where τ is the duration of the first and second exposures that are assumed to be equal to one another. After development of the photographic emulsion in accordance with relation (1.3), the photoplate becomes a transparency whose amplitude transmittance function is proportional to the exposure.

The second term in expression (4.68) represents a high-frequency carrier, the amplitude of which is modulated by low-frequency function $\cos(\delta/2)$. Hence, the expression describing a fringe pattern with bright fringes occurs when

$$\delta = 2n_s\pi, \qquad n_s = 1, 2, 3. \tag{4.69}$$

In order to transform variations of the carrier frequency into corresponding variations of the intensity (i.e. to receive an interference fringe pattern of the form (4.69)), a Fourier filtering has to be implemented through the use of the optical arrangement shown in Figure 4.41. The point source image is focused by the lens L_1 in the Fourier plane F. The photoplate FP is placed inside a convergent laser beam passing through the lens L_2. The lens L_2, the image plane of which is optically conjugated with the plane of the photoplate FP,

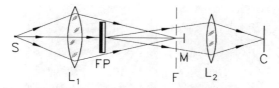

Figure 4.41. Arrangement for optical filtering of high spatial frequencies of light waves transmitted through a specklegram

reconstructs the object surface image in plane C with low spatial frequencies of the light wave passing through the double-exposure specklegram which must then be blocked with a non-transparent screen M. The object surface image thus reconstructed will be modulated by interference fringes.

For quantitative interpretation of interference fringes in the case concerned, the expression for the phase difference δ should be obtained in an explicit form. Let us represent the displacement vectors of points B_1 and B_2 by vectors $\vec{d}_1$ and $\vec{d}_2$, respectively

$$\vec{d}_1 = (w, u, v), \qquad \vec{d}_2 = (w + \Delta w, u + \Delta u, v + \Delta v). \qquad (4.70)$$

The required expression for phases δ_1 and δ_2 are given by equation (2.24):

$$\delta_1 = \frac{2\pi}{\lambda}(\vec{e}_1 - \vec{e}_s)\vec{d}_1$$

$$\delta_2 = \frac{2\pi}{\lambda}(\vec{e}_1 - \vec{e}_s)\vec{d}_2 \qquad (4.71)$$

where the vectors $\vec{e}_1$ and $\vec{e}_s$ are shown in Figure 4.40. Substituting equation (4.70) into (4.71) after some simple transformation yields the required expression for the relative phase shift:

$$\delta = \delta_2 - \delta_1 = \frac{2\pi}{\lambda}\left[\frac{\partial w}{\partial x_2}(1 + \cos\theta) + \frac{\partial u}{\partial x_2}\sin\theta\right]\Delta x_2 \qquad (4.72)$$

where Δx_2 is the distance between the points B_1 and B_2 on the object surface in the direction of the x_2-axis.

The combination of expressions (4.69) and (4.72) gives the relation describing the bright interference fringes:

$$\frac{\partial w}{\partial x_2}(1 + \cos\theta) + \frac{\partial u}{\partial x_2}\sin\theta = \frac{\lambda n_s}{\Delta x_2} \qquad (4.73)$$

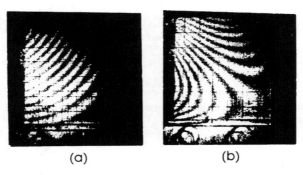

(a)   (b)

Figure 4.42. Interferograms of the slope obtained in two (a,b) diagonal directions of the square plate subjected to a bending

Analogous expressions can be obtained when the illuminating wave is propagated in the plane $(x_1 \, x_3)$:

$$\frac{\partial w}{\partial x_3}(1 + \cos \theta) + \frac{\partial v}{\partial x_3} \sin \theta = \frac{\lambda n_s}{\Delta x_3} \tag{4.74}$$

Equations (4.73) and (4.74) show that in speckle shearing interferometry the values of the first-order displacement derivatives with respect to spatial coordinates are responsible for the interference fringe pattern appearance. It should be noted that expressions (4.73) and (4.74) are valid in all cases for any image shearing type if the interferometer optical arrangement is designed so that the angle subtended by the imaging lens at the object surface is small.

If the bending strains are of main interest, the normal to the object surface illumination must be used. Substituting $\theta = 0$ into expressions (4.73) and (4.74) yields

$$\frac{\partial w}{\partial x_2} = \frac{\lambda n_s}{2\Delta x_2},$$

$$\frac{\partial w}{\partial x_3} = \frac{\lambda n_s}{2\Delta x_3}$$

Figure 4.42 illustrates the latter case. In this figure two interferograms of the regular square plate bent with a concentrated force are presented. These fringe patterns reveal distributions of the deflection derivatives with respect to diagonal directions of the plate. The value of image shearing in both directions was 5 mm. Note that the geometrical parameters of the object investigated and the loading scheme coincide with those shown in Figure 5.12.

4.5 Defocused speckle photography

The method of speckle photography is described in Sections 3.2 and 3.3 in reasonable detail. This is used to measure the tangential to the object surface displacement components and when implemented in general cases, the plane where a specklegram is recorded may not coincide with the plane of the focused image of the object surface. When this occurs, the relation between the displacement of speckles in an arbitrary plane in the image space and the displacement components of the object surface points is represented by expression (3.37).

Consider now the case when the image of the plane object surface subjected to bending. $\{\vec{d}_\tau(x_2, x_3) = 0, \ \vec{d} = [0, 0, w(x_2, x_3)]\}$ is recorded in the plane located at the distance Δl_1 from the plane of the focused image of the object surface (see Figure 4.43). Under these conditions, expression (3.37) describing the displacements of speckles in the recording plane takes the form

$$\vec{d}_s = -\frac{M\vec{r}w}{l_1} + \Delta l_1 \frac{1}{M} \nabla(\vec{m}_0 w \vec{e}_1) \tag{4.75}$$

where $\vec{r}$ is the radius-vector of the point under consideration in the object surface plane; w is the surface deflection at this point, $\vec{m}_0 = \vec{e}_s + \vec{e}_1$; and M is the transverse magnification of the recording set-up. The first term in equation (4.75) can be referred to as the so-called 'perspective' effect. That is, when the plane of the object surface does not coincide with the plane of the focused image, the normal displacement component of the object surface points generate fictitious tangential displacements of speckles in the recording plane. This may have a large influence on the speckle pattern formation whereas large deflections appear, and in most cases this influence can be neglected.

When investigating the bending of thin plates, the direction of illumination $\vec{e}_s$ and observation $\vec{e}_1$ coincide with the normal to the object surface. When this

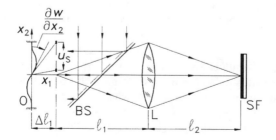

Figure 4.43. Optical setup for recording of a specklegram with defocusing

geometrical configuration of an optical arrangement is implemented, relation (4.75) can be rewritten in the following form:

$$\vec{d}_s = 2\Delta l_1 \frac{\text{grad } w}{M} \tag{4.76}$$

Therefore, the vector of the displacement of speckles in the recording plane $\vec{d}_s$ responds to the vector of the deflection gradient:

$$\text{grad } w = \frac{\partial w}{\partial x_2} \vec{j} + \frac{\partial w}{\partial x_3} \vec{k}$$

at each surface point under illumination. Expression (4.76) can be represented through the components of the two above-mentioned vectors:

$$u_s = \left(\frac{2\Delta l_1}{M}\right) \frac{\partial w}{\partial x_2}$$
$$v_s = \left(\frac{2\Delta l_1}{M}\right) \frac{\partial w}{\partial x_3} \tag{4.77}$$

where u_s and v_s are the components of the vector of the displacement of speckles in the recording plane, which can be determined by means of either whole-field Fourier filtering or pointwise procedure (see Section 3.3).

Expression (4.77) shows that the sensitivity of the defocused speckle photography (DSP) method depends on the magnification factor M and the distance between the recording plane and the plane of the focused object surface image Δl_1. The value of M, as a rule, is defined by the size of the object surface to be studied, and the size of the image can be constructed by the recording device used. It is evident that the sensitivity increases when the magnification becomes less than unity.

Defocusing results in a blurred surface image. This blurring is the main reason for the transformation of the slopes at surface points into tangential displacements of speckles in the recording plane. If the minimum value of the speckle displacement component $u_{s\text{min}}$ or ($v_{s\text{min}}$) in the recording plane that can be measured by the speckle photography technique is fixed, the sensitivity of the DSP method with respect to the deflection gradient increases with growth of the defocusing distance. However, the greater the defocusing distance, the greater the uncertainty of location of points on the object surface image. This may lead to significant errors, especially when a region of stress concentration has to be studied. Therefore, the defocusing distance should not exceed 15 per cent of the focal length of the recording lens. The basic principles of the method of defocused speckle photography were developed in some initial work by Tiziani [36], Gregory [37] and Chiang and Juang [38].

Consider now an influence of deflections of the object surface on the displacement of speckles when defocusing is introduced during recording of a specklegram. This means that the first and the second terms have to be retained in expression (4.75). If the directions of illumination and observation coincide with the normal to the object surface, expression (4.75) takes the form

$$\bar{d}_s = -M \frac{\vec{r}w}{l_1} + 2\Delta l_1 \frac{\text{grad } w}{M} \tag{4.78}$$

Expression (4.78) can be represented through the components of vectors contained in the left- and right-hand sides:

$$u_s = -M \frac{wx_2}{l_1} + \left(\frac{2\Delta l_1}{M} \right) \frac{\partial w}{\partial x_2}$$

$$v_s = -M \frac{wx_3}{l_1} + \left(\frac{2\Delta l_1}{M} \right) \frac{\partial w}{\partial x_3} \tag{4.79}$$

Expressions (4.78) and (4.79) show that the influence of deflections w on displacements of speckles in the recording plane depends on the radius-vector $\vec{r}$ of the surface point under consideration. Therefore, in order to determine the slope values $\partial w/\partial x_2$ and $\partial w/\partial x_3$ by using equation (4.79), information on the deflection field on the object surface is needed.

Now we shall describe the approach which is capable of considerably reducing the influence of the deflection field on the results of the deflection derivatives determination. This approach is based on the optical arrangement that allows us to record two defocused specklegrams SF1 and SF2 simultaneously (see Figure 4.44). This pair of specklegrams will be called a 'tandem'. Expression (4.78) can be written for each specklegram in tandem:

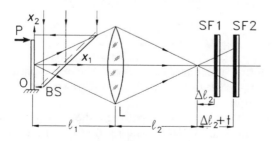

Figure 4.44. Optical setup for simultaneous recording of two defocused specklegram

$$\vec{d}_{s1} = -M_1 \frac{\vec{r}w}{l_1} + 2\Delta l_2 \frac{\text{grad } w}{M_1}$$

$$\vec{d}_{s2} = -M_2 \frac{\vec{r}w}{l_1} + 2(\Delta l_2 + t) \frac{\text{grad } w}{M_2} \qquad (4.80)$$

where

$$M_2 = \frac{l_2 + \Delta l_2 + t}{l_1} = M_1 + \frac{t}{l_1} = M_1 + \Delta M$$

The corresponding relations between vector components of equation (4.80) take the following form:

$$u_{s1} = -M_1 \frac{x_2 w}{l_1} + 2 \frac{\Delta l}{M_1} \frac{\partial w}{\partial x_2}$$

$$u_{s2} = -(M_1 + \Delta M) \frac{x_2 w}{l_1} + 2 \frac{\Delta l_2 + t}{M_1 + \Delta M} \frac{\partial w}{\partial x_2} \qquad (4.81)$$

Subtracting the first equation of (4.81) from the second equation of (4.81) yields

$$\Delta u_s = u_{s2} - u_{s1} = -\Delta M \frac{x_2 w}{l_1} + 2 \left(\frac{\Delta l_2 + t}{M_1 + \Delta M} - \frac{\Delta l_2}{m} \right) \frac{\partial w}{\partial x_2}$$

Taking into account that in practice $t \ll l_1$ and $\Delta M \ll M_1$, this expression can be simplified:

$$\Delta u_s = 2 \frac{t}{M_1} \frac{\partial w}{\partial x_2} \qquad (4.82)$$

Relation (4.82) does not contain the object surface deflection w. The value of the slope $\partial w / \partial x_2$ is defined by the distance between photoplates in the tandem and the difference between the displacements of speckles of the corresponding points of two photoplates. The expression for determination of other slope values can be written by analogy with equation (4.82):

$$\Delta v_s = 2 \frac{t}{M_1} \frac{\partial w}{\partial x_3}$$

It should be noted that the distance between the photoplates in tandem t contained in equation (4.82) can be established with a greater accuracy than the defocusing Δl_1 from equation (4.77).

In order to illustrate the difference between the two approaches to DSP implementation described above, the bending of the cantilever beam loaded with a concentrated force in its free short edge was used as an example. First,

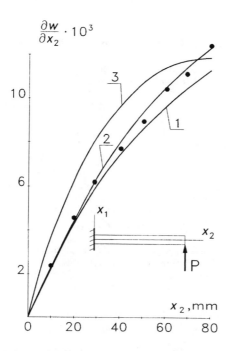

Figure 4.45. Distribution of the slope along the axis of a cantilever beam obtained by means of defocused speckle photography

the conventional DSP technique based on expression (4.77) was implemented to determine the slope $\partial w/\partial x_2$ distribution along the x_2-axis (see Figure 4.45). Curve 1 in Figure 4.45 describes the slope distribution obtained when the optical axis of the recording system passes through the beam segment near the beam clamping (near point O in Figure 4.44). Curve 2 indicates the data of the analytical solution of the corresponding elasticity problem. The comparison of curves 1 and 2 in Figure 4.45 reveals their good coincidence near beam clamping where the deflection values are small. The distinction between these two curves increases with the growth of the distance from beam clamping, i.e. with growth of the deflection value. Curve 3 corresponds to the case when the axis of the recording system is moved to the free edge of the beam, that is, the region of the maximum deflection values. This curve also differs from the analytical curve 2. Thus, we can affirm that the influence of the deflection field is capable of causing considerable distortions in a slope distribution.

Some results of the implementation of the tandem technique to slope measurement are shown in Figure 4.45 by dots. It is clearly seen that in this case experimental and analytical data are in good agreement.

References

1. Draper, N. and Smith, H. (1966) *Applied Regression Analysis*. John Wiley, New York.
2. Seber, G. A. F. (1977) *Linear Regression Analysis*. John Wiley, New York.
3. Vannik, V. N. (1979) *Dependences Reconstruction over Empiric Data*. Nauka, Moscow.
4. Rowlands, R. E., Liber, T., Daniel, I. M. and Rose, P. G. (1973) Differentiation of experimental information. *Exp. Mech.* **13**, 105–112.
5. Rowlands, R. E., Winters, K. D. and Jensen, J. A. (1978) Full field numerical differentiation. *J. Strain Anal.* **13**, 177–183.
6. Nosach, V. V. (1994) *Solution of Approximation Problems by Personal Computer*. MIKAP, Moscow.
7. Yakovlev, V. V., Shchepinov, V. P. and Odintsev, I. N. (1979) Circular plate bending studed by holographic interferometry. In *Physics and Mechanics of Deformations and Fracture*, pp. 133–139, Atomizdat, Moscow.
8. Timoshenko, S. P. and Woinovsky-Krieger, S. (1959) *Theory of Plates and shells*. McGraw-Hill, New York.
9. Bugrov, Ya. S. and Nikolsky, S. M. (1981) *Differential equations. Multiple integrals*. Series Complex variable functions. Nauka, Moscow.
10. Volkov, E. A. (1982) *Numerical methods*. Nauka, Moscow.
11. Timoshenko, S. P. and Goodier, J. N. (1970) *Theory of Elasticity*. McGraw-Hill, New York.
12. Grishin, V. I. and Donchenko, V. Yu. (1983) Investigation of stress concentration in thin and thick plates with a circular hole in elastoplastic deformation range. *Proc. TsAGI* **14**, 85–90.
13. Peterson, R. E. (1974) *Stress Concentration Factors*. Charts and relations useful in making strength calculations for machine parts. John Wiley, New York.
14. Van Dyke, P. (1965) Stress about a circular hole in cylindrical shell. *AIAA J.* **3**, 1733–1742.
15. Ahlberg, J. H., Nilson, E. N. and Walsh, J. L. (1967) *The Theory of Splines and their Application*. Academic Press, New York.
16. Sivokhin, A. V. and Yakovlev, V. V. (1988) Differentiation in hologram interferometry. In *Plasticity, Strength and Fracture Resistance of Materials and Nuclear Power Plants Elements*, pp. 50–55, Energoatomizdat, Moscow.
17. Korn, G. A. and Korn, T. M. (1968) *Mathematical Handbook*. McGraw-Hill, New York.
18. Hsu, T. R. (1974) Large deformation measurements by real-time holographic interferometry. *Exp. Mech.* **14**, 408–411.
19. Kupper, F. D. and Van Dijk, C. A. (1973) Reference fringes in holographic interferometry. *Opt. Laser Technol.* **5**, 69–74.
20. Wang, L., Hovanesian, J. D. and Hung, Y. Y. (1984) A new fringe carrier method for the determination of displacement derivatives in hologram interferometry. *Opt. Laser Engng* **5**, 109–120.
21. Abramson, N. (1975) Sandwich hologram interferometry. 2: Some practical calculations. *Appl. Opt.* **14**, 981–984.
22. De Larminat, P. M. and Wei, R. P. (1976) A fringe-compensaton technique for stress analysis by reflection hologram interferometry. *Exp. Mech.* **16**, 241–248.
23. Abramson, N. and Bjelkhagen, H. (1979) Sandwich hologram interferometry. 5: Measurement of in-plane displacement and comparison for rigid body motion. **18**, 2870–2880.

24. Kersch, L. A. (1971) Advanced concepts of holographic nondestructive testing. *Mater. Evaluat.* **29**, 125–129.
25. Klappert, W. R. (1976) The setting of double-exposure optical holographic NDT test parameters. **34**, 160–164.
26. Hung, Y. Y., Taylor, C. E. (1973) Speckle-shearing interferometric camera—a tool for measurement of derivatives of surface displacement. *Proc. SPIE* **44**, 169–175.
27. Hung, Y. Y. (1982) Shearography: a new optical method for strain measurement and nondestructive testing. *Opt. Engineer.* **21**, 391–395.
28. Sirohi, R. S., (ed.) (1993) *Speckle Metrology*. Marcel Dekker, New York.
29. Dainty, J. C., (ed.) (1975) *Laser Speckle and Related Phenomena*. Topics in Applied Physics, **9**. Springer Verlag, Berlin.
30. Hung, Y. Y., Rowlands, R. E., Daniel, I. M. (1975) Speckle-shearing interferometric technique: a full-field strain gauge. *Appl. Opt.* **14**, 618–622.
31. Rosenberg, A. and Politch, J. (1978) Fringe parameters in speckle shearing interferometry. *Opt. Commun.* **26**, 301–304.
32. Hung, Y. Y. and Liang, C. Y. (1979) Image shearing camera for direct measurement of surface strains. *Appl. Opt.* **18**, 1046–1051.
33. Hung, Y. Y. and Durelli, A. J. (1979) Simultaneous measurement of three displacement derivatives using a multiple image-shearing interferometric camera. *J. Str. Anal.* **14**, 81–88.
34. Iwahashi, Y., Iwata, K. and Nagata, R. (1984) Single-aperture speckle shearing interferometry with a single grating. *Appl. Opt.* **23**, 247–249.
35. Novikov, S. A., Pisarev, V. S., Fursov, A. N. and Shchepinov, V. P. (1992) Using an annular aperture to record speckle interferograms. *Opt. Spectrosc.* **72**, 741–746.
36. Tiziani, H. (1972) A study of the use of laser speckle to measure small tilts of optically rough surfaces accurately. *Opt. Commun.* **5**, 271–274.
37. Gregory, D. A. (1976) Basic physical principles of defocused speckle photography: a tilt topology inspection technique. *Opt. Laser Technol.* **8**, 201–213.
38. Chiang, F. P., Juang, R. M. (1976) Laser speckle interferometry for plate bending problems. *Appl. Opt.* **15**, 2199–2204.

5 STRAIN AND STRESS DETERMINATION IN THIN-WALLED STRUCTURES

The main feature of the holographic and speckle interferometry methods is that they enable non-contact measurements to be carried out of three-dimensional displacement fields on both a curved and a plane optically rough surface of opaque objects. The strain tensor used in solid and fracture mechanics is generally defined through three displacement components of the corresponding object's point. It should be noted that a transition from displacement components to strain values can be performed by means of geometrical relations only, without using any hypotheses concerning mechanical properties of the deformable medium [1]. This fact is of great importance from a standpoint of holographic and speckle interferometry implementation to strain or stress analysis, since the displacement vector components represent the physical values which can be directly measured on the object surface to be studied. Therefore, the application of these methods to a strain/stress analysis is most advantageous for thin-walled structures, in particular, of plate and shell structures which are widely used in various fields of engineering. In this case, strain distribution in the normal to the surface direction through the object thickness can be established by using the experimental data obtained on a single (usually external) surface of the structure.

In order to realize the above-mentioned capability, some theoretical hypotheses should be implemented [2]. Moreover, a calculation of the stresses or similar force parameters is connected with the need to make use of the mechanical model of deformable medium, for instance Hooke's law.

A determination of local strains and stresses near irregular zones of thin-walled structures, especially near holes and cut-outs of various shapes, where the displacement components are characterized with high gradients in all three directions, is of great importance from a strength analysis point of view. Analytical and numerical methods of stress concentration analysis in plates and curved shells with large cut-outs often represent a complicated problem [3–5].

The main topics of this chapter are the development, justification and practical implementation of the experimental approach to an investigation of local strains and stresses in thin-walled shells. This approach is based on three-dimensional displacement field measurement by means of holographic interferometry.

Different techniques of the local strain/stress investigation based on conventional methods of experimental mechanics have been developed [6–10]. However, it should be noted that the number of problems solved through the use of traditional experimental approaches is limited. There are some reasons for this. The first is associated with the high value of the surface curvature of the object to be studied. The second reason resides in the fact that the three-dimensional displacement field in an irregular region of a curved shell is characterized by a high gradient of all three displacement components. The need to determine the strain distribution in the normal to the surface direction through the object thickness is the third serious difficulty that has to be overcome to ensure a positive result of a local stress investigation in curved shells.

The first two above-mentioned problems to be solved prove that holographic interferometry can be effectively implemented in this field. The capability of holographic interferometry for local strain analysis by measuring three-dimensional displacement fields in an irregular zone of a curved shell has been demonstrated through the use of an optical set-up based on a Fresnel hologram recording [11–13], a curved overlay interferometer [14] and a numerical-experimental approach [15].

However, only the combination of the optimal reflection hologram interferometer system with the absolute fringe order counting technique of quantitative interferogram interpretation is capable of ensuring the required accuracy of displacement field measurement of the curved surface which is sufficient for a reliable local strain and stress determination [16,17]. This approach will be used below for local strain and stress analysis in curved shells.

5.1 Principal relations between displacements, strains and stresses

Equations describing calculus on curved surface and the deformation of an object in the most general tensor form are given, for instance, in the work of Schumann, Zurcher and Cuche [18]. In order to gain a better understanding of the experimental results presented below, a more simple approach to the strain/ stress analysis of thin-walled structures is considered in this section.

Measurements of the deformation characteristics of a strained body implementing methods of holographic and speckle interferometry can be performed on an opaque object surface only. Mathematically, a surface in three-dimensional Euclidean space where the Cartesian coordinate system x_i ($i=1,2,3$) is defined can be given by a vector valued function of two parameters

$$\vec{r} = \vec{r}(\alpha_1, \alpha_2) \tag{5.1}$$

or three scalar valid functions

$$x_i = x_i(\alpha_1, \alpha_2) \qquad i = 1, 2, 3 \tag{5.2}$$

where α_1, α_2 are curvilinear coordinates on the surface. We will assume that functions (5.2) or (5.1) are continuous, as well as the first and second derivatives. Under this condition the lines $\alpha_1 = \text{const}$ and $\alpha_2 = \text{const}$ can be considered as the coordinate lines on the surface. In the theory of thin plates and shells, as a rule, the coordinate lines α_1 and α_2 coincide with the lines of principal curvature of the surface [2].

At each point of the surface which is referred to the line of principal curvature of the surface, a base of three orthogonal unit vectors can be defined:

$$\vec{e}_1 = \frac{1}{A_1}\frac{\partial \vec{r}}{\partial \alpha_1}, \ \vec{e}_2 = \frac{1}{A_2}\frac{\partial \vec{r}}{\partial \alpha_2}, \ \vec{e}_3 = [\vec{e}_1 \times \vec{e}_2] \tag{5.3}$$

where $\vec{r} = \vec{r}(\alpha_1, \alpha_2)$ is a vector equation of the basic surface (5.1);

$$A_1 = \left|\frac{\partial \vec{r}}{\partial \alpha_1}\right|, \ A_2 = \left|\frac{\partial \vec{r}}{\partial \alpha_2}\right|$$

are Lame's coefficients, which represent the local scaling factors in the directions α_i ($i = 1, 2$); vectors $\partial\vec{r}/\partial\alpha_i$ ($i = 1, 2$) are line tangent to the coordinate lines α_1 and α_2.

Usually, a trio of vectors $\{\vec{e}_1, \vec{e}_2, \vec{e}_3\}$ should be chosen in accordance with the right-hand rule and the positive direction of the vector $\vec{e}_3$ coincides with the surface convexity. If α_3 is the distance from the surface, then $\vec{e}_3 \equiv \vec{n}$ becomes the unit normal (see Figure 5.1), as determined by the third term in equation (5.3). In this case the coordinate lines α_3 represent straight lines ($A_3 = 1$). Therefore, a radius-vector of a point M whose position on the surface corresponds to curvilinear coordinate values (α_1, α_2, α_3) can be expressed in the following form:

$$\vec{R} = \vec{r}(\alpha_1, \alpha_2) + \alpha_3\vec{n}. \tag{5.4}$$

Let's assume that a point M lies on the surface which can be described by equation $\alpha_3 = \alpha_3^0 = \text{const}$ and usually called a basic surface. When the body under consideration is deformed, the arbitrary point M, describing the

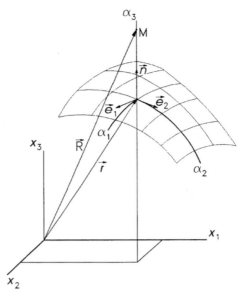

Figure 5.1. Cartesian and curvilinear co-ordinate systems used for an arbitrary surface definition

radius-vector $\vec{R}$ (5.4), is placed in a new position which corresponds to radius-vector $\vec{R} + \vec{U}$ ($\vec{U}$ denotes a vector valued function of a displacement):

$$\vec{U} = u_1\vec{e}_1 + u_2\vec{e}_2 + u_3\vec{e}_3 \tag{5.5}$$

where u_i ($i = 1, 2, 3$) are the displacement vector components on a local basis (5.3).

Each family of coordinate lines belonging to the deformed surface will be distorted. The segments of the coordinate lines will be varied both in length and in direction. The differential relations between the parameters of the coordinate mesh in an initial state and final state represent, as known, a quantitative description of some degree of the body's deformation.

For orthogonal curvilinear coordinate system α_i ($i = 1, 2, 3$), in which the lines α_1 and α_2 coincide with the lines of the surface principal curvature, a square of a total differential of an arc length has a form

$$(\mathrm{d}s)^2 = \sum_{i=1}^{3} A_i^2(\mathrm{d}\alpha_i)^2 \qquad i = 1, 2, 3. \tag{5.6}$$

If $\mathrm{d}s_i$ denotes an arc length along a coordinate line α_i, expression (5.6) gives the following result:

$$A_i = \frac{\mathrm{d}s_i}{\mathrm{d}\alpha_i} \qquad i = 1, 2, 3 \tag{5.7}$$

where A_i are Lame's coefficients from expression (5.3).

A partial derivative of a vector valued displacement function $\vec{u}$ with respect to an independent variable s_i calculated on a local vector basis (5.3) can be expressed as

$$\frac{\partial \vec{u}}{\partial s_i} = \frac{\partial \vec{u}}{\partial \alpha_i} \frac{\partial \alpha_i}{\partial s_i} = \frac{1}{A_i} \frac{\partial \vec{u}}{\partial \alpha_i} = a_{ij}\vec{e}_j. \tag{5.8}$$

It is well known that within limitations of 'small' deformations the components of decomposition (5.8) allow us to construct the components of a symmetrical strain tensor [19]. The diagonal elements of decomposition (5.8) correspond to the linear dilatations or usual strains of an elemental distance $\mathrm{d}s_i$ in the direction $\vec{e}_i$:

$$\varepsilon_{ii} = a_{ii} \qquad i = 1, 2, 3. \tag{5.9'}$$

The non-diagonal elements of the strain tensor represent the angular dilatations or shearing strains considering two perpendicular directions $\vec{e}_i$ and $\vec{e}_j$:

$$\gamma_{ij} = a_{ij} + a_{ij} \equiv \omega_{ij} + \omega_{ji}$$
$$i,j = 1, 2, 3; \qquad i \neq j. \tag{5.9''}$$

To express strain tensor component (5.9) in an explicit form through the geometrical parameters of the surface to be considered, it is necessary to apply the formulae of Gauss–Veingarten for derivatives calculation of an arbitrary vector $\vec{U}$ referred to local basis (5.3) and the relations of Codazzi–Gauss which connect the parameters of the surface curvature and Lame's coefficients [2,4,20]:

$$\varepsilon_{11} = \frac{1}{A_1} \frac{\partial u_1}{\partial \alpha_1} + \frac{1}{A_1 A_2} \frac{\partial A_1}{\partial \alpha_2} u_2 + k_1 u_3$$
$$\varepsilon_{22} = \frac{1}{A_2} \frac{\partial u_2}{\partial \alpha_2} + \frac{1}{A_1 A_2} \frac{\partial A_2}{\partial \alpha_1} u_1 + k_2 u_3 \tag{5.10'}$$
$$\varepsilon_{33} = \frac{\partial u_3}{\partial \alpha_3}$$

$$\omega_{12} = \frac{1}{A_1}\frac{\partial u_2}{\partial \alpha_1} - \frac{u_1}{A_1 A_2}\frac{\partial A_1}{\partial \alpha_2}$$

$$\omega_{13} = \frac{1}{A_1}\frac{\partial u_3}{\partial \alpha_1} - \frac{u_1}{A_1 A_3}\frac{\partial A_1}{\partial \alpha_3}$$

$$\omega_{23} = \frac{1}{A_2}\frac{\partial u_3}{\partial \alpha_2} - \frac{u_2}{A_2 A_3}\frac{\partial A_2}{\partial \alpha_3}$$

$$\omega_{21} = \frac{1}{A_2}\frac{\partial u_1}{\partial \alpha_2} - \frac{u_2}{A_2 A_1}\frac{\partial A_2}{\partial \alpha_1} \qquad (5.10'')$$

$$\omega_{31} = \frac{\partial u_1}{\partial \alpha_3}$$

$$\omega_{32} = \frac{\partial u_2}{\partial \alpha_3}$$

$$\gamma_{12} = \frac{A_1}{A_2}\frac{\partial}{\partial \alpha_2}\left(\frac{u_1}{A_1}\right) + \frac{A_2}{A_1}\frac{\partial}{\partial \alpha_1}\left(\frac{u_2}{A_2}\right)$$

$$\gamma_{13} = A_1\frac{\partial}{\partial \alpha_3}\left(\frac{u_1}{A_1}\right) + \frac{1}{A_1}\frac{\partial u_3}{\partial \alpha_1} \qquad (5.10''')$$

$$\gamma_{23} = A_2\frac{\partial}{\partial \alpha_3}\left(\frac{u_2}{A_2}\right) + \frac{1}{A_2}\frac{\partial u_3}{\partial \alpha_2}$$

where

$$k_i = -\vec{n}\,\frac{\partial^2 \vec{r}(\alpha_1, \alpha_2)}{A_i \partial \alpha_i A_i \partial \alpha_i} \qquad i = 1, 2, 3 \qquad (5.11)$$

are principal curvatures of the basic surface in the directions α_i.

In the Cartesian coordinate system ($\alpha_i \equiv x_i$, $A_i = 1$, $k_i = 1$, $i = 1, 2, 3$) the components of tensor (5.10) are identical to the usual strains and shearing strains, which result from the well-known Cauchy equations [21]:

$$\varepsilon_{ii} = \frac{\partial u_i}{\partial x_i} \qquad i = 1, 2, 3, \qquad (5.12')$$

$$\gamma_{ij} = \omega_{ij} + \omega_{ji} = \frac{\partial u_i}{\partial x_j} + \frac{\partial u_j}{\partial x_i} \qquad i, j = 1, 2, 3; \; i \neq j. \qquad (5.12'')$$

The known strain values (5.10) or (5.12) allow us to calculate the stresses acting in the deformed body by using some physical law. For instance, if a homogeneous and isotropic material submits to Hooke's law, the following relations are valid:

$$\varepsilon_{ii} = \frac{\sigma_{ii} - \mu(\sigma_{jj} + \sigma_{kk})}{E} \tag{5.13'}$$

$i = 1, j = 2, k = 3$ and then the indices must be cyclically removed,

$$\gamma_{ij} = \frac{2\sigma_{ij}(1 + \mu)}{E} \qquad i, j = 1, 2, 3; \; i \neq j, \tag{5.13''}$$

where σ_{ij} are the stress tensor components; E is an elasticity modulus of the material; and μ is Poisson's ratio. Finally, the unit vectors which lie tangent to the lines of the distorted coordinate mesh due to usual and shearing strains (5.10) have the following form:

$$\begin{aligned}
\vec{e}_1' &= \vec{e}_1 + \omega_{12}\vec{e}_2 + \omega_{13}\vec{e}_3 \\
\vec{e}_1' &= \omega_{21}\vec{e}_1 + \vec{e}_2 + \omega_{23}\vec{e}_3 \\
\vec{e}_3' &= \omega_{31}\vec{e}_1 + \omega_{32}\vec{e}_2 + \vec{e}_2.
\end{aligned} \tag{5.14}$$

It should be noted that equations (5.10) or (5.12) which define the strain tensor components and relations (5.13) which connect the strain and stress components have a general form, since a point M, whose position in space is described by the radius-vector (5.4), may lie on both the object's surface element and the volume element underneath it. For various specified classes of bodies which are characterized by some definite geometrical parameters the above-mentioned relations can be represented in a more simple and convenient form. One of these classes is a family of thin-walled plates and shells of constant thickness. Such a thin-walled structure represents a physical body limited by two sizes with coordinate surfaces, for instance, $\alpha_3 = h_1 = \text{const}$ and $\alpha_3 = h_2 = \text{const}$. The value $h = h_2 - h_1$ (the thickness of a plate or shell) must be considerably less than the maximum characteristic dimension of the structure R_0 in another two directions. The usual condition, which should be satisfied to consider a structure as thin-walled, is $h/R_0 \leqslant 0.05$. The surfaces $\alpha_3 = h_1$ and $\alpha_3 = h_2$ are usually called face surfaces or faces.

If analytical or numerical methods are applied to strain/stress analysis of plates and shells, the surface which lies at an equal distance from the faces (the middle surface) is taken as the basic surface. Since optical interferometric techniques allow measurements to be carried out on faces of an opaque object, we shall conveniently use the external face of a body $h = h_2$ as the basic surface.

Expressions (5.10), describing the strain tensor components, are valid on any basic surface $\alpha_3 = \text{const}$. In order to obtain the analogous relations which will be able to express the strains and stresses on an arbitrary plate or shell surface through the strains and stresses on the basic surface and geometrical parameters of the basic surface, it is necessary to use relations between principal curvature values (5.11) and the magnitude of α_3-coordinate and analogous relations for Lame's coefficients (5.7), which can be expressed in the following forms respectively [2]:

$$k_1(\alpha_3) = \frac{k_1}{1 + k_1\alpha_3}, \quad k_2(\alpha_3) = \frac{k_2}{1 + k_2\alpha_3} \tag{5.15}$$

$$A_1(\alpha_3) = (1 + k_1\alpha_3)A_1, \quad A_2(\alpha_3) = (1 + k_2\alpha_3)A_2 \tag{5.16}$$

where k_i and A_i ($i = 1, 2, 3$) are the values of corresponding functions on the external basic face. Specific deformation properties of thin-walled structures provide us with the possibility to simplify relations (5.10) and (5.13) which define the strain and stress tensor components. We shall consider the simplification procedure which leads to the so-called technical theory of plates and shells. In this case, the corresponding transformations of the above-mentioned relations are based on the hypotheses which have an evident physical sense, and are known as the Kirchhoff–Love hypotheses [2–4].

The first, called kinematic hypothesis, can be formulated in the following form:

The straight segments of the coordinate lines α_3 between two shell or plate faces maintain their initial length during deformation and remain as straight lines which are perpendicular to the deformed coordinate surface $\alpha_3 = $ const.

Mathematically, a constancy in a length of the coordinate lines between two faces means that the strain tensor component ε_{33} from equation (5.10') is equal to zero. In this case equation (5.10') becomes

$$\varepsilon_{33} = \frac{\partial u_3}{\partial \alpha_3} \cong 0 \tag{5.17'}$$

and, hence,

$$u_3(\alpha_3) = \text{const} \tag{5.17''}$$

where u_3 is the displacement component from equation (5.5) normal to a shell or plate surface which is defined on the external face.

The first Kirchhoff–Love hypothesis also allow us to consider that the unit normal vector $\vec{e}_3'$ defined in the deformed basis (5.14) is independent of the α_3 coordinate value. This means that rotation angles ω_{31} and ω_{32} are also independent of the α_3 coordinate value—see the third equation in (5.14). In this case the last two equations from (5.10'') become

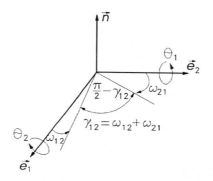

Figure 5.2. Scheme of sign definition of rotation angles θ_1 and θ_2

$$\omega_{31} \equiv \theta_1 = \frac{\partial u_1}{\partial \alpha_3} = \text{const}$$

$$\omega_{32} \equiv \theta_2 = \frac{\partial u_2}{\partial \alpha_3} = \text{const.}$$

(5.18)

The positive directions of the rotations θ_1 and θ_2 are shown in Figure 5.2. By integrating the relations in equation (5.18) along α_3 coordinate between two faces we can obtain

$$u_1(\alpha_3) = \theta_1 \cdot \alpha_3 + u_1$$

$$u_2(\alpha_3) = \theta_2 \cdot \alpha_3 + u_2$$

(5.19)

where u_1 and u_2 are the displacement components along the coordinate lines α_1 and α_2, respectively, defined on the basic surface of a thin-walled structure. Remember that in the presented analysis, the external face is considered as the basic surface of a shell or plate. Expressions (5.19) denote that displacement components u_1 and u_2 which lie tangent to the coordinate surface $\alpha_3 = \text{const}$ represent linear functions of the α_3 coordinate.

A change in the cosine of the angle between each couple of coordinate lines characterizes the shearing strain values, in particular the values of γ_{31} and γ_{32} from equation (5.10'''). The fact that according to the first Kirchhoff–Love hypothesis the unit vector $\vec{e}_3'$ in (5.14) is perpendicular to the coordinate surface $\alpha_3 = \text{const}$ in a deformed thin structure results in

$$\gamma_{31} = \gamma_{32} = 0.$$

(5.20)

Using equations (5.20) and (5.10''') we can find that the following relations are valid for the rotation angles:

$$\omega_{31} = -\omega_{13}, \qquad \omega_{32} = -\omega_{23}. \tag{5.21}$$

By substituting the corresponding rotation angle values in equation (5.21) with those of equation (5.10″), we obtain

$$\theta_1 = -\frac{1}{A_1}\frac{\partial u_3}{\partial \alpha_1} + k_1 u_1$$
$$\theta_2 = -\frac{1}{A_2}\frac{\partial u_3}{\partial \alpha_2} + k_2 u_2 \tag{5.22}$$

where u_i ($i = 1,2,3$) are the components of the displacement vector (5.5) defined on the external face.

Relations (5.17) and (5.19) allow us (taking into account equation (5.22)) to express the displacement components of an arbitrary point of a thin-walled structure through the displacement components and its partial derivatives along the coordinate lines α_1 and α_2 defined on the external face (basic surface) of a plate or shell. As mentioned above, these parameters represent the deformation characteristics which can be measured by various holographic and speckle interferometric techniques. This remarkable feature means that experimental study of a deformation of a thin-walled structure, within the technical theory of plates and shells, consists of the investigation of a deformation of its single basic surface, usually the external face.

Substitution of expressions (5.17), (5.20) and (5.22) into the corresponding relations in (5.10) leads to the following strain tensor components:

$$\varepsilon_{11}(\alpha_3) = \varepsilon_{11} + \alpha_3 \ae_1$$
$$\varepsilon_{22}(\alpha_3) = \varepsilon_{22} + \alpha_3 \ae_2 \tag{5.23}$$
$$\gamma_{12}(\alpha_3) = \gamma_{12} + 2\alpha_3 \ae_{12}$$

where ε_{11}, ε_{22}, γ_{12} must be determined on the basic surface in accordance with expression (5.10), and

$$\ae_1 = \frac{1}{A_1}\frac{\partial \theta_1}{\partial \alpha_1} + \frac{\theta_2}{A_1 A_2}\frac{\partial A_1}{\partial \alpha_2}$$
$$\ae_2 = \frac{1}{A_2}\frac{\partial \theta_2}{\partial \alpha_2} + \frac{\theta_1}{A_1 A_2}\frac{\partial A_2}{\partial \alpha_1} \tag{5.24}$$
$$\ae_{12} = \frac{1}{A_1}\frac{\partial \theta_2}{\partial \alpha_1} - \frac{\theta_1}{A_1 A_2}\frac{\partial A_1}{\partial \alpha_2} + k_1 \omega_{12} = \frac{1}{A_2}\frac{\partial \theta_1}{\partial \alpha_2} - \frac{\theta_2}{A_1 A_2}\frac{\partial A_2}{\partial \alpha_1} + k_2 \omega_{21}$$

where ω_{12} or ω_{21} is derived from expression (5.10″) and k_i ($i = 1, 2$) are defined by (5.11). The values $\ae_1$ and $\ae_2$ are the increments of the principal curvature values of basic surface due to deformation. The meaning of $\ae_{12}$ is the increment of torsion of the same surface. In order to derive the equations (5.23) and (5.24) it is necessary to apply the Codazzi–Gauss relations and to neglect all members containing the products of $\alpha_3 k_i$ ($i = 1, 2$), since, in accordance with the definition of a thin plate or shell, $\alpha_3 k_i \ll 1$.

Now, in order to obtain the complete set of the strain tensor components we must more accurately determine the usual strain ε_{33} in the normal to the basic surface direction only, taking into account that two shearing strain components γ_{13} and γ_{23} (5.20) are equal to zero due to the first Kirchhoff–Love hypothesis.

Estimation of the strain tensor component ε_{33} can be obtained proceeding from the second hypothesis of Kirchhoff–Love whose formulation may be represented in the following form:

The normal to the basic surface stress tensor component can be neglected in comparison with other stress components:

$$\sigma_{33} = 0. \tag{5.25}$$

Note that combination (5.25) with the corresponding equation in (5.13) gives us

$$\varepsilon_{33} = -\frac{\mu}{E}(\sigma_{11} + \sigma_{22}).$$

This result is in contradiction to expression (5.17′). That is why a sign of an approximated equality is used in relation (5.17′). This fact illustrates that the technical theory of thin plates and shells is not free from internal contradictions. Other contradictions of the theory involved which can be revealed by means of application of holographic and speckle interferometry to local strain determination in plates and curved shells with cut-outs will be discussed in Sections 5.2 and 5.3.

As seen from equations (5.23), the strain tensor components $\varepsilon_{ii}(\alpha_3)$ ($i = 1, 2$) and $\gamma_{12}(\alpha_3)$ at an arbitrary point of a thin-walled structure (plate or shell) can be represented through the principal curvature values of the basic surface and six parameters ε_{11}, ε_{22}, γ_{12}, $æ_1$, $æ_2$ and $æ_{12}$, which are the functions of two variables α_1 and α_2 and defined on the basic surface. It has been established that these parameters completely characterize a deformation of the basic surface of a thin-walled structure and, therefore, can be adopted as the actual strain tensor components [22].

Therefore, an assumption about the validity of the Kirchhoff–Love first hypothesis, in common with the subsequent simplification of the corresponding relations by means of a member's neglecting of an order of magnitude of $k_i h$ ($i = 1, 2$; h is thickness of shell or plate) with respect to unity, leads to the linear distribution of the strain tensor components in the normal to surface direction through a structure's thickness.

Taking into account relations (5.17′), (5.20) and (5.25), Hooke's law in the form of expression (5.13) may be rewritten for the case concerned:

$$\varepsilon_{11} = \frac{1}{E}(\sigma_{11} - \mu\sigma_{22})$$

$$\varepsilon_{22} = \frac{1}{E}(\sigma_{22} - \mu\sigma_{11}) \qquad (5.26)$$

$$\gamma_{12} = \frac{2(1 + \mu)}{E}\sigma_{12}.$$

Solving equations (5.26) in relation to the stress tensor components gives

$$\sigma_{11} = \frac{E}{1 - \mu^2}(\varepsilon_{11} + \mu\varepsilon_{22})$$

$$\sigma_{22} = \frac{E}{1 - \mu^2}(\varepsilon_{22} + \mu\varepsilon_{11}) \qquad (5.27)$$

$$\sigma_{12} = \frac{E}{2(1 + \mu)}\gamma_{12}.$$

Substituting equation (5.23) into (5.27) yields

$$\sigma_{11}(\alpha_3) = \sigma_{11} + \alpha_3 \frac{E}{1 - \mu^2}(\mathfrak{æ}_1 + \mu\mathfrak{æ}_2)$$

$$\sigma_{22}(\alpha_3) = \sigma_{22} + \alpha_3 \frac{E}{1 - \mu^2}(\mathfrak{æ}_2 + \mu\mathfrak{æ}_1) \qquad (5.28)$$

$$\sigma_{12}(\alpha_3) = \sigma_{12} + \alpha_3 \frac{E\mathfrak{æ}_{12}}{1 + \mu}$$

where $\sigma_{ii}(i = 1, 2)$ and σ_{12} are defined on the basic surface by relations (5.27).

If a stress analysis of a shell is carried out by analytical or numerical methods, formulae (5.28) are usually represented in another form. In these cases the stresses acting on the middle surface of a structure $\alpha_3 = -h/2$ should be conveniently selected:

$$\sigma_{ii}^{M} = \sigma_{ii}\left(\alpha_3 = -\frac{h}{2}\right) \qquad i = 1, 2$$

$$\sigma_{12}^{M} = \sigma_{12}\left(\alpha_3 = -\frac{h}{2}\right). \qquad (5.29)$$

Then, instead of equation (5.28), we can write

$$\sigma_{ii}(\alpha_3) = \sigma_{ii}^M + \left(\frac{h}{2} - \alpha_3\right)\frac{2}{h}\sigma_{ii}^B \qquad i = 1, 2$$

$$\sigma_{12}(\alpha_3) = \sigma_{12}^M + \left(\frac{h}{2} - \alpha_3\right)\frac{2}{h}\sigma_{12}^B \qquad \text{(5.30)}$$

where

$$\sigma_{11}^B = \frac{h}{2}\frac{E}{(1-\mu^2)}(\ae_1 + \mu\ae_2)$$

$$\sigma_{22}^B = \frac{h}{2}\frac{E}{(1-\mu^2)}(\ae_2 + \mu\ae_1) \qquad \text{(5.31)}$$

$$\sigma_{12}^B = \frac{h}{2}\frac{E}{(1+\mu)}\ae_{12}.$$

Stresses (5.29) are usually called membrane stresses and stresses (5.31) are bending stresses. Notations (5.29) and (5.31) will be used below for a comparison of analytical and experimental data.

We shall now consider an application of the above-presented analysis in strain and stress determination of typical engineering structures. This is the subject of the following experimental investigations.

A CYLINDRICAL SHELL SUBJECTED TO ANY TYPE OF LOADING

A circular cylindrical shell of a constant thickness h and external radius R is shown in Figure 5.3. The direction of the coordinate line α_1 coincides with the shell generatrix. The α_1-line and α_2-line, which represent a circle line generated by an intersection of the external shell face and plane perpendicular to a shell axis, coincide with the lines of the shell principal curvature. The corresponding values of the principal curvatures

$$k_1 = 0 \qquad k_2 = \frac{1}{R} \qquad \text{(5.32')}$$

and Lame's scaling factors

$$A_i = 1 \qquad i = 1, 2, 3 \qquad \text{(5.32'')}$$

are derived from equations (5.11) and (5.7) respectively. In the cylindrical coordinate system the components u_i ($i = 1, 2, 3$) of the displacement vector $\vec{u}$ (5.5) are usually denoted as u, v and w respectively—see Figure 5.3.

The strain tensor components (5.10'), (5.10'') and rotation angles (5.22) taking into account (5.32) can be rewritten in the following form:

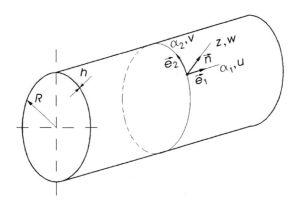

Figure 5.3. Scheme of a regular circular cylindrical shell with notation used for strain determination

$$\varepsilon_{11} = \frac{\partial u}{\partial \alpha_1}, \qquad \varepsilon_{22} = \frac{\partial v}{\partial \alpha_2} + \frac{w}{R}, \qquad \gamma_{12} = \frac{\partial u}{\partial \alpha_2} + \frac{\partial v}{\partial \alpha_1} \qquad (5.33)$$

$$\theta_1 = -\frac{\partial w}{\partial \alpha_1}, \qquad \theta_2 = -\frac{\partial w}{\partial \alpha_2} + \frac{v}{R}. \qquad (5.34)$$

Using relations (5.32) and (5.34) allows us to rewrite the expressions for the curvature and torsion increments:

$$æ_1 = -\frac{\partial^2 w}{\partial \alpha_1^2}, \qquad æ_2 = \frac{\partial^2 w}{\partial \alpha_2^2} - \frac{1}{R}\frac{\partial v}{\partial \alpha_2}, \qquad æ_{12} = -\frac{\partial^2 w}{\partial \alpha_1 \partial \alpha_2} + \frac{1}{R}\frac{\partial v}{\partial \alpha_1}. \qquad (5.35)$$

The corresponding dependencies between strains and stresses are defined by relations (5.26)–(5.31).

The deformation characteristics (5.33)–(5.35) in common with equations (5.23) describe a deformed state of a shell in terms referred to the curvilinear cylindrical coordinate system, two coordinate lines of which (α_1 and α_2) coincide with the lines of the principal curvature. This approach is usually applied in strain or stress analysis of regular shells without holes. But solving the most interesting problems of a strain or stress concentration determination is connected with the analysis of strain distributions along an arbitrary closed and smooth contour (loop, curve) on a shell surface. This contour, in a general case, does not coincide with the lines of the principal curvature and are defined on the basic shell surface and, also, on a line surface formed by moving of the coordinate line α_3 along this contour.

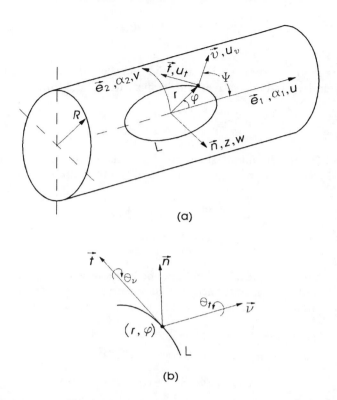

Figure 5.4. General (a) and local (a,b) coordinate system referring to the closed contour L on the surface of a cylindrical shell

Let us consider a closed smooth contour L which represents a boundary of an arbitrary form cut-out on the external shell face—see Figure 5.4(a). The semi-geodetic coordinate system (r, z, φ) is connected with a geometric centre of the cut-out. This coordinate system corresponds to the polar coordinate system on the development of a shell. Both normal $\vec{v}$ and tangential $\vec{t}$ to the contour directions are defined at each point of the line L with coordinates (r, φ). The angle between directions $\vec{v}$ and α_1 is denoted Ψ—see Figure 5.4(a). Displacement component values along directions α_1, α_2, on the one hand, and $\vec{v}, \vec{t}$ on the other, and between the corresponding rotations angles $\theta_1\ \theta_2$ and θ_v, θ_t (see Figure 5.4(b)) are connected through the following relations [4]:

$$u_v - iu_t = e^{i\Psi}(u - iv) \qquad (5.36')$$

$$\theta_v - i\theta_t = e^{i\Psi}(\theta_1 - i\theta_2) \tag{5.36''}$$

where $i^2 = -1$.

The analogue to relation (5.36) is valid between derivatives along the directions α_1 and α_2 and derivatives with respect to ∂s_v and ∂s_t:

$$\frac{\partial}{\partial s_v} - i\frac{\partial}{\partial s_t} = e^{i\Psi}\left(\frac{1}{A_1}\frac{\partial}{\partial \alpha_1} - \frac{i}{A_2}\frac{\partial}{\partial \alpha_2}\right) \tag{5.37'}$$

where ∂s_v and ∂s_t are the infinitesimal arc segments in the directions $\vec{v}$ and $\vec{t}$ respectively. From equation (5.37'), taking (5.32'') into account, we can obtain

$$\begin{aligned}
\frac{\partial}{\partial \alpha_1} &= -\sin\Psi\frac{\partial}{\partial s_t} + \cos\Psi\frac{\partial}{\partial s_v} \\
\frac{\partial}{\partial \alpha_2} &= \cos\Psi\frac{\partial}{\partial s_t} + \sin\Psi\frac{\partial}{\partial s_v}.
\end{aligned} \tag{5.37''}$$

Strains on the basic shell surface along the line L can be expressed by means of known relations of the elasticity theory which describe a transformation of the strain tensor components due to a rotation of coordinate axes [20,21]:

$$\begin{aligned}
\varepsilon_{tt} &= \varepsilon_{22}\cos^2\Psi - \sin\Psi\cos\Psi\gamma_{12} + \varepsilon_{11}\sin^2\Psi \\
\varepsilon_{vv} &= \varepsilon_{22}\sin^2\Psi - \gamma_{12}\sin\Psi\cos\Psi + \varepsilon_{11}\cos^2\Psi \\
\varepsilon_{tv} &= (\varepsilon_{22} - \varepsilon_{11})\sin\Psi\cos\Psi + \frac{1}{2}(\cos^2\Psi - \sin^2\Psi)\gamma_{12}
\end{aligned} \tag{5.38}$$

where ε_{11}, ε_{22}, γ_{12} are derived from equation (5.33), ε_{tt}, ε_{vv} are relative dilatations (usual strains) of an infinitely small element belonging to the line L, ε_{tv} is a half of a shear angle between the unit vectors $\vec{v}$ and $\vec{t}$.

The strain component ε_{tt} is of a great interest from a strength analysis viewpoint especially if a boundary of an open hole is load-free. In the last case the values ε_{tt} and ε_{vv} are related through Poisson's ratio. The value of ε_{tt}-strain from equation (5.38) can be conveniently represented through partial derivatives of the displacement components taken along the line L. Substituting equation (5.33) into (5.38) and taking (5.37) into account yields

$$\varepsilon_{tt} = \cos\Psi\frac{\partial v}{\partial s_t} - \sin\Psi\frac{\partial u}{\partial s_t} + \cos^2\Psi\frac{w}{R}. \tag{5.39}$$

Now, to complete an analysis of a deformed state of a load-free hole edge in a circular cylindrical shell, it is necessary to establish a law of change of the ε_{tt}-strain in the normal to surface direction through a thin-walled object thickness. In order to find this law, one can represent the displacement vector $\vec{u}$ (5.5) at each point of a line L in a base of three unit vectors $(\vec{v}, \vec{t}, \vec{n})$ referred to a contour line L ($\vec{v}$ is the normal direction to the line L, $\vec{t}$ is the tangential direction to the line L, $\vec{n}$ is the normal direction to the basic shell surface, see Figure 5.4(b)):

$$\vec{U} = u_v \cdot \vec{v} + u_t \cdot \vec{t} + w \cdot \vec{n}. \tag{5.40}$$

The displacement vector components in the right-hand side of equation (5.40) are connected to those in the cylindrical coordinate system through relations (5.36').

Using the displacement components (5.40), expression for the strain ε_{tt} (5.39) becomes

$$\varepsilon_{tt} = \frac{\partial u_t}{\partial s_t} + æ_t u_v + \frac{w}{R} \cos^2 \Psi \tag{5.41}$$

where

$$æ_t = \partial \Psi / \partial s_t \tag{5.42}$$

is a geodetic curvature of a contour L.

The combination of expressions (5.17), (5.19) and (5.36) allows us to receive relations between displacement component values on the external (u_v, u_t) and internal (u_v', u_t') shell faces:

$$
\begin{aligned}
u_v' &= u_v - h\theta_v \\
u_t' &= u_t - h\theta_t \\
w' &= w
\end{aligned}
\tag{5.43}
$$

where θ_t and θ_v are the angles of the surface normal $\vec{n}$ rotations in planes $(\vec{t} \cdot \vec{n})$ and $(\vec{t} \cdot \vec{v})$, respectively, and h is a shell thickness. Rotation angles θ_v and θ_t can be expressed through the following equations:

$$\theta_v = -\frac{\partial w}{\partial s_v} + \frac{u_v}{R} \sin^2 \Psi + \frac{u_t}{2R} \cos 2\Psi = -\frac{\partial w}{\partial s_v} + \frac{v}{R} \sin \Psi \tag{5.44}$$

$$\theta_t = -\frac{\partial w}{\partial s_t} + \frac{u_t}{R} \cos^2 \Psi + \frac{u_t}{2R} \sin 2\Psi = -\frac{\partial w}{\partial s_t} + \frac{v}{R} \cos \Psi. \tag{5.45}$$

The second equalities in equations (5.44) and (5.45) are obtained by using expressions (5.34), (5.36) and (5.37).

Relations (5.41) and (5.43)–(5.45) allow us to obtain differences between strain ε_{tt} on the external face and ε_{tt}' on the internal face of the shell. To implement this possibility it is assumed that geodetic curvature $æ_t$ (5.42) along the line L does not depend on the α_3-coordinate value, and rotation angles θ_v and θ_t are defined on the basic surface (external face) of the shell:

$$(\varepsilon_{tt} - \varepsilon_{tt}') = \frac{\partial}{\partial s_t}(u_t - u_t') + æ_t(u_v - u_v') = h\left[-\frac{\partial^2 w}{\partial s_t^2} + \frac{\partial v}{\partial s_t} \frac{\cos \Psi}{R} - \frac{\partial w}{\partial s_v} æ_t \right]. \tag{5.46}$$

If the coordinate axes α_1 and α_2 are rigidly turned around from the normal to a shell surface direction $\vec{n}$ to a new position $\vec{v}$, $\vec{t}$ the total stresses (5.30) referred to the line L can be transformed by means of known relations of the elasticity theory which are analogous to equation (5.38):

$$\sigma_{vv}(z) = \sigma_{vv}^M + \left(\frac{h}{2} - z\right)\frac{2}{h}\sigma_{vv}^B$$

$$\sigma_{tt}(z) = \sigma_{tt}^M + \left(\frac{h}{2} - z\right)\frac{2}{h}\sigma_{tt}^B \qquad (5.47)$$

$$\sigma_{tv}(z) = \sigma_{tv}^M + \left(\frac{h}{2} - z\right)\frac{2}{h}\sigma_{tv}^B$$

where $z \equiv \alpha_3$ is the coordinate in the normal to a shell surface direction, $z = 0$ and $z = -h$ on the external and internal face, respectively.

In accordance with the basic ideas of the classical elasticity theory, the following relations are valid along a load-free boundary of an open hole in a thin-walled structure:

$$\sigma_{vv}(z) = 0, \qquad \sigma_{tv}(z) = 0. \qquad (5.48)$$

Then, expressions (5.48) and (5.27) reveal that the stress stage along the load-free hole edge can be represented in the following form:

$$\sigma_{tt}(z) = E\varepsilon_{tt}(z). \qquad (5.49')$$

According to equation (5.49'), the stresses on the external face σ_{tt} and internal face along a hole edge can be expressed as

$$\sigma_{tt} = E\varepsilon_{tt} \qquad \sigma_{tt}' = E\varepsilon_{tt}'. \qquad (5.49'')$$

Bending stresses from (5.47) which are related to the changing of the corresponding curvature values through relations of form (5.31) are given by

$$\sigma_{tt}^B = \frac{E}{2}(\varepsilon_{tt} - \varepsilon_{tt}'). \qquad (5.50)$$

Membrane stresses on the middle plane of a shell are determined as a difference between any stresses of equation (5.49'') and bending stresses (5.50). If the stresses on the external are used, one can obtain

$$\sigma_{tt}^M = \sigma_{tt} - \sigma_{tt}^B. \qquad (5.51)$$

Now we represent the above-presented relations for the strain and stress tensor components along the closed hole edge in a circular cylindrical shell in the explicit form for a circular and elliptical form of the line L.

A circular hole of radius r_0

In this case the angles φ and Ψ coincide one with another (see Figure 5.4(a)), $\partial/\partial s_t = \partial/r\partial\varphi$, $\partial/\partial s_v = \partial/\partial r$ and geodetic curvature (5.42) $\ae_t = 1/r$. Relations (5.41) and (5.46) can be rewritten in the form

$$\varepsilon_{tt} \equiv \varepsilon_\varphi = \frac{1}{r_0}\left[\frac{\partial v}{\partial\varphi}\cos\varphi - \frac{\partial u}{\partial\varphi}\sin\varphi\right] + \frac{w\cos^2\varphi}{R} \qquad (5.52)$$

$$(\varepsilon_\varphi - \varepsilon_\varphi') = \frac{h}{r_0}\left(-\frac{\partial^2 w}{r_0\partial\varphi^2} + \frac{\partial v}{\partial\varphi}\frac{\cos\varphi}{R} - \frac{\partial w}{\partial r}\right). \qquad (5.53)$$

An elliptical cut-out

We assume that a semi-axis a lies along a shell generatrix and another semi-axis in a perpendicular direction is denoted b, $a > b$. In this case the elliptical curvilinear coordinate system must be defined on the basic external surface of a shell to analyse a strain or stress distribution along a hole edge [4]. A conformal mapping of the contour of the unit circle onto the contour of an elliptical form is given by

$$x + iy = 2r_0\sqrt{e}\,\text{ch}\,(\xi + i\gamma) \qquad (5.54)$$

where the axes x and y are the coordinate lines α_1 and α_2 respectively on the development of a shell, $2r_0 = a + b$, $e = (a - b)/(a + b)$, $a > b$. The coordinate lines $\xi = \text{const}$ are the ellipses with the same focus point. The semi-axes of these ellipses are described by the equations

$$a = 2r_0\sqrt{e}\,\text{ch}\xi \qquad b = 2r_0\sqrt{e}\,\text{sh}\xi. \qquad (5.55)$$

The law of the coordinate transformation corresponding to equation (5.54) is

$$\begin{aligned} x \equiv \alpha_1 = 2r_0\sqrt{e}\,\text{ch}\xi\cos\gamma \\ y \equiv \alpha_2 = 2r_0\sqrt{e}\,\text{sh}\xi\sin\gamma. \end{aligned} \qquad (5.56)$$

The second equalities in equation (5.56) result from (5.55).

An angle between the direction of the normal $\vec{v}$ to the contour L and α_1 axis (the x-axis in the case concerned) is defined by the following equations [4]:

$$\cos\Psi = \frac{A_2}{B_L}\frac{\partial\alpha_2}{\partial\gamma} \qquad \sin\Psi = \frac{A_1}{B_L}\frac{\partial\alpha_1}{\partial\gamma} \qquad (5.57)$$

where A_i ($i = 1, 2$) are Lame's scaling factors (5.7) and

$$B_{\mathrm{L}} = \sqrt{\left(A_1 \frac{\partial \alpha_1}{\partial \gamma}\right)^2 + \left(A_2 \frac{\partial \alpha_1}{\partial \gamma}\right)^2} = \sqrt{a^2 \sin^2 \gamma + b^2 \cos^2 \gamma}. \qquad (5.58)$$

The second equality in equation (5.58) is valid for a circular cylindrical shell accordingly to expressions (5.32″), (5.56) and (5.57).

The most convenient approach to an experimentally measured displacement component representation along an elliptical hole edge consists of using the polar (semi-geodetic) coordinate system (r, φ)—see Figure 5.4a. The relation between these coordinates and elliptical ones results from equation (5.56):

$$\frac{y}{x} = \mathrm{tg}\varphi = \frac{b}{a}\mathrm{tg}\gamma, \qquad (5.59)$$

$$r^2 = \alpha_1^2 + \alpha_2^2 \equiv x^2 + y^2 = (2r_0\sqrt{e})^2 \, (\mathrm{ch}^2\xi\cos^2\gamma + \mathrm{sh}^2\xi\sin^2\gamma).$$

Using relations (5.56)–(5.59), one can obtain the relations between current parameters describing an elliptical cut-out boundary and the current values of polar angle φ in the following form:

$$\mathrm{tg}\Psi = \frac{a}{b}\mathrm{tg}\gamma = \frac{a^2}{b^2}\mathrm{tg}\varphi \qquad (5.60)$$

$$B_{\mathrm{L}} = \sqrt{\frac{a^4\sin^2\varphi + b^4\cos^2\varphi}{a^2\sin^2\varphi + b^2\cos^2\varphi}}. \qquad (5.61)$$

Formulae (5.60) and (5.61) are necessary to calculate the derivatives $\partial/\partial s_t$ and $\partial/\partial s_v$ and other function derivatives, contained in the expressions of strain (5.41) and (5.46), through the polar coordinates (r, φ). The required relations for partial derivatives yield from equations (5.37′) and (5.57)

$$\frac{\partial\phi}{\partial s_t} = \frac{1}{B_{\mathrm{L}}}\frac{\partial\phi}{\partial\gamma} = \frac{1}{B_{\mathrm{L}}}\frac{\partial\phi}{\partial\varphi}\frac{\partial\varphi}{\partial\gamma} \qquad (5.62)$$

$$\frac{\partial\phi}{\partial s_v} = \frac{1}{B_{\mathrm{L}}}\frac{\partial\phi}{\partial\xi} = \frac{1}{B_{\mathrm{L}}}\frac{\partial\phi}{\partial r}\frac{\partial r}{\partial\xi} \qquad (5.63)$$

where ϕ represent a generalized function describing a distribution of the corresponding displacement component.

In addition to expressions (5.62) and (5.63), the geodetic curvature $\mathfrak{æ}_t$ (5.42)

$$\mathfrak{æ}_t = \frac{\partial\Psi}{\partial s_t} = \frac{1}{B_{\mathrm{L}}}\frac{\partial\Psi}{\partial\gamma} = \frac{1}{B_{\mathrm{L}}}\frac{\partial\Psi}{\partial\varphi}\frac{\partial\varphi}{\partial\gamma} \qquad (5.64)$$

and the second-order derivatives of the generalized deflection ϕ must be used to calculate the bending strains. The last function can be derived by applying the rule of complex function differentiation to formula (5.62):

$$
\frac{\partial^2 \phi}{\partial s_t^2} = \frac{1}{B_L^2} \left\{ \frac{\partial^2 \phi}{\partial \varphi^2} \left(\frac{\partial \varphi}{\partial \gamma} \right)^2 + \frac{\partial \phi}{\partial \varphi} \frac{\partial^2 \varphi}{\partial \gamma^2} - \frac{1}{B_L} \frac{\partial B_L}{\partial \gamma} \frac{\partial \phi}{\partial \varphi} \frac{\partial \varphi}{\partial \gamma} \right\}. \tag{5.65}
$$

The derivatives $\partial \Psi / \partial \varphi$, $\partial B_L / \partial \gamma$, $\partial \varphi / \partial \gamma$, and $\partial^2 \varphi / \partial \gamma^2$ which are also contained in relations (5.64) and (5.65), can be obtained using (5.59) and

$$
\frac{\partial \Psi}{\partial \gamma} = \frac{a^2 b^2}{b^4 \cos^2 \varphi + a^4 \sin^2 \varphi} \tag{5.66}
$$

$$
\frac{\partial B_L}{\partial \gamma} = \frac{ab \sin 2\varphi (a^2 - b^2)}{2 B_L (a^2 \sin^2 \varphi + b^2 \cos^2 \varphi)} \tag{5.67}
$$

$$
\frac{\partial \varphi}{\partial \gamma} = \frac{a^2 \sin^2 \varphi + b^2 \cos^2 \varphi}{ab} \tag{5.68}
$$

$$
\frac{\partial^2 \varphi}{\partial \gamma^2} = \frac{\sin 2\varphi (a^2 - b^2)}{ab} \cdot \frac{\partial \varphi}{\partial \gamma}. \tag{5.69}
$$

Now relations (5.60), (5.62) and (5.68) enable us to rewrite expression (5.39) for strain $\varepsilon_{tt} = \varepsilon_\varphi$ in the case of an elliptical cut-out:

$$
\varepsilon_\varphi = \sqrt{\frac{a^2 \sin^2 \varphi + b^2 \cos^2 \varphi}{B_L^2 ab}} \left\{ \frac{\partial v}{\partial \varphi} b^2 \cos \varphi - \frac{\partial u}{\partial \varphi} a^2 \sin \varphi \right\}
$$
$$
+ \frac{b^4 \cos^2 \varphi}{(b^4 \cos^2 \varphi + a^4 \sin^2 \varphi)} \frac{w}{R}. \tag{5.70}
$$

In order to find the ε_φ-strain distribution in the normal to a surface direction $\vec{n}$ through a shell thickness, we need to use an explicit form of the geodetic curvature (5.64) which can be obtained by using expressions (5.66) and (5.68):

$$
\text{æ}_t = \frac{ab}{B_L^3}. \tag{5.71}
$$

An explicit form of the partial derivatives $\partial \phi / \partial s_v$ (5.63) comes from the second expression in equation (5.59):

$$\frac{\partial \phi}{\partial s_v} = \frac{1}{B_L} \frac{\partial \phi}{\partial r} \frac{ab}{r}. \tag{5.72}$$

Now the required expression for the bending strains along an elliptical hole edge can be obtained from relations (5.46), (5.60)–(5.62), (5.65), (5.67)–(5.69), (5.71) and

$$
\begin{aligned}
(\varepsilon_\varphi - \varepsilon_\varphi') = & -\frac{h}{B_L^2} \left\{ \frac{a^2\sin^2\varphi + b^2\cos^2\varphi}{a^2b^2} \left[(a^2\sin^2\varphi + b^2\cos^2\varphi)\frac{\partial^2 w}{\partial \varphi^2} \right. \right. \\
& \left. + \frac{\partial w}{\partial \varphi} \frac{\sin 2\varphi(a^2 - b^2)}{2} \cdot \frac{2a^4\sin^2\varphi + 2b^4\cos^2\varphi - a^2b^2}{(a^4\sin^2\varphi + b^4\cos^2\varphi)} \right] \\
& \left. -\frac{b}{a} \frac{\sqrt{a^2\sin^2\varphi + b^2\cos^2\varphi}}{R} \frac{\partial v}{\partial \varphi} + \frac{a^2b^2}{B_L^2 r} \frac{\partial w}{\partial r} \right\}.
\end{aligned}
\tag{5.73}
$$

When $a = b = r_0$, formulae (5.70) and (5.73) are transformed into the corresponding relations (5.52) and (5.53) for a shell with a circular hole. The stresses along a load-free circular or elliptical hole boundary must be calculated in accordance with the relations (5.49)–(5.51).

DEFORMATION OF THIN PLATES

Thin plates represent a broad family of thin-walled structures with widespread engineering application. A plate is a physical body which is limited to plane faces $z = \pm h/2$ and cylindrical or prismatic surfaces whose generatrices lie parallel to the z-axis (see Figure 5.5). In order to describe strain and stress distributions in regular plates the Cartesian coordinate system with the usual notation, $\alpha_1 \equiv x$, $\alpha_2 \equiv y$, $\alpha_3 \equiv z$ is used. It is evident that the values of principal curvatures of plate (5.11) are equal to zero and all Lame's coefficients (factors) (5.7) are equal to unity. Note that the displacement components along the axes x, y and z are denoted below as u, v and w.

As generally accepted in the classic elasticity theory, we will distinguish two main types of plate deformation in accordance with a character of applied loading. A deformation caused by forces and moments acting in the planes which are perpendicular to the middle surface of a plate is usually classified as a bending. In this case the strain and stress values mainly depend on a magnitude of the normal to plate surface displacement component w. A small contribution in strains and stresses resulting from the in-plane displacement components u and v can be neglected.

If a plate is loaded with forces and moments acting in the middle plane of a plate, this type of a deformation is usually called the plane stress state. In this case strains and stresses depend on the in-plane displacement components u and v only.

Relations between the displacement vector components and the strain and stress tensor components for both the above-mentioned classes of problems can be derived as particular cases of thin-shell theory.

Bending of plates

As was noted above, the vector valued displacement functions (5.5) or (5.40) in the case concerned consists of one term only:

$$\vec{U} = w\vec{n}. \tag{5.74}$$

Also, in accordance with relations (5.17) the normal displacement component w does not depend on a value of the z-coordinate for each point of a plate with fixed in-plane coordinates (x, y). Moreover, as was mentioned, the following relations are valid:

$$k_1 = k_2 = 0, \qquad A_i = 1 \qquad (i = 1, 2, 3) \tag{5.75}$$

where k_i ($i=1,2$) are the principal curvatures of the basic surface (5.11) and A_i ($i=1,2,3$) are Lame's scaling factors.

Taking relations (5.74) and (5.75) into account, it is convenient to define the middle plane of a bending plate as the basic surface—see Figure 5.5. Then, rotation angles (5.22) and increments of curvature and torsion of the basic surface (5.24) take the form

$$\theta_1 \equiv \theta_x = -\frac{\partial w}{\partial x}, \qquad \theta_2 \equiv \theta_y = -\frac{\partial w}{\partial y},$$

$$\text{æ}_1 \equiv \text{æ}_x = -\frac{\partial^2 w}{\partial x^2}, \qquad \text{æ}_2 \equiv \text{æ}_y = -\frac{\partial^2 w}{\partial y^2}, \tag{5.76}$$

$$\text{æ}_{12} \equiv \text{æ}_{xy} = -\frac{\partial^2 w}{\partial x \partial y}.$$

It should be noted that relations (5.76) to within sign coincide with the notation of the rotation angles and curvatures (4.29), which were introduced in Section 4.3.

The strain tensor components corresponding to relation (5.76) are given by

$$\varepsilon_x(z) = z\left(-\frac{\partial^2 w}{\partial x^2}\right), \qquad \varepsilon_y(z) = z\left(-\frac{\partial^2 w}{\partial y^2}\right),$$

$$\varepsilon_{xy}(z) = z\left(-\frac{\partial^2 w}{\partial x \partial y}\right). \tag{5.77}$$

The generalized Hooke's law remain the form (5.26) where the strain values must be expressed in form (5.77).

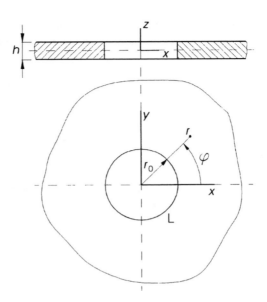

Figure 5.5. Scheme of thin plate with notation used for strain determination

Membrane stresses, evidently, become equal to zero for any z-coordinate values in the range $-h/2 \leqslant z \leqslant h/2$. Bending stresses (5.31) are given by

$$
\sigma_x^B = \frac{E}{1 - \mu^2} \left[\frac{\partial^2 w}{\partial x^2} + \mu \frac{\partial^2 w}{\partial y^2} \right] \frac{z}{2},
$$

$$
\sigma_y^B = \frac{E}{1 - \mu^2} \left[\frac{\partial^2 w}{\partial y^2} + \mu \frac{\partial^2 w}{\partial x^2} \right] \frac{z}{2}, \qquad (5.78)
$$

$$
\sigma_{xy}^B = \frac{E}{1 + \mu} \frac{\partial^2 w}{\partial x \partial y} \frac{z}{2}.
$$

The maximum bending stress values occur on the plate faces and equal the bending stresses (5.78) when $z = \pm h/2$.

Solving a stress concentration problem in a bending plate with a circular hole can be conveniently performed by using a strain and stress representation in the polar coordinate system (r, φ)—see Figure 5.5. The distribution of the ε_φ-strain along a circular hole edge which contains the spatial deflection derivatives taken with respect to the Cartesian coordinates x and y as the input data can be derived from equation (5.52):

$$\varepsilon_\varphi = \pm \frac{h}{2r_0} \left[\frac{\partial}{\partial y} \left(\frac{\partial w}{\partial y} \right) \cos\varphi - \frac{\partial}{\partial \varphi} \left(\frac{\partial w}{\partial x} \right) \sin\varphi \right]. \tag{5.79}$$

In order to obtain relation (5.79), it should be taken into account that, in accordance with expression (5.19) and (5.56′), the following relations are valid in a bending plate:

$$u = \pm \frac{h}{2} \frac{\partial w}{\partial x}, \qquad v = \pm \frac{h}{2} \frac{\partial w}{\partial y}. \tag{5.80}$$

Sometimes an interpretation of holographic and speckle interferometry measured data obtained near a circular hole can be effectively carried out by using the expression of the ε_φ-strain containing the partial derivatives of the deflection w with respect to polar coordinates r and φ. The required relation results from formula (5.53):

$$\varepsilon_\varphi = \pm \frac{h}{2r_0} \left[\frac{\partial^2 w}{r_0 \partial \varphi^2} + \frac{\partial w}{\partial r} \right]. \tag{5.81}$$

As in the case of an open load-free cut-out in a cylindrical shell, it is assumed that a boundary of an open load-free hole in a bending plate is subjected to the one-axis stress stage condition due to equation (5.48). Therefore, relation (5.50) describing the bending stresses at a load-free hole edge is valid in the case of a plate bending, taking into account that $\varepsilon_\varphi = -\varepsilon_\varphi'$.

Plates under plane stress conditions

In this case, taking into account the specific features of optical interferometric techniques, one of the plate faces is defined as the basic surface. The vector valued displacement function (5.5) can be represented in the following form:

$$\vec{U} = u\vec{e}_1 + v\vec{e}_2 + \tilde{w} \cdot \vec{n}. \tag{5.82}$$

The normal displacement component $\tilde{w}$ in equation (5.82) is marked by a tilde due to the following reason. In accordance with the assumed hypotheses about the deformation character of a plate, the normal stress component σ_z (5.25) on the plate face must be equal to zero. Then, the corresponding relation in (5.13) yields

$$\varepsilon_{33} \equiv \varepsilon_z = -\frac{\mu}{E}(\sigma_x + \sigma_y). \tag{5.83}$$

On the other hand, the corresponding Cauchy's relation (5.12′) gives

$$\varepsilon_z = \frac{\partial w}{\partial z}. \tag{5.84}$$

Substituting equation (5.84) into (5.83) and then integrating the result over a plate thickness, yields

$$\tilde{w} = \frac{1}{2} \int_{-h/2}^{+h/2} \frac{\partial w}{\partial z} \, dz = -\frac{h}{2} \frac{\mu}{E} (\sigma_x + \sigma_y). \tag{5.85}$$

Expression (5.85) defining the normal displacement component $\tilde{w}$ on the plate face has been obtained on the assumption that stresses σ_x and σ_y do not change through the plate thickness at a point with fixed coordinates x and y. The adequacy of this hypothesis to the conditions of a real deformation process lessens with growth of the plate's thickness. A more detailed analysis of the influence of a specimen thickness on surface strains in the case of plane stress conditions validity will be presented in Section 8.1

Relations (5.83)–(5.85) show that determination of the strain components (5.12) may be performed by using the first two displacement components in the decomposition (5.82) only. The required results have the following form:

$$\varepsilon_x = \frac{\partial u}{\partial x}, \qquad \varepsilon_y = \frac{\partial v}{\partial y}, \qquad \varepsilon_z \cong \frac{\tilde{w} \cdot 2}{h},$$

$$\gamma_{xy} = \frac{\partial u}{\partial y} + \frac{\partial v}{\partial x}. \tag{5.86'}$$

Sometimes it is necessary to write the plane strain components in the polar coordinate system

$$\varepsilon_r = \frac{\partial u_v}{\partial r}, \qquad \varepsilon_\varphi = \frac{1}{r} \frac{\partial u_t}{\partial \varphi} + \frac{u_v}{r},$$

$$\gamma_{r\varphi} = \frac{1}{r} \frac{\partial u_v}{\partial \varphi} + \frac{\partial u_t}{\partial r} - \frac{u_t}{r}. \tag{5.86''}$$

Relations between the strains and stresses can be obtained by substituting equation (5.86) into (5.26) or (5.27).

The distribution of the circumferential strain ε_φ along a circle of a constant radius r, where the input data are expressed through the displacement components and u and v in the Cartesian coordinate system, are derived from equation (5.52):

$$\varepsilon_\varphi = \frac{1}{r} \left[\frac{\partial v}{\partial \varphi} \cos\varphi - \frac{\partial u}{\partial \varphi} \sin\varphi \right]. \tag{5.87}$$

An analogue of equation (5.87) in the polar coordinate system is the second equation in expression (5.86″). The displacement component in the Cartesian (u, v) and polar (u_v, u_t) coordinate systems are connected through relation (5.36′). Circumferential stress σ_φ along a hole boundary can be obtained from equation (5.49).

5.2   Thin plates under plane stress conditions

Investigations of strain or stress concentration in plates subjected to tension–compression without any bending or torsion are widely discussed in this book in order to illustrate various applications of holographic and speckle interferometry from metrological and methodological points of view.

The most common task of dealing with the stress concentration determination in a plane specimen with a central circular open hole under tension is described in Section 2.5 where a detailed analysis of the accuracy of the displacement component measurement is also given. The solving of the same problem in the elasto-plastic deformation range is presented in Sections 7.4 and 8.1 The study of the local cyclic deformation process of plane specimens with both open and filled holes is the subject of the analysis described in Sections 8.2–8.4. This section will consider two examples which represent an interest from a strength analysis point of view and an illustration of the measurement capability of holographic and speckle interferometric techniques.

SHEARING STRESS CONCENTRATION IN THIN-WALLED STRUCTURAL ELEMENTS

This problem demonstrates the white-light speckle photography application for stress concentration determination in metallic models of structures which are most characteristic in certain fields of engineering. These investigations are often carried out in combination with various numerical methods in order to choose an optimal discrete modelling of the structure element to be studied, and to verify that the model adopted for calculations is in a good agreement with actual loading and boundary conditions.

In particular, aircraft structures such as ribs and spar webs are essentially subjected to shearing loads. Most of these structures have cut-outs or holes of various shape, which are conditioned by their design and manufacturing purpose. The stress concentration factor on the edge of these cut-outs can reach relatively large magnitudes [3]. An accurate determination of local stress values is an essential condition for the optimal design of these structures.

The objects investigated were two specimens of a special design made of aluminium alloy. The symmetrical half of each specimen with a central open hole of radius $r_0 = 7.5$ mm and $r_0 = 8$ mm for specimens 1 and 2 respectively is

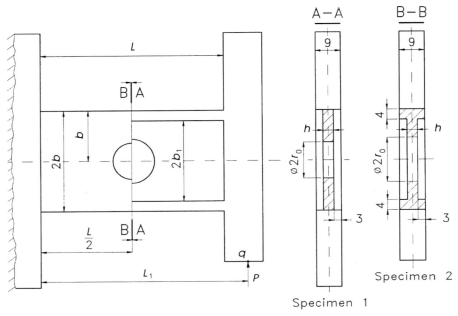

Figure 5.6. Side-on and cross-sectional view of two plane specimens with central circular hole

shown in Figure 5.6. The adopted loading scheme allows us to model the real conditions of operation of these structures.

The white-light speckle-photography technique was used for in-plane displacement component determination along the hole edges (see Section 3.8). The sensitivity threshold of the coating used was 20–25 μm.

Interpretation of a double-exposure specklegram was performed by means of a pointwise filtering procedure. The loading increment values needed for a reliable quantitative interpretation of the negatives at all points of interest on the hole edge lie within the range of 2 to 3 kN. The corresponding magnitudes of the circumferential strain are in the range of 0.3×10^{-3} to 2×10^{-3}.

The in-plane displacement components were obtained in 16 points of interest along the hole edge by a circular division into 16 equal parts. The origin point lies on the neutral axis of the specimens. The parameters needed for calculations with relation (3.36) have the following meanings: $l = 2700$ mm, $M = 0.65$, $\lambda = 0.633$ μm.

The values of the in-plane displacement components u and v in the Cartesian coordinate system with its origin at the centre of the circular hole obtained for three different loading increments were averaged and normalized to the force meaning $P = 10$ N. The displacement component distributions along the hole edge thus obtained for two specimens are presented in Figure 3.36.

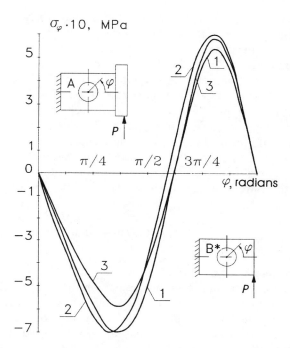

Figure 5.7. Distributions of the circumferential stress along the hole edge in specimen 1

The discrete experimental data sets were approximated by means of a Fourier series according to the procedure described in Section 4.1.

Distributions of the circumferential stress σ_φ along the hole boundary for specimens 1 and 2 calculated with equations (5.87) and (5.49) for the load $P = 10$ N and elasticity modulus $E = 72\,000$ MPa are shown in Figures 5.7 and 5.8 by curves 1. The parts of both curves in the angular range $\pi < \varphi < 2\pi$ have an opposite symmetry.

The same distribution of stress σ_φ in specimen 1 obtained by means of the finite element method for the same load and elasticity modulus values and Poisson's ratio $\mu = 0.32$ is represented in Figure 5.7 by curve 2.

It is also interesting to compare the experimental results obtained with the data of the analytical solution of the problem concerned. To realize this approach for specimen 1, one can use the solution of Savin for the bending of a plate with an open hole stressed with concentrated force lying in its plane [3]:

$$\sigma_\varphi = -\frac{P\left(L_1 - \dfrac{L}{2}\right)}{J}\,(\sin\varphi - \sin3\varphi) - \frac{P}{J}\left[(2b^2 - r_0^2)\sin2\varphi + r_0^2\sin4\varphi\right] \quad (5.88)$$

where r_0 is the hole radius; $L = 35$ mm is the distance between the clamped outer plate edge and the centre of the hole; $2b = 40$ mm is the height of the plate; $L_1 = 80$ mm is the distance between the clamped plate edge and the line of the force action; and $J = (2b)h^3/12$ (where $h = 3$ mm is the plate thickness) is the moment of inertia of the net cross-section of the plate. Distribution of the circumferential stress σ_φ resulting from equation (5.88) is shown in Figure 5.7 by curve 3 (load P is applied in accordance with scheme B*).

The geometrical parameters, design and loading system of specimen 2 were chosen so that its thin-walled plane part (rectangular plate thickness $h = 3$ mm) is stressed according to pure shear condition. Distribution of the circumferential stress σ_φ along the hole boundary which corresponds to the solution of Savin [3] for the pure shearing of the plate of width $2b_1 = 32$ mm and hole radius $r_0 = 8$ mm

$$\sigma_\varphi = -\tau[7.56\sin2\varphi - 0.87\sin4\varphi + 0.12\sin6\varphi]$$

is presented in Figure 5.8 by curve 2.

The magnitude of the uniformly distributed shear stress τ acting along the plate edges is calculated as a ratio of the total sum of forces resulting from shear stress acting on the specimen cross-section part of thickness $h = 3$ mm which are determined from Zhuravsky relations [21] to the square of this cross-section part, s:

$$\tau = \frac{0.7P}{s}$$

where $P = 10$ N and $s = b_1 \cdot h = 96$ mm^2.

Comparison of the experimental results presented with the data of the analytical and numerical decisions of the same problem proves that using white-light speckle photography for solving the stress concentration determination problem ensures the accuracy of the results, which is sufficient for engineering applications.

A possibility of elasto-plastic strain investigation in irregular areas of the structure using the white-light speckle photography technique is also of great interest. To illustrate this possibility, the determination of the residual displacement caused by plastic strains of the hole vicinity in specimen 2 was carried out.

To achieve this objective, the original and final states of the specimen surface after loading with force $P = 10\,500$ N and subsequent unloading to zero were recorded on a double-exposure white-light speckle photography negative. This value of the acting force results in the maximum circumferential stress magnitude on the hole edge $\sigma_\varphi^{\max} = 1.3\sigma_{0.2}$ (where $\sigma_{0.2}$ is the yield limit of the

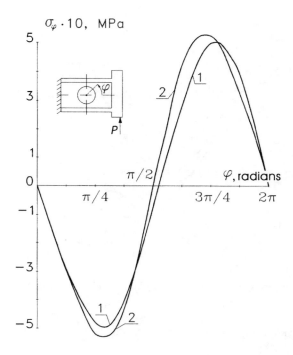

Figure 5.8. Distributions of circumferential stress along the hole edge in specimen 2

specimen material). It should be noted that a pointwise filtering procedure provides interference fringes of high quality at all points of interest on the hole boundary.

Distributions of the residual in-plane displacement components u (curves 1 and 2) and v (curves 3 and 4) along the hole edge (curves 1 and 3) and circle line $r = 20$ mm (curves 2 and 4) caused by plastic deformation of the hole neighbourhood between the original and final state of the object surface are shown in Figure 5.9. The direction of the u displacement component is transversal to the direction of force P (see Figure 5.6).

The most noted advantage of the white-light speckle photography technique is its simple procedure of speckle patterns recording in comparison with the usual laser-light speckle photography. This is due to the unlimited power of white-light illumination sources, and its relatively small dimensions. This is why white-light speckle photography can be effectively used for stress concentration study in different real structures.

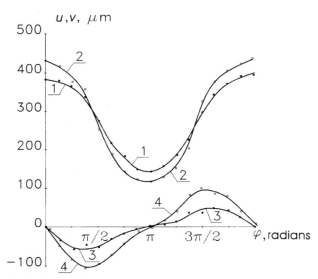

Figure 5.9. Distributions of the residual in-plane displacement components near the hole in specimen 2

A STRAIN CONCENTRATION STUDY OF A LAMINATED COMPOSITE SPECIMEN

The approach to a local strain/stress investigation based on the use of an overlay interferometer which was described in Section 2.5 can be effectively implemented in a strain concentration study of plane composite structures subjected to tension or compression.

Determination of circumferential strain distribution along an open circular hole edge in a plane composite specimen under tension is used below to illustrate a corresponding experimental procedure. The object investigated was a plane specimen made of a laminated graphite-epoxy material of length $L = 260$ mm, width $b = 60$ mm, and thickness $h = 5.52$ mm with a central open hole of radius $r_0 = 6$ mm. These geometrical parameters (except for a slight difference in thickness) coincide with those for a plane metallic specimen shown in Figure 2.24.

The graphite-epoxy laminate of which the specimen to be investigated was made consists of 46 graphite fibre layers. Twenty-two layers are directed along the tension direction (x-axis), 20 layers have an inclination $\pm 45°$ to the tension direction and 4 fibre layers lie in the transversal to tension direction $(22\text{-}0°/20\text{-}(\pm 45°)/4\text{-}90°)$.

The specimen was subjected to tensile loading by means of a computer-aided closed-loop servohydraulic testing machine. The holographic experimental procedure was the same as described in Section 2.5.

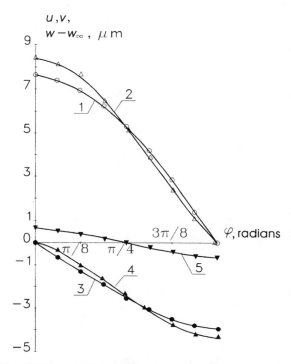

Figure 5.10. Distributions of the displacement components near the hole in the composite specimen subjected to tension

Distributions of the in-plane displacement components u (curve 1) and v (curve 3) along the hole edge ($r_0 = 6$ mm) and analogous distributions (curves 2, 4) along the circle line $r = 20$ mm are presented in Figure 5.10. The analogous distribution of relative normal to the object surface displacement component $w - w_\infty$ is also shown in this figure (curve 5). The above-mentioned dependencies were obtained for the net stress increment $\sigma_0 = 25$ MPa.

A circumferential strain ε_φ along the hole boundary is plotted against the polar angle φ in Figure 5.11 by curve 1. This distribution was derived from equation (5.87). The analogous distribution along the circle line $r = 20$ mm is shown in the same figure by curve 2. A circumferential strain ε_φ distribution resulting from the well-known solution of Lekhnitsky [23] has the following form:

$$\varepsilon_\varphi = \frac{\sigma_0}{E_1} \left[-k \cos^2\varphi + (1 + n)\sin^2\varphi \right] \qquad (5.89)$$

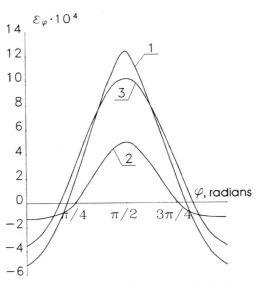

Figure 5.11. Distributions of circumferential strain near holes in analogous composite and aluminum specimens subjected to tension

For the specimen concerned, $E_1 = 70\,630$ MPa is the elasticity modulus in the tension direction; $k = 1.53$ and $n = 2.52$ are the elastic constants which characterize an orthotropy of the laminate.

The φ angle is shown in Figure 2.24. The distribution in equation (5.89) is found to coincide notably with the experimental results described by curve 1 in Figure 5.11.

This demonstrates the remarkable possibilities of holographic interferometry application in local strain study of composite structures deformed according to plane stress conditions. It is interesting to compare the circumferential strain distributions along a hole edge obtained for both metallic and composite specimens. The aluminium specimen considered in Section 2.5 and the composite specimen dealt with in this section have almost the same geometrical parameters and elasticity modulus in the tension direction. Distribution of the circumferential strain ε_φ along the hole boundary in the above-mentioned aluminium specimen, which corresponds to the net stress increment $\sigma_0 = 25$ MPa, is shown in Figure 5.11 by curve 3.

5.3 Bending of thin plates

Structures formed by thin plates and subjected to bending loading are widely used in different engineering fields. Analytical and numerical methods of local strain and stress investigations in bending plates are often too complex and sometimes do not have the capacity to take into account the actual loading schemes and boundary conditions with a high degree of accuracy. For this reason, experimental approaches to stress concentration study in plates under bending are of both scientific and practical interest.

As mentioned in Section 5.1, the plates are characterized by two face plane surfaces. This allows the application of both holographic and speckle interferometry to obtain the input data for subsequent local strain and stress calculation. Note that in the case concerned the bending stresses are the subject of main interest from a strength analysis point of view. The membrane stresses are too small.

Methods of holographic and speckle interferometry allow non-contact measurements to be carried out of deflection fields on rough surface of opaque plates and displacement derivative fields along the spatial coordinates lying in a plate plane—see Chapters 2–4. These data can be effectively used for local strain or stress study in irregular zones of thin-walled plates characterized by the high gradients of the deflection and its derivatives. It should be noted that a high sensitivity of holographic and speckle interferometric techniques ensures the validity of different hypotheses which are usually used to establish a theoretical description of the plates' deformation.

For instance, the experimental data obtained can be used for a determination of the stiffness parameters of complex shape plates with various support schemes that enhance the reliability of the results. This information represents, usually, a direct interest from a strength analysis viewpoint and is necessary for the verification of various discrete schemes and computer routines used for numerical strain and stress analysis of plates. The latter circumstance is of great importance for the design of different types of special finite elements, which can be used for accurate determination of local stress in bending plates.

On the other hand, the approaches presented in this section are capable of being used as a direct powerful tool of local strain or stress analysis in bending plates and as a basis for different combined experimental-numerical techniques to solve the above-mentioned problems.

In this section some characteristic examples are presented to illustrate the possibilities of holographic and speckle interferometry for local strain analysis in bending plates. A procedure for its practical application is also described. Determination of the bending strains and stress are considered only at the points belonging to a hole edge where the magnitudes of the stresses reach maximum values.

Some general comments should be made before the description of the experimental procedure used and data obtained. The holographic interferometry method has a remarkable capability of visualizing and measuring fields of small surface deflection of bending plates—see Sections 2.4 and 2.5. However, the relatively simple procedure of recording such fringe patterns became a necessity for second-order numerical differentiation of discrete experimental data sets in the angular direction and first-order differentiation in the radial direction—see expression (5.81). This approach may lead to significant errors of strain and stress determination especially in irregular zones of structures. Additional computational difficulties result from the fact that in the case of a radial differentiation, the points of main interest at a hole edge coincide with the end points of the segment where radial derivatives must be determined.

One method of determining the deflection derivatives in two orthogonal directions along the hole edge is described in Section 4.2.

It appears that the use of directly measured displacement derivatives along spatial coordinates (so-called slopes) as input data for strain and stress calculation in plates are preferable. The first-order derivatives must be used in this case only for the subsequent strain and stress calculation—see relation (5.79). These slope values can be measured, for instance, by means of defocused speckle photography (DSP, see Section 4.5), shearing speckle interferometry (SSI, see Section 4.4), and finite width fringe holographic interferometry (FWFHI, see Section 4.1, 4.2).

BENDING OF A SQUARE PLATE WITH A CENTRAL OPEN CIRCULAR HOLE SUBJECTED TO A CONCENTRATED FORCE

Methods of holographic and speckle interferometry provide us with very important and interesting capabilities of analysing the correctness and validity of some kinematic and force hypotheses on which different theories of plate bending are based. It is well known, for instance, that the foundation of the technical theory of plate bending, from the Kirchhoff–Love hypotheses, leads to contradictions. One consists of the fact that the three main boundary conditions used for elasticity bending problem solution, namely, an equality to zero of transversal force, bending, and torsional moments, cannot be satisfied simultaneously on load-free plate edges [24]. As plate edges, prismatic or cylindrical cut-outs or holes faces and external lateral boundary faces of plate are usually considered. The above-mentioned situation can be illustrated by the example which will be described in this section.

The investigated object was an aluminium square plate of width $L = 140\,mm$, thickness $h = 4\,mm$ and with a central open hole of radius $r_0 = 18\,mm$. The elasticity modulus of plate material was $E = 73\,000\,MPa$, and its Poisson's ratio was $\mu = 0.34$.

Two neighbouring sizes of the plate were rigidly clamped and two others were free-of-fixing. The plate was loaded with a concentrated force P applied at the point of the crossing of the two free-of-fixing plate sizes. A diagram showing the plate fixing, loading and coordinate system used is presented in

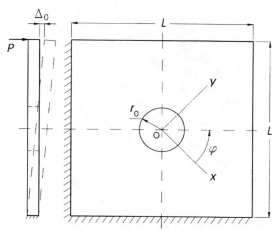

Figure 5.12. Schematic diagram of the bending of a square plate with central circular hole

Figure 5.12. Due to an absence of an axial symmetry, the loading scheme used led to plate stressing with bending moment, torsional moment and transversal force simultaneously. The strain stage along a hole edge resulting from this type of plate loading is the most unfavourable from a metrological point of view.

To determine a strain–stress stage along a hole boundary the slope values $\partial w/\partial x$ and $\partial w/\partial y$ in the Cartesian coordinate system were measured by means of a defocused speckle photography method—see Section 4.5. Circumferential strain and stress distributions were derived through equations (5.79) and (5.49) respectively.

A comparison of experimental data (obtained by different methods) with the results of numerical solution of an analogous bending problem (by means of the finite element method) was used to control the accuracy of the measurement procedure implemented. Finite element calculations were carried out by using a software package called FITING [25]. Figure 5.13 presents a double-exposure interferogram of a hole vicinity, illustrating the lines of equal deflection values. This interferogram corresponds to initial deflection $\Delta_0 = 15\,\mu m$ applied at the free plate corner. The * symbol in the figure indicates the zero-motion fringe which is an origin of absolute fringe order counting.

The deflection w distribution along a hole boundary, which is presented in Figure 5.14 by curve 1, was constructed using equation (2.74). The corresponding distribution of the deflection derivative in the angular direction $\partial w/\partial \varphi$ obtained by means of a Fourier series approximation of experimental data is shown in the same figure by curve 3. The analogous deflection and

Figure 5.13. Interferogram of deflection of bent square plate with central circular hole

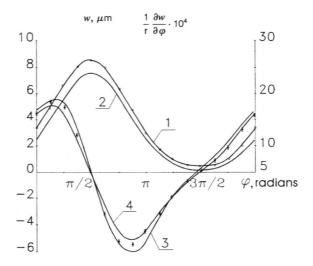

Figure 5.14. Deflection and angular deflection derivative distributions along the hole edge in bending square plate

derivatives distributions obtained by using a finite element technique are shown in Figure 5.14 by curves 2 and 4 respectively. The dots • indicate the derivative value $\partial w/\partial \varphi$ directly measured by means of a defocused speckle photography technique. All data presented in Figure 5.14 are normalized to initial deflection magnitude $\Delta_0 = 100 \ \mu m$.

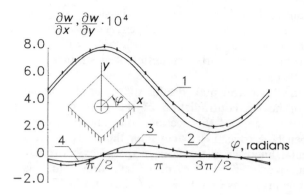

Figure 5.15. Distribution of slopes along the hole edge in square plate under bending

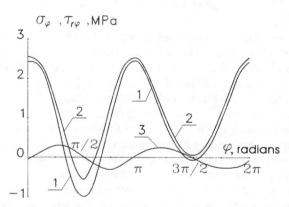

Figure 5.16. Stress distributions along the hole edge in square plate under bending

One can find that all corresponding experimental results which are shown in Figure 5.14 are in good agreement. This fact is of great importance because it demonstrates a coincidence of the data obtained by numerical and two optical interferometric methods. It should also be noted that the sensitivity of the two experimental techniques differs by an order of magnitude.

Figure 5.15 illustrates the results of a determination of the slope values $\partial w/\partial y$ (curves 1, 2) and $\partial w/\partial x$ (curves 3, 4) along a hole edge in the Cartesian coordinate system by using a finite element technique (curves 2, 4) and a defocused speckle photography method (curves 1, 3). The distribution of the circumferential stress σ_φ along a hole edge, plotted by using the experimentally measured slope values as the input data in expressions (5.79) and (5.49), is shown in Figure 5.16 by curve 1. Curve 2 describes the same stress obtained

through a Fourier series approximation of the slope values calculated by means
of a finite element method. Both dependencies correspond to an initial
deflection of the point q on the plate surface $\Delta_0 = 100\ \mu m$ (see Figure 5.12).
The actual values of measured circumferential strain ε_φ lie within the range
from (-10^{-5}) to 3×10^{-5}.

The obtained experimental results allow us to present a quantitative
illustration of how a contradiction of the plate bending technical theory,
noted at the beginning of this section, may influence the stress distribution
along an open hole edge. Let us consider the shearing stress $\tau_{r\varphi}$ distribution
along a circular line $r = \mathrm{const}$ written in the polar coordinate system through
the Cartesian partial derivatives of a deflection. This distribution can be
derived from expressions (5.33), (5.37), (5.38) and (5.47):

$$\tau_{r\varphi} = \frac{Eh}{2(1 + \mu)r}\left[\frac{\partial}{\partial\varphi}\left(\frac{\partial w}{\partial x}\right)\cos\varphi + \frac{\partial}{\partial\varphi}\left(\frac{\partial w}{\partial y}\right)\sin\varphi\right] \qquad (5.90)$$

These stresses cannot really exist along a load-free open hole boundary
$(r = r_0)$, at least on the external face of a plate, where displacement
measurements are carried out, due to the rule of coupling of shearing stresses.
Nevertheless, a calculation by means of equation (5.90) using the same slopes
magnitudes as for σ_φ calculation yields non-zero values of the shearing stresses
$\tau_{r\varphi}$—see curve 3 in Figure 5.16. The values of the circumferential σ_φ and
shearing $\tau_{r\varphi}$ stress have the same order of magnitude and the existence of the
latter distribution along an open hole edge cannot account for an influence on
the measurement errors only. It is interesting to point out the characteristic
feature of the shearing stress distribution which consists of the fact that its
integral taken around a closed line of a hole edge is equal to zero. Therefore, a
hypothetical total load caused by the shearing stresses $\tau_{r\varphi}$ is self-balanced.

The results presented in this section clearly demonstrate the non-traditional
and quite unexpected capabilities of holographic and speckle interferometric
techniques. These possibilities allow us to analyse the validity of some
hypotheses which are the basis of different theories of deformation of thin
plates and shells.

Note, in conclusion, that questions concerning conformity of the Kirchhoff–
Love hypotheses to a deformation process of an open hole edge in a curved
cylindrical shell will be discussed in Section 5.4.

BENDING OF A CIRCULAR PLATE WITH AN ECCENTRIC CIRCULAR HOLE UNDER UNIFORM PRESSURE

Difficulties resulting from the necessity of an accurate discrete modelling of a
load acting at the nearest vicinity of various cut-outs represent the main

problem which must be overcome to ensure a correct numerical determination of a strain or stress concentration in bending plates subjected to distributed loading. This special feature may sometimes lead to significant errors in the calculation of local strains and stresses if a numerical model is chosen incorrectly. In order to establish the criteria of an appropriate numerical model construction, experimental data describing a local deformation character for different loading schemes can be effectively used. While type of loading (uniform pressure, concentrated force, bending or torsional moments etc.) is not of major importance for holographic and speckle interferometric techniques for local strain determination, its application in the case of a distributed load can be very powerful.

In order to demonstrate this, we shall again consider the bending of a circular plate clamped along the external edge with an eccentric circular open hole. The plate is loaded with uniformly distributed pressure along the whole face with the exception of the hole area—see Figure 4.19.

Figure 4.23 presents distributions of the deflection derivatives $\partial w/\partial t$ and $\partial w/\partial r$ along the hole edge plotted against polar angle φ. We must remember that derivative $\partial w/\partial t$ was obtained by means of deflection w approximation with a Fourier series and subsequent differentiation of the corresponding analytical dependence. Radial derivative $\partial w/\partial r$ was derived by using the semi-geometrical approach described in Section 4.2. In addition, the spatial derivatives in the Cartesian coordinate system (slopes) $\partial w/\partial x$ and $\partial w/\partial y$ can be determined by using the fringe patterns shown in Figures 4.20(b) and (c) according to relation (4.26).

A special kind of a fringe pattern, which represents a straight-line system, and can be observed on the surface of a witness placed inside the hole (see Figures 4.20 (b) and (c)), allows us to measure uniform fringe spacing values $\bar{p}_x$ and $\bar{p}_y$ resulting from linear phase shift only. These magnitudes are necessary for slope calculation in accordance with equation (4.26). It should be noted that the known direction of an illumination wave displacement provides the possibility to determine physical signs of the slopes $\theta_x = \partial w/\partial x$ and $\theta_y = \partial w/\partial y$ with certainty. Distributions of the slopes θ_x and θ_y along a hole edge obtained by the FWFHI technique are shown in Figure 5.17 by curves 1 and 3 respectively. The analogous slope distributions which were determined by using DSP technique are presented in the same figure by curves 2 and 3.

The circumferential strain ε_φ along a hole boundary plotted against a polar angle φ is shown in Figure 5.18. The curve presented to within experimental error corresponds to slope distributions obtained by the semi-geometrical approach (Section 4.2), FWFHI (Section 4.1) and DSP (Section 4.5).

A good coincidence of all experimental data obtained, which proves a high accuracy and reliability of all measurement techniques used, should be noted in conclusion.

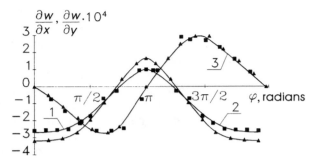

Figure 5.17. Slope distributions along the edge of eccentric hole in circular plate subjected to an internal pressure

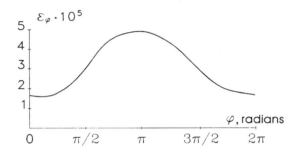

Figure 5.18. Circumferential strain distribution along the edge of eccentric hole in circular plate subjected to an internal pressure

INTERACTION OF HOLES IN A BENDING PLATE

A problem concerning the interaction of several holes in a thin-walled structural element subjected to bending is of great practical interest since the stress concentration factor values in this case depend on both the hole diameter and the relative distance between the centres of the holes. Numerical solution of such problems may be accompanied by great difficulties resulting from the necessity of a detailed discretization of a large part of the structure to be studied.

As an example of holographic interferometry implementation for solving the above-mentioned problem, consider the determination of local strains in a circular bending plate with four open holes. The set of holes consists of two pairs of holes of equal diameter, the centres of which are located symmetrically with respect to the plate centre (see Figure 5.19).

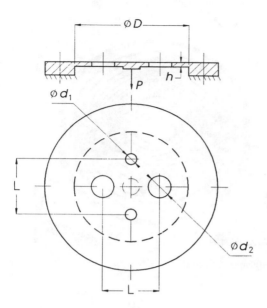

Figure 5.19. Drawing of circular plate with four circular holes

The plate diameter was $D = 120$ mm, and plate thickness $h = 4$ mm. The diameter of the lesser hole was $d_1 = 18$ mm, the diameter of the greater hole $d_2 = 32$ mm, and the distance between the centres of each hole pair $L = 58$ mm. The plate was bent by a concentrated force P applied at the centre of the plate.

The two-exposure interferogram of the plate obtained through the use of the optical arrangement shown in Figure 2.19 for the load increment $P = 0.5$ N is presented in Figure 5.20. The dark interference fringes in this case are the lines of equal levels of the normal to the surface displacement component w with spacing $\lambda/2$. The results of interpretation of this infinite width fringes interferogram are given in Figure 5.21. The deflection distributions along the hole edges are plotted in two local polar coordinate systems, the origins of which coincide with the hole centres, and the polar angle φ originates from the point of the hole boundary which is nearest to the plate centre. The experimental deflection distributions are approximated with a trigonometrical series in the following form:

$$w = C_0 + C_1\cos\varphi + C_2\cos2\varphi + C_3\cos3\varphi$$

The approximation procedure used completely coincides with that described in Section 4.1.

Figure 5.20. Interferogram of deflection of the plate with four circular holes

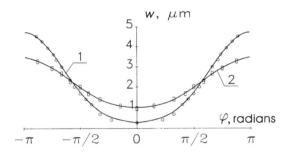

Figure 5.21. Distributions of deflection along the edge of hole of diameter 32 mm (curve 1) and 18 mm (2) in the circular plate with four holes

The results of the first- and second-order numerical differentiation of the experimental deflection w distributions with respect to the angular variable φ for two holes of different diameters are presented in Figures 5.22(a) and (b) respectively.

In order to determine the radial deflection derivatives, which are needed to calculate the circumferential strains along the hole edges in accordance with equation (5.81), the semi-geometrical approach described in Section 4.2 is implemented. Analysis of the fringe pattern shown in Figure 5.20 demonstrates that the principal equation (4.37) that must be solved to determine the first-order derivatives becomes degenerate at the points of greater interest

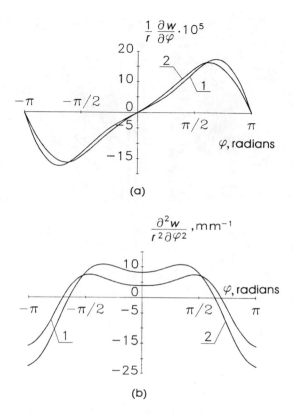

Figure 5.22. Distributions of the first (a) and second (b) order deflection derivatives with respect to the angular variable φ along the edge of holes of different diameter in circular plate with four holes

corresponding to the values of angular coordinate $\varphi = 0$ and $\varphi = 180°$. Therefore, interferograms of plate bending obtained by interference fringes of finite width were used for implementation of the direct geometrical approach. These interferograms which allow us to determine the distributions of the radial first-order derivative of the deflection w along the edge of holes of diameter 32 mm and 18 mm are presented in Figures 5.23(a) and (b), respectively.

The discrete set of radial derivative values $\partial w/\partial r$ obtained at the points of intersection of the dark interference fringes and the hole boundaries is approximated with the segment of trigonometrical series. The initial experimental data and the corresponding approximating curves for two hole edges are shown in Figure 5.24. Figure 5.25 depicts the distributions of the

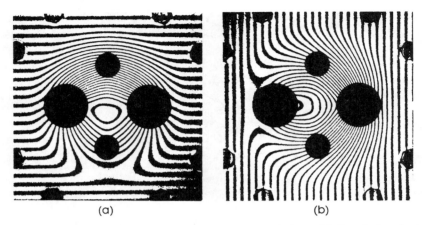

(a)   (b)

Figure 5.23. Finite width interferograms of deflection of plate with four holes obtained for the first order radial deflection derivative determination along the edge of holes of diameter 32 mm (a) and 18 mm (b)

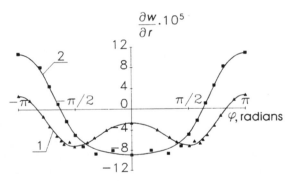

Figure 5.24. Distributions of the first order radial deflection derivative along the edge of holes of different diameter in plate with four holes

circumferential strains ε_φ along the edges of the holes of different diameter which result from introduction of the experimental data in Figures 5.22(b) and 5.24 into relation (5.81).

5.4 Displacement fields along the boundary of circular and elliptical cut-outs in a circular cylindrical shell subjected to tension and torsion

In this section we shall present some basic experimental information necessary for the determination of strain and stress concentration in a curved cylindrical

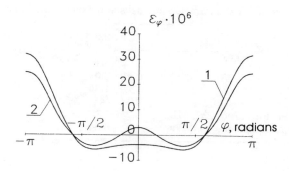

Figure 5.25. Distributions of the circumferential strain along the edge of holes of different diameter in plate with four holes subjected to a bending

shell. These experimental data represent 3-D displacement fields along the edge of open circular holes of different diameters and along the elliptical hole boundary in a circular cylindrical shell. Data presented have been obtained for the first time by means of reflection hologram interferometry.

The displacement component distributions given below are of both scientific and practical interest. In particular, the results of the precise measurement of the displacement components on a curved shell surface can be used for a verification of various numerical programs, estimation of the adequacy between a numerical model and an actual object under investigation, etc.

All displacement component distributions presented in this section will be used in the rest of this chapter for a description of the local strain analysis procedure. Remember that the components of the displacement vector $\vec{U}$ (equation (5.5)) in the cylindrical coordinates will be denoted by u, v and w (the u and v components are tangential, and w is normal to the surface of the shell under study).

All four aluminium specimens investigated had the form of a circular cylindrical shell of external radius $R = 30\,\text{mm}$, wall thickness $h = 1.5\,\text{mm}$ and length of thin-walled cylindrical part $L = 100\,\text{mm}$—see Figure 2.27. Three of the specimens had an open circular hole of different radius r_0. The relations between the geometrical characteristics of the shell can be combined to define the so-called shell parameters [4,5]:

$$\beta_1 = \frac{r_o}{\sqrt{Rh}}$$

$$\beta = r_0 \left(\frac{\sqrt{12(1 - \mu^2)}}{8Rh} \right)^{1/2} \tag{5.91}$$

Table 5.1 The geometrical parameters of shells and experimental characteristics

Parameters	Circular Hole			Elliptical Hole
Hole radius r_0, mm	9	14	18	15.5
Shell parameter β_1	1.34	2.08	2.68	2.31
Shell parameter β	0.87	1.36	1.75	1.48
Observation angle ψ, degrees	50	48	30	30
Net stress increment for tension, σ_0, MPa	5.5	3	2	2.4
Net stress increment for torsion τ_0, MPa	1.4	0.64	0.35	0.5

where $\mu = 0.32$ is Poisson's ratio of the shell material. The parameters of the shell used are listed in Table 5.1. It should be noted that the value $\beta = 0.87$ for the external radius $R = 30$ mm lies at the boundary of validity of the theory of non-curved shells, and the value $\beta = 1.75$ is far off this boundary.

The fourth analogous cylindrical specimen had an elliptical open hole with semi-axes a directed along the shell generatrix (the α_1-axis in Figure 5.3) and b which lies in the perpendicular direction, $a > b$. The values $a = 18$ mm and $b = 13$ mm represent the dimensions of the elliptical hole edge corresponding to the shell development. The parameter r_0 (see Table 5.1), which is an analogue to a circular hole radius, can be defined for the elliptical hole in the following way in accordance with equation (5.54):

$$r_0 = \frac{a + b}{2}$$

All specimens were subjected to tension and then to torsion by means of special high-precision loading devices. The results of experimental testing of these devices are partially presented in Section 2.5. Reflection holograms were used to record information concerning the displacement component distributions at hole vicinities. Holographic interferograms were recorded by using both the two-exposure method and combined technique that is capable of visualizing bright zero-motion fringes (see Sections 2.1 and 2.2).

In the case of shell tension, the increment of the net stresses acting in regular cross-sections of the shell far from the central part of the shell containing the hole can be expressed as

$$\sigma_0 = \frac{P}{2\pi R_0 h} \tag{5.92}$$

where P is tensile force increment between two exposures or during the combined procedure and $R_0 = R - h/2$, R is an external radius of the shell and h is the wall thickness.

Figure 5.26. Three interferograms of shell with the circular hole subjected to a tension obtained for different viewing directions: (a) $\alpha = 0$, (b) $\alpha = 120$, (c) $\alpha = 240$ degrees

In the case of shell torsion by a moment M, acting in the plane which lies perpendicular to the shell axis, the regular cross-sections of the shell undergo an influence of shearing net stresses:

$$\tau_0 = \frac{M}{2\pi R_0^2 h} \tag{5.93}$$

The symbols in the denominator of equation (5.93) coincide with those of equation (5.92).

The net stress increment values used for hologram recording and listed in Table 5.1 were chosen consequent to the need to ensure an optimal fringe spacing along the hole edges for all viewing directions used for quantitative interpretation of the interferograms.

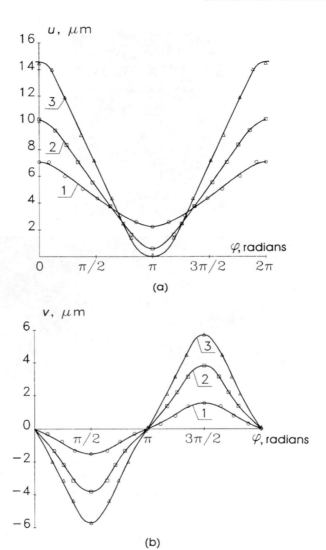

Figure 5.27(a) and (b)

The process of the quantitative interpretation of holographic interferograms completely coincides with that described in Section 2.5. The optimal systems of the holographic interferometers had parameters $\theta = 60°$ (2.61) and $m = 4$ (2.65). The values of the observation angles ψ for different magnitudes of the hole radius r_0 are listed in Table 5.1.

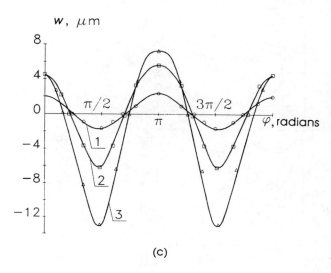

(c)

Figure 5.27. Distributions of the displacement components u (a), v (b) and w (c) along the circular hole edge of radius 9mm (1), 14mm (2) and 18mm (3) in shell under tension

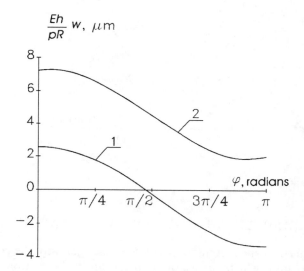

Figure 5.28. Distribution of the relative deflection along the circular hole edge in a shell under tension obtained by experimental (1) and analytical (2) way

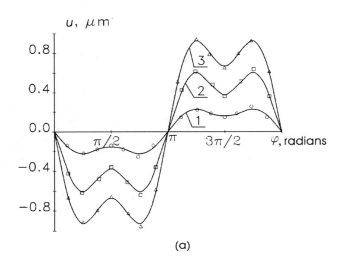

(a)

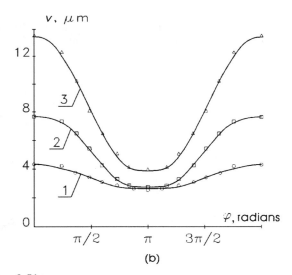

(b)

Figure 29(a) and (b).

TENSION OF SHELLS WITH AN OPEN CIRCULAR HOLE

Three typical two-exposure interferograms of the shell with the circular hole of radius $r_0 = 14\,\text{mm}$, which corresponds to the optimal equation system (2.67) with the number $j = 1$ are presented in Figure 5.26. The reflection hologram that is a source of interferograms shown in the figure was recorded for net

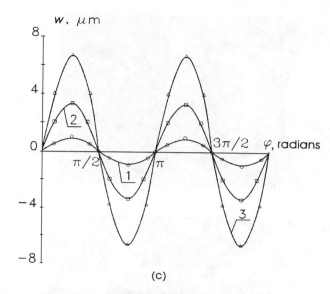

(c)

Figure 5.29. Distributions of the displacement components u (a), v (b) and w (c) along the circular hole edge of radius 9 mm (1), 14 mm (2) and 18 mm (3) in shell under torsion

stress increment value (5.92) $\sigma_0 = 3$ MPa. Identification of the fringe orders is made by direct counting from a zero-motion fringe on the unstrained part of the cylindrical specimen.

Distributions of the displacement components u, v and w on the external face of shells along the hole edge for different values of hole radius r_0 are shown in Figure 5.27. For a clear comparison, all data presented are normalized to the net stress increment value $\sigma_0 = 5$ MPa. Attention should be paid to the asymmetrical character of the deflection w distributions (Figure 5.27(b)(c)) which increases with the growth of the hole radius r_0. This clearly results from an influence of the clamped shell part on the local deformation of the hole vicinity.

It is interesting to compare the deflection distribution along the hole edge obtained by experimental data with that corresponding to the analytical solution for an infinite shell with an open circular hole [4]. Experimental (curve 1) and analytical (curve 2) distributions of the relative deflection $\bar{w}$ along a hole edge

$$\bar{w} = \frac{Ehw}{\sigma_0 R}$$

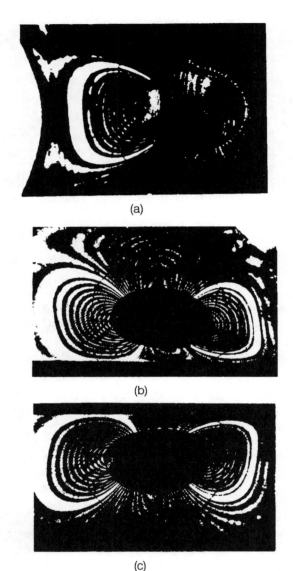

(a)

(b)

(c)

Figure 5.30. Three interferograms of shell with the elliptical hole subjected to a tension obtained for different viewing directions: (a) $\alpha = 0$, (b) $\alpha = 120$, (c) $\alpha = 240$ degrees

(where E is the elasticity modulus of the shell material, h is the shell thickness, R is the external radius of the shell and σ_0 is the net stress increment) obtained for the hole radius $r_0 = 14$ mm are shown in Figure 5.28. The analytical dependence, corresponding to the parameter (5.91) $\beta_1 = 2$, differs from experimental dependence by a constant value for all magnitudes of the polar angle φ. The difference between the two curves in Figure 5.28 reveals an influence of the actual boundary conditions in the finite-length shell on local deformation of the hole vicinity.

TORSION OF A SHELL WITH AN OPEN CIRCULAR HOLE

The displacement component distributions which are analogous to those presented in Figure 5.27 are shown in Figure 5.29. All the data in this figure are normalized to the net stress increment (5.93) $\tau_0 = 0.65\,\text{MPa}$. In the case concerned, as for shells subjected to tension, the asymmetry of the deflection distributions, resulting from the difference of the boundary conditions on two sides of the cylindrical part of the specimen used, can be clearly seen.

TENSION OF A SHELL WITH AN OPEN ELLIPTICAL CUT-OUT

Three typical interferograms of the shell obtained by combining the two-exposure and time-average techniques are shown in Figure 5.30. These fringe patterns provide the optimal equation system (2.67) with the number $j = 1$. The displacement components on the external face of the shell along the hole boundary are plotted against the polar angle φ and are presented in Figure 5.31.

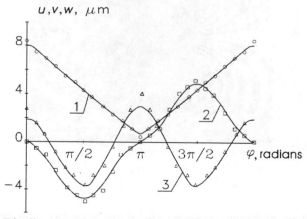

Figure 5.31. Distributions of the displacement components u (1), v (2) and w (3) along the elliptical hole edge in shell under tension

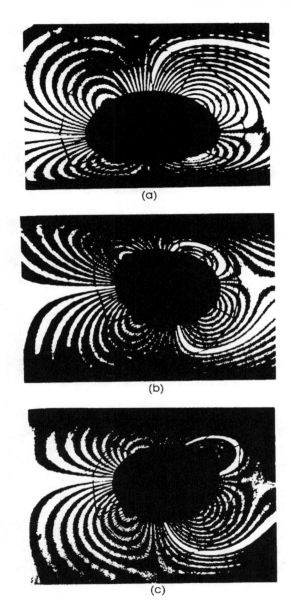

Figure 5.32. Three interferograms of shell with the elliptical hole subjected to a torsion obtained for different viewing directions: (a) $\alpha = 90$, (b) $\alpha = 210$, (c) $\alpha = 330$ degrees

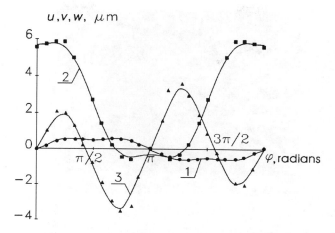

Figure 5.33. Distributions of the displacement components u (1), v (2) and w (3) along the elliptical hole edge in shell under torsion

TORSION OF A SHELL WITH AN OPEN ELLIPTICAL CUT-OUT

Three typical interferograms with a bright zero-order fringe, corresponding to the optimal equation system (2.67) with the number $j=4$ are presented in Figure 5.32. The displacement component distributions along the hole edge on the external shell face are illustrated in Figure 5.33.

The most characteristic displacement component distributions along a hole boundary only are shown in the figures presented in this section. However, the interferograms obtained allow us to construct the displacement component distributions in the vicinity of a hole which can be observed simultaneously from at least three viewing directions.

5.5   Membrane and bending stress determination in a shell with a hole

Three-dimensional displacement fields obtained on the external face near a hole edge in a circular cylindrical shell (some of which are presented in Section 5.4) suffice as input data to solve a strain/stress concentration problem, as follows from the theoretical relations described in Section 5.1. However, the local strain and stress analysis is founded upon two Kirchhoff–Love hypotheses. Therefore, before practical implementation of the relations between the displacement components (strain and stresses in an irregular zone of a curved shell) we must be sure that the assumptions adopted reliably describe a real deformation process. Special attention should be given to a

Figure 5.34. Multi-exposure interferogram of the shell with the circular hole under tosion

correct and reliable transition from the displacement components measured to the strain and stress distributions at the vicinity of the large cut-out in a curved cylindrical shell, where displacement component distributions are characterized by notable gradients, both in the tangential and normal to shell surface directions.

To obtain experimental information for direct confirmation of the main hypotheses of the thin shell theory, a special investigation was carried out. The object studied was an aluminium cylindrical shell with a circular open hole subjected to a torsion. A diagram of this specimen is analogous to that shown in Figure 2.27. The geometrical dimensions of the thin-walled shell are: length

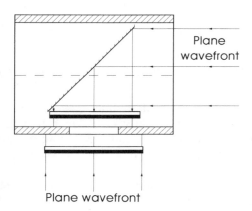

Figure 5.35. Reflection hologram setup for an inspection of two shell faces near a hole

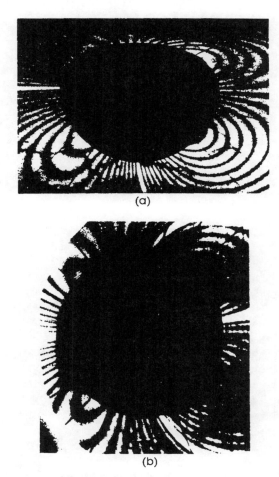

(a)

(b)

Figure 5.36. Reconstructed image of the shell with the circular hole subjected to a torsion obtained on the external (a) and internal (b) face

$L = 320$ mm, external radius $R = 60$ mm, hole radius $r_0 = 25$ mm, and wall thickness $h = 3$ mm ($h/R = 0.05$). The corresponding value of the shell parameter β (5.91) is equal to 1.2. This meaning for the external radius of a shell $R = 60$ mm lies far off the boundary of the validity of the non-curved shell theory [5].

The specimen was loaded with a torsional moment by means of a special precision loading device. An equivalence of the strained state of the object under study with the corresponding theoretical strained state of a thin shell under torsion (which is a necessary basis of comparison of the experimental and

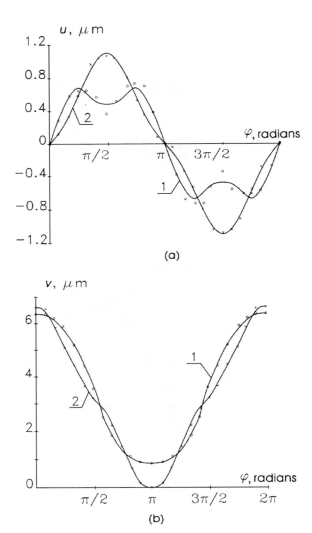

(a)

(b)

Figure 5.37(a) and (b)

analytical results of a strain or stress concentration determination) was proved to
be within the displacement measurement accuracy established in Section 2.5 by
testing an analogous regular shell. Holding accuracy, linearity and reproduci-
bility of the shell loading was well established by means of the multi-exposure
technique (see Section 2.1). Typical multiple-exposure interferograms obtained

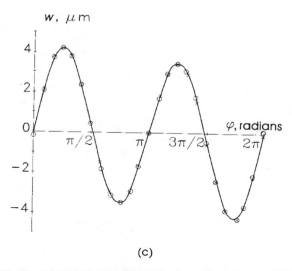

(c)

Figure 5.37. Distributions of the displacement components u (a), v (b) and w (c) along the circular hole edge in the cylindrical shell under torsion on the external (1) and internal (2) face

on the external face of the shell for the torsional moment increment $M = 7\,\text{Nm}$ and the exposure number $N = 5$ is shown in Figure 5.34.

Reflection holograms that are required to determine the displacement fields along the hole edge were recorded simultaneously on both the external and internal faces of the shell. The hologram recording was carried out by means of combining the two-exposure and time-average methods in order to visualize a bright zero fringe (see Section 2.2). Figure 5.35 is a diagram of the optical set-up. Typical interferograms of the shell obtained by the above-mentioned method for the external and internal faces are shown in Figure 5.36(a) and (b) respectively. These fringe patterns correspond to the torsional moment increment $M = 24.5\,\text{Nm}$. Note that the experimental technique and the procedure of quantitative interpretation of the fringe patterns are the same as those described in Section 2.5.

Distributions of the tangential to the shell surface displacement components u and v along the hole boundary corresponding to the net stress increment $\tau_0 = 0.36\,\text{MPa}$ are presented in Figures 5.37(a) and (b) for the external and internal shell faces by curves 1 and 2 respectively. The analogous distribution of the normal to the shell surface displacement component w, which has the same form on the external and internal faces, is shown in Figure 5.37(c).

First, these data allow us to calculate the circumferential strains ε_φ and ε'_φ on the external and internal faces of the shell in accordance with relation (5.52).

Table 5.2 The coefficients of approximating series (5.94) for displacement component
distributions along the circular hole edge in the shell under torsion

Displacement component	Shell face	Series coefficient				
		$k = 1$	$k = 2$	$k = 3$	$k = 4$	$k = 5$
u, A_k, μm	External	–	0.72	–	0.25	–
	Internal	–	0.95	–	−0.15	–
v, B_k, μm	External	3.6	2.94	–	−0.2	–
	Internal	3.3	2.96	–	0.38	–
w, C_k, μm	External	–	0.41	3.85	0.1	−0.15
	Internal	–	0.54	3.80	–	–

Note that first- and second-order numerical differentiation of the discrete sets
of the measured displacement components which are necessary for strain
determination are performed by means of the Fourier approximation
procedure (see Section 4.1). The trigonometrical series of the following form
is used for this purpose:

$$u = A_1 + \sum_{k=2}^{p} A_k \sin[(k - 1)\varphi]$$

$$v = \sum_{k=1}^{q} B_k \cos[(k - 1)\varphi] \tag{5.94}$$

$$w = C_1 + \sum_{k=2}^{s} C_k \sin[(k - 1)\varphi]$$

The values of series (5.94) coefficients which are used for the following strain
and stress calculation are listed in Table 5.2. The criterion on which a choice of
the number of coefficients (5.94) is based is described in detail in Section 4.1.

Distributions of the relative circumferential stress σ_φ / τ_0 (where net stress
increment τ_0 is calculated according to equation (5.93)) obtained by using
expressions (5.49) and (5.52) and series (5.94) are presented in Figure 5.38 by
curves 1 and 2 for the external and internal faces respectively. The dots indicate
the analogous analytical results of Van Dyke [5]. In this section the value of the
elasticity modulus $E = 72\,000$ MPa is used for stress calculation.

Before we consider the membrane and bending stress in accordance with
formulae (5.50) and (5.51), let us analyse the values of the rotation angles (5.44)
and (5.45) along the hole edge obtained from the experimental data. For a
circular hole in a circular cylindrical shell the latter relations take the following
form:

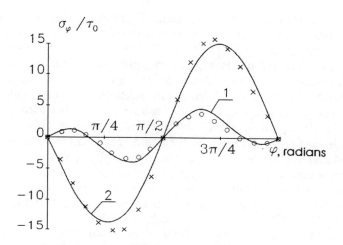

Figure 5.38. Distributions of the relative circumferential stress along the circular hole edge in the shell under torsion obtained on the external (1) and internal (2) face

$$\theta_v \equiv \theta_r = -\frac{\partial w}{\partial r} + \frac{v}{R}\sin\varphi \qquad (5.95)$$

$$\theta_t \equiv \theta_\varphi = -\frac{\partial w}{r_0\partial\varphi} + \frac{v}{R}\cos\varphi \qquad (5.96)$$

where r is the hole radius and R is the radius of an external (internal) surface of the shell. It is evident that to within an accuracy of the thin shell theory, which is a value of an order of a magnitude of h/R, the second terms in (5.95) and (5.96) can be neglected. On the other hand, the rotation angle values can be expressed, proceeding from equation (5.43), through the corresponding displacement components in the following way:

$$\theta_r = \frac{u_v - u_v'}{h} \qquad (5.97)$$

$$\theta_\varphi = \frac{u_t - u_t'}{h} \qquad (5.98)$$

where displacement components u_t, u_t' and u_v, u_v' are related to the displacement components in the cylindrical coordinate system u and v by expression (5.36').

Therefore, the rotation angle magnitudes θ_r and θ_φ can be calculated from both the normal displacement component w measured on any face of a shell according to equations (5.95) and (5.96) and the difference between the corresponding tangential displacement components measured on the external and internal faces along a hole edge in accordance with equations (5.97) and (5.98). It should, however, be pointed out that experimental data describing the displacement fields near a hole boundary enable us to obtain with a high and comparable accuracy three values only: θ_φ from equation (5.96) and θ_φ and θ_r from equations (5.97) and (5.98). Numerical or geometrical determination of the radial derivative $\partial w/\partial r$ at the end of a segment lying on a curved surface, which is characterized by a high gradient of deflection w distribution, is always accompanied by both a considerable growth of the input experimental data volume and a calculation problem. In other words, any procedure of radial deflection derivative determination through the deflection fields only cannot ensure the required accuracy of the final result which must be suitable for a reliable local strain analysis. That is why the rotation angle value θ_r (equation (5.95)) will not be taken into consideration below.

Distributions of the rotation angle values θ_φ (5.96) (curve 1) and θ_φ (5.98) (curve 2) θ_r (5.95) (curve 3) are presented in Figure 5.39. Rather unexpectedly, Figure 5.39 reveals a notable difference between rotation angle values θ_φ obtained by using two methods, namely, relations (5.96) and (5.98). An analysis of the accuracy of the displacement components determination (Section 2.4) and the procedure of numerical differentiation along a closed contour (Section 4.1) denotes that the observed non-coincidence lies far off the level of the maximum possible total error in determining the final result. Now we have to establish what consequences this will cause from the standpoint of membrane and bending stress calculation:

(1) The first method of determining bending stress (strain) results from relations (5.52) and (5.53) and requires information about displacement component distributions on both faces of a shell. The corresponding bending stress σ_φ^B/τ_0 distribution along the hole edge in the circular cylindrical shell is presented in Figure 5.40 by curve 1. The different shaped dots, again, indicate the data of Van Dyke [5].

(2) The second method to calculate bending stress consists of the application of the relation (5.53), which contains the radial deflection derivative $\partial w/\partial r$. A combination of expressions (5.97) and (5.95) is the only reliable approach to obtain the value of this derivative:

$$\frac{\partial w}{\partial r} = \frac{u_v - u_v'}{h} + \frac{v}{R}\sin\varphi \qquad (5.99)$$

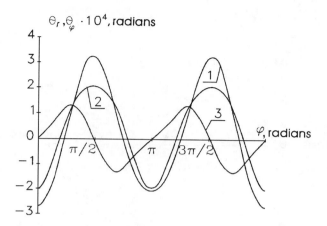

Figure 5.39. Distributions of rotation angles of the surface normal along the circular edge of hole in the shell under torsion

Substituting (5.99) into formula (5.53), and then substituting the obtained result into relation (5.50) allows us to obtain the other version of the bending stress σ_φ^B distribution, which is shown in Figure 5.40 by curve 2. To construct this distribution, it is necessary, as in the previous case, to combine the displacement components measured on both shell faces along the hole edge.

The difference between the bending stress values obtained by the two above-mentioned methods is apparently the same as the non-coincidence of the rotation angle θ_φ values resulting from relations (5.96) and (5.98). Today, it is difficult for us to indicate accurately the main reason for this. One can assume, however, that the contradictions observed are connected with the complex space form of a large cut-out edge in a curved shell.

The results obtained can be considered from two viewpoints. On the one hand, the information clearly demonstrates the efficiency of the holographic interferometry application in obtaining new data which describes a deformation character in irregular zones of curved thin-walled structures. Note that an analysis of the validity of various assumptions on which the thin shell theory is founded, and its further development, is perhaps the most interesting implementation of the approach considered.

On the other hand, the question is raised concerning which of the two above methods of bending stress determination should be adopted as the most correct approach to the deformation process. We shall attempt to provide an answer to this question by proceeding from relatively simple considerations without taking into account complex aspects related to the derivation of different thin shell theories.

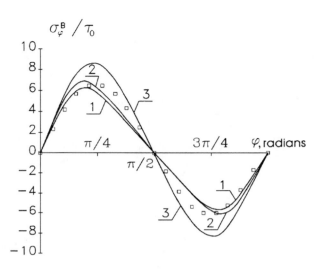

Figure 5.40. Distributions of the relative bending stress obtained by different ways along the circular hole boundary in the shell under torsion

The experimental data presented in Figures 5.37–5.40 allow us to draw some conclusions. First, the constancy of the normal to the surface displacement component w through a thin-walled shell thickness in the normal to the surface direction has been proven by means of a direct physical measurement carried out near the irregular zone of a metallic object—see Figure 5.37. Therefore, the validity of the Kirchhoff–Love hypothesis (5.17) is confirmed, in fact, near a large hole in a circular curved cylindrical shell where the displacement component and strain distributions are characterized by high gradients. The most important corollary of this hypothesis is the linear distribution of the strains (5.23) and stresses (5.28) through the shell thickness in the normal to surface direction. Thus, relation (5.50) for bending stress determination has direct experimental evidence obtained in an irregular zone of a curved shell. This is of great importance, since the expression for the circumferential strain (5.41) (or (5.42)) results from the single assumption that deformations are small which is valid in the corresponding curvilinear coordinate system, and not related to the Kirchhoff–Love hypotheses.

These facts lead to the following conclusions. First, the approach to the determination of bending stress based on using relations (5.52) and then (5.50) (curve 1 in Figure 5.40) gives us with a high probability the result which correctly represents the real situation. Second, taking into account the coincidence of the dependencies presented in Figure 5.38 and curve 1 in Figure 5.40 with the corresponding analytical results, we can reach a non-trivial conclusion that data obtained by Van Dyke [5] in the range of the non-curved shell theory can be used in the considerably broad range of shell radius magnitudes for the corresponding values of shell parameters β (5.91).

Therefore, a reliable method for analysis of strain/stress distributions in the normal to surface direction through a shell thickness has been found. This approach requires the displacement components measured on both shell faces along an open hole edge to be used for the determination of bending stress.

However, from the optical method application point of view, the most effective and convenient technique to determine bending stress should be based on the use of experimental data obtained on a single (usually external) shell face. This problem can be overcome in a rather unexpected way. The difference between rotation angle values θ_φ (5.96) and (5.98), which is represented in Figure 5.39, gives rise to the consideration that to calculate the bending stress (5.46) (or, in our particular case, (5.53)) the rotation angle θ_v (or θ_r) (5.44) may be neglected. Mathematically, this assumption means

$$\theta_v = -\frac{\partial w}{\partial s_v} + \frac{v}{R}\cos\Psi \equiv 0 \tag{5.100}$$

Substituting (5.100) into (5.46) yields

$$(\varepsilon_{tt} - \varepsilon'_{tt}) = h\left\{ -\frac{\partial^2 w}{\partial s_t^2} + \frac{\partial v}{\partial s_t}\frac{\cos\Psi}{R} - \frac{v}{R}\frac{\sin\Psi}{r_0} \right\} \tag{5.101}$$

For a circular hole in a circular cylindrical shell equation (5.101) takes the form

$$(\varepsilon_\varphi - \varepsilon'_\varphi) = -\frac{h}{r_0}\left\{ \frac{\partial^2 w}{r_0\partial^2\varphi} - \frac{\partial v}{\partial\varphi}\frac{\cos\varphi}{R} + \frac{v\sin\varphi}{R} \right\} \tag{5.102}$$

The result of the bending stress calculation σ_φ^B using equation (5.102) for the experimental data obtained above is presented in Figure 5.40 by curve 3, which almost coincides with curve 1 in the same figure. Thus, the bending stress distribution can be obtained by using the displacement components and their first and second derivatives along a hole boundary measured on a single (external) face of a shell. In what follows, expression (5.101) (or (5.102)) will be used for bending stress calculation only, since the experimental foundation of this approach can be considered as reliable confirmation of the validity of expression (5.101).

In conclusion, note that the distribution of the relative circumferential stress σ_φ/τ_0 along the hole edge obtained through the use of expressions (5.102) and (5.52) on the internal shell face almost coincides with curve 2 in Figure 5.38.

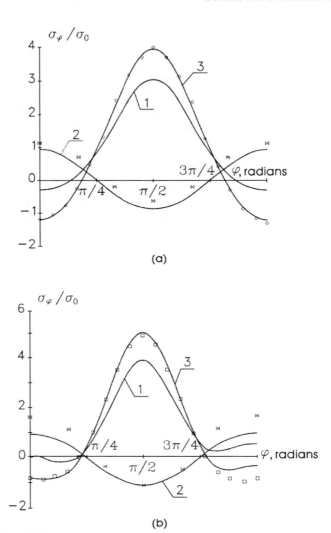

(a)

(b)

Figure 5.41(a) and (b)

5.6 Influence of boundary conditions on stress concentration

In many cases, a mathematically correct description of the boundary conditions exerts its main influence on the results of a local strain/stress determination by analytical and numerical methods. In fact, modelling a

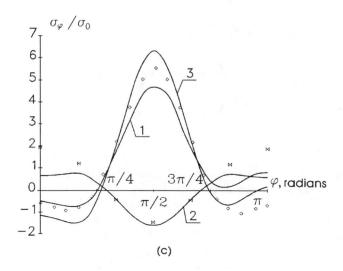

(c)

Figure 5.41. Distributions of the relative circumferential total stress (1), its bending (2) and membrane (3) components along the edge of centrally located circular hole of radius 9 mm (a), 14 mm (b) and 18 mm (c) in a shell under tension

structure's supports cannot reproduce the actual boundary conditions with a high degree of accuracy. The reflection hologram technique of a local strain investigation is free from these limitations, since its practical implementation requires only the use of local displacement component distributions as input data. The accuracy of the determination of these local parameters depends on the geometrical parameters of the interferometer optical system. This accuracy is not connected with the geometrical parameters of the structure, its fixing conditions or loading scheme.

In order to demonstrate the capability of the technique presented in Section 5.5 we shall consider the influence of a hole radius value and a hole centre position on the local stress distribution in a circular cylindrical shell. Distributions of the displacement components u, v and w along the hole boundary when the hole centre is located at the central cross-section of the shell subjected to tension are shown in Figure 5.27 for different values of the hole radius r_0.

A numerical differentiation of the displacement component distributions, which is necessary for stress determination in accordance with expressions (5.52), (5.102) and (5.49)–(5.51), is carried out by approximation of the discrete sets of the experimental data with a Fourier series of the following form:

$$u = \sum_{k=2}^{p} A_k \cos[(k-1)\varphi]$$

$$v = B_1 - \sum_{k=2}^{q} B_k \sin[(k-1)\varphi] \qquad (5.103)$$

$$w = \sum_{k=2}^{r} C_k \cos[(k-1)\varphi]$$

The procedure for selection of the coefficients of series (5.103) is completely analogous to that described in Section 4.1. Distributions of the relative circumferential stress σ_φ/σ_0 on the external shell face obtained from equations (5.52) and (5.49) (σ_0 is defined according to expression (5.92)), its bending component $\sigma_\varphi^B/\sigma_0$ (relations (5.102) and (5.50)) and membrane component $\sigma_\varphi^M/\sigma_0$ (relation (5.51)) along the hole edge resulting from the experimental data corresponding to the symmetrical hole position are presented in Figure 5.41 by curves 1, 2, and 3 respectively. Figures 5.41(a)–(c) correspond to the hole radius value $r_0 = 9$, 14 and 18 mm. The different shaped dots indicate the data of Van Dyke [5].

A comparison of the experimental and analytical distributions in Figure 5.41(a) (hole radius $r_0 = 9$ mm) proves that for the shell, in which the distance between the hole edge and the end face can be considered as infinity from a mechanical point of view, these data almost coincide. Some differences between analogous dependencies presented in Figures 5.41(b) and (c) (hole radius $r_0 = 14$ and 18 mm), apparently, are due to a non-equivalence of the deformation character of the shell under study and an abstract model of an infinite shell with a hole. The influence of the boundary conditions on the local strained state near the hole is the main reason of this non-equivalence.

The influence of the clamped shell edge on local strains at a hole vicinity are revealed more strongly if the distance between the hole and shell edges are reduced. In order to illustrate this, we shall now consider a deformation of the cylindrical shell with a non-symmetrical position of the hole of radius $r_0 = 14$ mm in relation to the shell end face. The distance between the symmetrical cross-section of the shell and the hole centre in the case involved is $L_1 = 22$ mm. The hole is displaced towards the left more rigid end of the shell (see Figure 2.27). Figure 5.42 shows a comparison between the displacement component distributions along the hole edge in the shell and the central and displaced hole. Curves 1, 2 and 3 describe the components u, v and w in the former case, curves 4, 5 and 6 correspond to the same components in the latter case. Note that preparation of the discrete experimental data sets for a numerical differentiation procedure are performed by using series (5.103) in both cases.

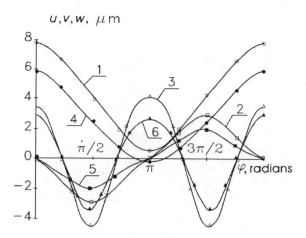

Figure 5.42. Comparison of distributions of the displacement components along the edge of centrally located and displaced hole in a shell under tension

Distributions of the relative circumferential stress σ_φ/σ_0 on the external face and σ_φ'/σ_0 on the internal face along the boundary of the central hole are shown in Figure 5.43 by curves 1 and 2 respectively. Curves 3 and 4 represent the analogous dependencies in the case of a non-symmetrical hole position. The stresses on the external face are calculated from expressions (5.52) and (5.49) using series (5.103). The stresses on the internal face can be defined by

$$\sigma_\varphi' = \sigma_\varphi - 2\sigma_\varphi^B \tag{5.104}$$

where σ_φ are the stresses on the external face and σ_φ^B are the bending stresses obtained by using (5.102) and (5.50).

The differences between the corresponding couples of dependencies presented in Figure 5.43 can be considered as a characteristic of the sensitivity of the holographic interferometric technique used to reveal the influence of boundary conditions on local strains or stresses.

5.7 Stress concentration in a shell with an elliptical cut-out

The application of the reflection hologram interferometry technique to a local strain or stress investigation near a hole with a non-circular boundary can be illustrated by determination of stress concentration along an elliptical hole edge, the parameters of which are listed in Table 5.1.

Distributions of the displacement components along the elliptical hole boundary, which are shown in Figures 5.31 and 5.33, are approximated with a

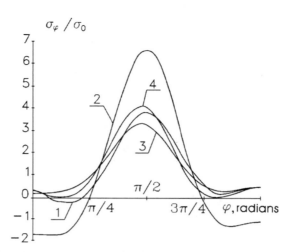

Figure 5.43. Comparison of distributions of the relative circumferential stress along the edge of centrally located and displaced hole in a shell under tension

trigonometrical series in the polar coordinate system (r, φ) the centre of which coincides with the centre of the cut-out. These series have form (5.103) in the case of shell tension and (5.94) in shell torsion.

Stresses on the external face of the shell are defined by using expressions (5.70) and (5.49). The bending strains, in accordance with the method established in Section 5.5, can be calculated by analogy with relation (5.102). The corresponding formula in the case of an elliptical hole can be derived by combining expressions (5.101), (5.72) and (5.73). Expressions (5.50), (5.51) and (5.104) give the bending stresses, membrane stresses and total circumferential stresses on the internal shell face along the elliptical hole edge respectively. The corresponding normalizing factors needed for calculation of stress concentration are obtained according to expressions (5.92) and (5.93) for tension and torsion respectively.

Distributions of the relative circumferential stress on the external shell face along the hole edge σ_φ/σ_0 for tension of the shell and σ_φ/τ_0 in the case of shell torsion are indicated by curves 1 in Figures 5.44 and 5.45. The dependencies for the bending stress $\sigma_\varphi^B/\sigma_0$ (curves 2) and the membrane stress $\sigma_\varphi^M/\sigma_0$ (curves 3) are shown in Figure 5.44. Curve 2 in Figure 5.45 reveals the total relative stresses on the internal face of the shell. The different shaped points in Figure 5.44 represent the data of Gus' and others [4] obtained for shell parameter (5.91) $\beta_1 = 2.5$ and semi-axes ratio $a/b = 1.4$. The corresponding parameters of the shell with an elliptical hole under study are: $\beta_1 = 2.31$, $a/b = 1.4$.

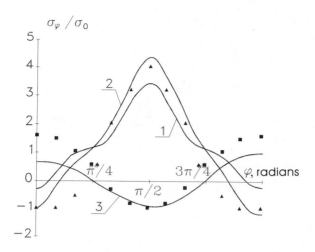

Figure 5.44. Distributions of the relative circumferential total stress (1), its bending (2) and membrane (3) components along the elliptical hole edge in a shell under tension

A comparison of the experimental and analytical data in the case of a torsion of a circular cylindrical shell with an open elliptical hole can also be performed by using the data of Gus' and others [4]. In order to find the required correspondence between two types of data, we have to consider a special parameter λ_0 [4]:

$$\lambda_0 = \sqrt[4]{\frac{3(1-\mu^2)}{4}} \cdot \frac{a}{\sqrt{Rh}}$$

where a is the greater semi-axis of the ellipse. The work of Gus' and others [4] contains, in particular, the bending and membrane stress distributions for $\lambda_0 = 2.4$ and ratio $b/a = 0.6$ and 0.8. The specimen and elliptical hole parameters are: $\lambda_0 = 2.5$, $b/a = 0.72$.

A comparison of the experimental stress distributions and the two above-mentioned analytical results reveals a good agreement of the maximum values both of the bending and membrane stresses. The same is true also in relation to the angular coordinate where maximum stress values are reached.

5.8   Deformation of a composite shell with a circular hole

The deformed state in irregular zones of thin-walled structures made of composite materials is normally described by highly complex relations

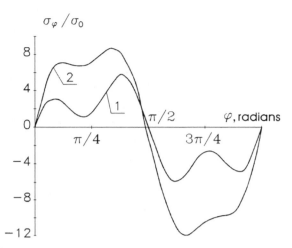

Figure 5.45. Distributions of the relative circumferential total stress along the elliptical hole edge in a shell under torsion obtained on the external (1) and internal (2) face

compared to that of metallic structures. These relations are founded upon various assumptions, which do not have such an obvious practical meaning as the Kirchhoff–Love hypotheses [4]. Moreover, one can argue that an experimental justification of these hypotheses, especially in a stress concentration region, is insufficient to enable a reliable local strain analysis. Consequently this is where an investigation of a strain or stress concentration in thin-walled composite structures by holographic interferometry becomes of scientific and practical interest.

A deformation of a circular hole vicinity in a thin-walled circular cylindrical shell made of glass-fibre reinforced laminate is given as a brief example which illustrates the capability of the reflection hologram interferometry in the experimental field. The geometrical parameters of this composite shell are analogous to those of the metallic shells listed in Table 5.1: the length of the shell $L = 280$ mm; external radius of shell $R = 30$ mm; wall thickness $h = 1.5$ mm; and radius of hole $r_0 = 14$ mm.

Typical fringe patterns reconstructed with a reflection double-exposure hologram, whose set-up is described in the optimal equation system (2.67) with the number $j = 4$ and corresponds to the tensile force increment $P = 400$ N, are presented in Figure 5.46. Distributions of the displacement components in the cylindrical coordinate system u, v and w along the hole edge, corresponding to the interferograms in Figure 5.46, are shown in Figure 5.47(a) by curves 1, 2 and 3 respectively.

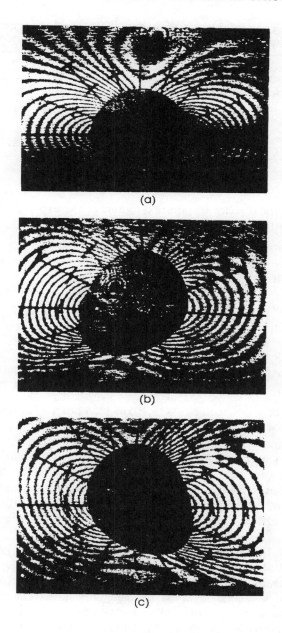

Figure 5.46. Three interferograms of the composite shell with the circular hole subjected to a tension obtained for different viewing directions: (a) $\alpha = 90$, (b) $\alpha = 210$ and (c) $\alpha = 330$ degrees

Table 5.3 The coefficients of approximating series (5.105) for displacement component distributions along the circular hole edge in the composite shell under tension

Displacement component	Series coefficient	k = 1	k = 2	k = 3
u	A_k	3.4	2.2	–
v	B_k	0.93	0.4	0.1
	C_k	−2.6	–	0.15
w	D_k	1.9	−0.21	2.9

The trigonometrical series for approximation of the experimental data can be represented in the following form:

$$u = \sum_{k=1}^{p} A_k \cos[(k-1)\varphi]$$

$$v = \sum_{k=1}^{q} B_k \cos[(k-1)\varphi] + \sum_{k=1}^{a} C_k \sin[(k-1)\varphi] \qquad (5.105)$$

$$w = \sum_{k=1}^{r} D_k \cos[(k-1)\varphi]$$

The magnitudes of series coefficients (5.105) which are used for calculation of the circumferential strains are listed in Table 5.3.

In order to obtain the circumferential strain distribution along a circular hole boundary in a composite circular cylindrical shell, we can use relation (5.52), as in the case of a metallic shell. The corresponding dependence is represented in Figure 5.47(b) by curve 1. The analogous dependence for the aluminium shell of the same geometrical parameters and hole radius (see Table 5.1) normalized to the tensile force increment $P = 400$ N is shown in Figure 5.47(b) by curve 2.

In conclusion, it should be noted that an analysis of the strain change through a shell thickness in the normal to surface direction cannot be performed by means of the approach described in Section 5.5 which allows determination of the bending stress by using the displacement component distributions measured along an open hole edge on the external shell face. The solution to this problem requires additional experimental and numerical resources which lie outside the scope of this book. This also applies to problems concerning transition from strains to stresses in irregular zones of thin-walled composite structures.

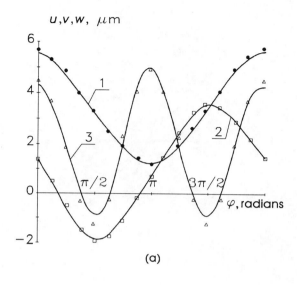

(a)

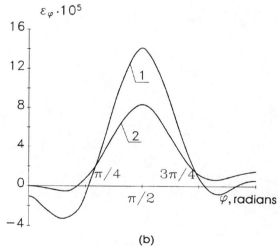

(b)

Figure 5.47. Distributions of the displacement components (a) and circumferential strain (b) along the edge of circular hole in a composite shell under tension

5.9   Stress concentration in a shell with multiple cut-outs

The capability of reflection hologram interferometry in the field of a local strain/stress investigation in thin-walled structures with several irregular zones can be clearly illustrated by solving the problem of determination of stress

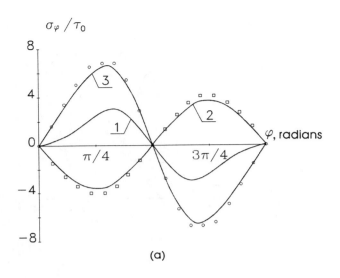

(a)

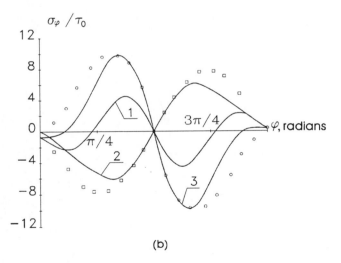

(b)

Figure 5.48(a) and (b)

redistribution in a shell with two co-axial circular holes of different radius subjected to tension and torsion.

The object investigated, whose geometrical parameters completely coincide with those indicated in Figure 2.27 and listed in Table 5.1. Two co-axial holes of radius $r_0 = 9$ mm and $r_0 = 18$ mm were made in the specimen.

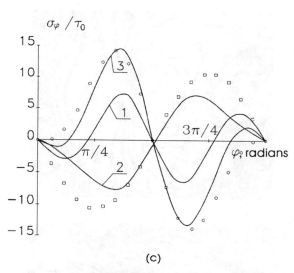

(c)

Figure 5.48. Distributions of the relative circumferential total stress (1), its bending (2) and membrane (3) components along the edge of circular hole of radius 9 mm (a), 14 mm (b) and 18 mm (c) in the shell under torsion

The data describing the displacement component distributions along the edge of a single hole of radius $r_0 = 9$ mm and $r_0 = 18$ mm in the circular cylindrical shell are shown in Figures 5.27 and 5.29 for tension and torsion respectively. Figures 5.41(a) and (c) illustrate the stress concentration in the case of a single hole in the shell under tension for the hole radius values $r_0 = 9$ mm and $r_0 = 18$ mm, respectively. Distributions of the relative circumferential stress σ_φ / τ_0 on the external shell face along the single circular hole boundary in the circular cylindrical shell subjected to torsion with bending $\sigma_\varphi^B / \tau_0$ and membrane $\sigma_\varphi^M / \tau_0$ components are presented in Figures 5.48(a)–(c) for hole radius $r_0 = 9$, 14 and 18 mm, respectively. The notation used in Figure 5.48 is similar to that in Figure 5.41. The stresses in Figure 5.48 correspond to the displacement components in Figure 5.29 and are obtained by means of the procedure described in Section 5.5. The different shaped points indicate the data of Van Dyke [5]. Note that in the case of shell torsion, to construct the graphs plotted in Figure 5.48 we take into account that the holes in the actual shell are made by milling with a cylindrical cutter. Due to this manufacturing process the distance between two shell faces measured in the direction of the generatrix of the lateral cylindrical surface of the hole h' (see Figure 2.27) always exceeds, except for the point of intersection of a hole edge and shell generatrix coinciding with the α_1-axis (see Figure 5.3), the wall thickness in the

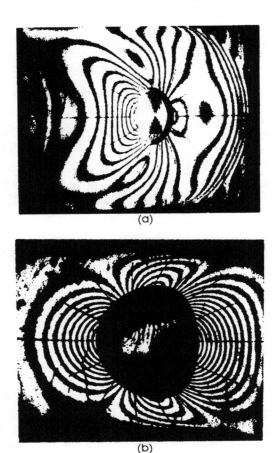

(a)

(b)

Figure 5.49. Interferograms of shell with two co-axial holes under tension obtained for the hole of radius 9 mm (a) and 18 mm (b)

normal to surface direction h, which is used for analytical determination of stress:

$$h' = \frac{h}{\cos\omega_0}, \quad \text{where } \omega_0 = \arcsin\left[\frac{r_0\sin\varphi}{R}\right]$$

Therefore, the value of h' is used in relation (5.102) instead of h.

Now we shall note some of the special features of the holographic interferometric experiment in the case of a shell with two co-axial holes with different radii. The reflection holograms were recorded using the opposite

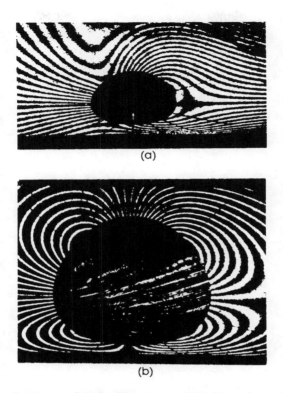

(a)

(b)

Figure 5.50. Interferograms of shell with two co-axial holes under torsion obtained for the hole of radius 9 mm (a) and 18 mm (b)

beam optical arrangement shown in Figure 5.26, but in this case two photoplates were mounted simultaneously from two external and opposite sides of the shell perpendicular to the hole axis. To ensure the correctness of the holographic recording procedure, it should be taken into account that the optimal fringe density is reached on an interferogram recorded near the hole of radius $r_0 = 18$ mm at a load value less than that for the hole of radius $r_0 = 9$ mm. Therefore, in the case of shell tension, two double-exposure holograms, corresponding to the same initial state but two different final states, were recorded simultaneously. Typical interferograms obtained for the net stress increment $\sigma_0 = 5.4$ MPa (hole radius $r_0 = 9$ mm) and $\sigma_0 = 1.8$ MPa (hole radius $r_0 = 18$ mm) are presented in Figures 5.49(a) and (b) respectively.

When the shell was subjected to torsion, the multiple-exposure hologram was recorded near the hole with radius $r_0 = 18$ mm with a torsional moment increment of $M = 3$ Nm in each of the four loading steps, and a second double-

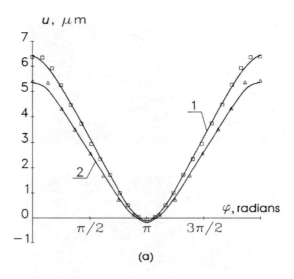

(a)

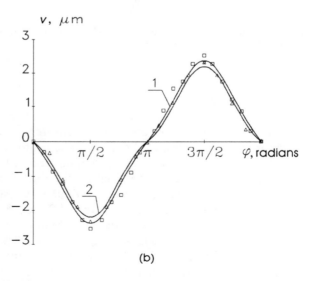

(b)

Figure 5.51(a) and (b)

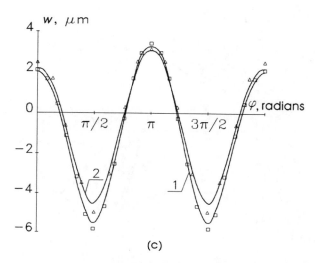

(c)

Figure 5.51. Comparison of distributions of the displacement components u (a), v (b) and w (c) along the edge of hole of radius 18mm in the shell with a single hole (1) and two co-axial holes (2) under tension

exposure hologram was recorded near the hole of radius $r_0 = 9$ mm. The latter hologram corresponds to the total torsional moment increment of the hologram obtained by multi-exposure technique $M = N \times M = 12$ Nm (N is the exposure number for the first hologram). Typical interferograms thus obtained are presented in Figures 5.50(a) (double-exposure hologram) and 5.50(b) (multiple-exposure hologram). This experimental design is necessary to minimize the errors made in a load holding. This circumstance is very important to achieve a correct comparison of the experimental results obtained on the opposite sides of the shell.

The technique of quantitative fringe patterns interpretation to obtain the displacement components and subsequent strain and stress distributions along the hole edges and its distributions in the normal to surface direction through the shell thickness completely coincides with that described in Sections 2.5 and 5.5. Note only that the absolute fringe orders were identified by observing the unstrained area near the rigidly clamped end face of shell.

Distributions of the displacement components along the hole edge of radius $r_0 = 18$ mm for tension and torsion are shown in Figures 5.51 and 5.52 respectively. Parts (a), (b) and (c) correspond to components u, v and w. Curve 1 indicates the data for the shells with a single hole and curve 2 denotes the results obtained in the case of two co-axial holes. The displacement

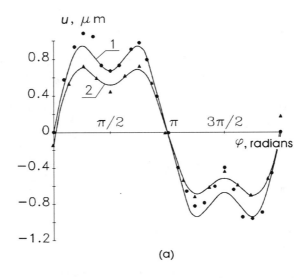

(a)

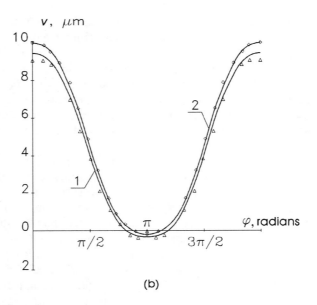

(b)

Figure 5.52(a) and (b)

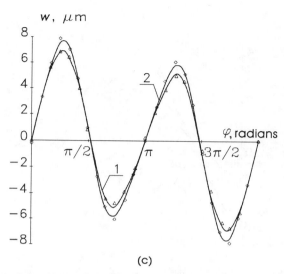

(c)

Figure 5.52. Comparison of distributions of the displacement components u (a), v (b) and w (c) along the edge of hole of radius 18mm in a shell with a single hole (1) and two co-axial holes (2) under torsion.

components corresponding to the shell tension and torsion are normalized to the net stress increment $\sigma_0 = 5$ MPa and $\tau_0 = 0.65$ MPa, respectively.

Distributions of the relative circumferential stress σ_φ/σ_0 and σ_φ/τ_0 along the edges of two holes in a single shell are shown in Figures 5.53 and 5.54 for a tension and torsion of the shell respectively. Parts (a) and (b) correspond to the hole radius values $r_0 = 9$ mm and $r_0 = 18$ mm. Numbers 1 and 2 denote the stresses on the external and internal shell faces respectively. For comparison, analogous stresses for two shells with a single hole of the same radius are presented in these figures by curves 3 and 4.

The results obtained demonstrate that when the shell with two co-axial holes of different diameters is subjected to tension or torsion, the hole of greater diameter is additionally stressed in comparison with the case of the single hole of the same diameter, and an opposite situation takes place for the hole of smaller diameter. The corresponding graphs presented in Figures 5.53 and 5.54 describe the quantitative characteristics of the stress redistribution observed.

In conclusion, we should once again call our attention to the important aspects of the experimental investigations, which mainly consist of a combined implementation of various hologram recording techniques in order to reduce the influence of errors of a load holding on the final results.

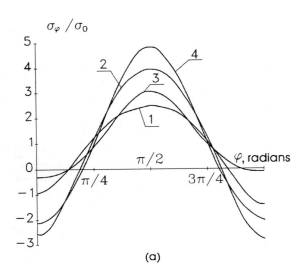

(a)

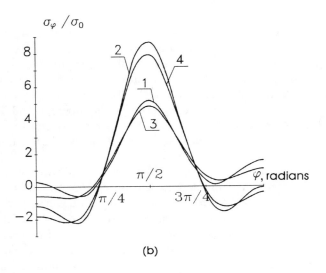

(b)

Figure 5.53. Comparison of distributions of the relative circumferential stress on two faces along the edge of hole of radius 9mm (a) and 18mm (b) in the case of shell tension

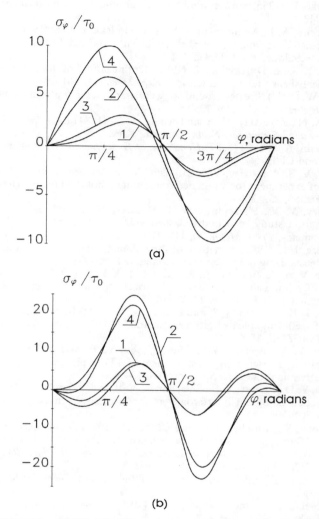

Figure 5.54. Comparison of distributions of the relative circumferential stress on two faces along the edge of hole of radius 9 mm and 18 mm in the case of shell torsion

References

1. Rabotnov, Yu. I. (1979) *Mechanics of Solid Deformable Body.* Nauka, Moscow.
2. Novozhilov, V. V. (1951) *The Theory of Thin Shells.* Sudpromgiz., Leningrad.
3. Savin, G. N. (1951) *Stress Concentration Near Holes.* GITTL, Moscow.
4. Guz', A. N., Chernyshenko, I. S., Chekhov, V. N. *et al.* (1980) *The Theory of Thin Shells with Holes. The methods of shells calculation 1.* Nauk. Dumka, Kiev.

5. Van Dyke, P. (1965) Stresses about a circular hole in a cylindrical shell. *AIAA J.* **3**, 1733–1742.
6. Alexandrov, A. J., Akhmetzianov, M. Kh. and Rakin, A. S. (1966) Investigation of elastic-plastic deformation of shells with cut-outs and fasteners by photoelastic coating technique. *Prikl. Mekh.* **2**, 1–9.
7. Rajaih, K. and Durelli, A. J. (1981) Optimization of hole shapes in circular cylindrical shells under axial tension. *Exp. Mech.* **21**, 201–204.
8. Bull, J. W. (1982) Stresses around large circular holes in uniform circular cylindrical shells. *J. Strain Anal.* **17**, 9–12.
9. Guz', A. N., Zarutskiy, V. A. and Amiro, I. Ja. (1984) *Experimental Investigations of Thin-walled Structures.* Nauk. Dumka., Kiev.
10. Kobayashi, A. S. (ed.) (1987) *Handbook of Experimental Mechanics.* Prentice Hall, Englewood Cliffs, NJ.
11. Pisarev, V. S., Yakovlev, V. V., Indisov, V. O. and Shchepinov, V. P. (1983) The design of experiment for strain determination by holographic interferometry. *Zhur. Tekh. Fiz.* **53**, 292–300.
12. Yakovlev, V. V., Shchepinov, V. P., Pisarev, V. S. and Indisov, V. O. (1984) Deformation study of cylindrical shell with rectangular cut-out by holographic interferometry. *Prikl. Mekh.* **20**, 117–120.
13. Goldberg, J. L. (1983) A method of three-dimensional strain measurement on non-ideal objects using hologram interferometry. *Exp. Mech.* **23**, 59–73.
14. Zhylkin, V. A., Ustimenko, A. P. and Boryniak, L. A. (1986) Investigation of strain stage of thin-walled circular cylindrical shells by means of panoramic interferometer. *Prikl. Mekh.* **22**, 79–84.
15. Begeev, T. K., Grishin, V. I. and Pisarev, V. S. (1984) Investigation of stress–strain stage of shell with hole by finite element and holographic interferometry techniques. *Proc. TsAGI.* **15**, 85–96.
16. Balalov, V. V., Pisarev, V. S., Shchepinov, V. P. and Yakovlev, V. V. (1990) Holographic interferometry measurement of 3-D displacement fields and its use for stress determination. *Opt. Spectrosc.* **68**, 134–139.
17. Indisov, V. O., Pisarev, V. S., Shchepinov, V. P. and Yakovlev, V. V. (1986) The use of reflection hologram interferometers for local strain investigation. *Zhur. Tekh. Fiz.* **56**, 701–707.
18. Schumann, W., Zurcher, J.-P. and Cuche, D. (1985) *Holography and Deformation Analysis.* Springer Series in Optical Sciences 46. Springer-Verlag, Berlin.
19. Lurie, A. I. (1955) *Three-dimensional Problems of the Elasticity Theory.* Gostekhizdat, Moscow.
20. Chernykh, K. F. and Novozhilov, V. V. (1991) *Linear Theory of Shells.* Nauka, St. Peterburg.
21. Timoshenko, S. P. and Goodier, J. N. (1970) *Theory of Elasticity.* McGraw-Hill, New York.
22. Podstrigach, Ya. S. and Shvets, R. N. (1978) *Thermoelasticity of Thin Shells.* Nauk. Dumka, Kiev.
23. Lekhnitsky, S. G. (1977) *The Elasticity Theory of an Anisotropic Body.* Nauka, Moscow.
24. Timoshenko, S. P. and Woinovsky-Kriger, S. (1959) *Theory of Plates and Shells.* McGraw-Hill, New York.
25. Baryshnicov, V. I., Grishin, V. I., Donchenko, V. Yu. and Tikhonov, Yu. V. (1983) Implementation of the finite element method to local strength investigation of aircraft structural elements. *Proc. TsAGI.* **14**, 66–73.

6 DETERMINATION OF FRACTURE MECHANICS PARAMETERS FOR SHELLS WITH SURFACE FLAWS AND THROUGH CRACKS

The problem of pressure vessel and pipeline failure due to fatigue growth in surface flaws is a frequent occurrence today. Through-crack propagation in thin-walled shell structures is the main reason for numerous accidents. For predictive analysis of a structure's safety from a fracture mechanics point of view, engineering estimates of criterial parameters should be obtained. A variety of parameters and criteria, such as the stress intensity factor (SIF) K, crack driving force (or strain energy release rate) G, invariant J-integral, crack opening displacement (COD) δ, weight function t, etc. are usually determined in practice to analyse fracture initiation, stable crack growth and unstable fracture.

Some stress intensity solutions for typical shell elements geometries with surface or through-cracks are available in the literature [1–4]. However, in practical problems of fracture analysis, where gross mechanics and loadings of engineering structures are often so complex, the solutions currently available are found to be rather inadequate.

In order to analyse cracks located in non-regular shell structure zones, for instance near holes, nozzles or rigid elements, it is neccessary to apply three-dimensional numerical modelling or to develop experimental methods, which have reasonable accuracy and are easy to use.

The displacement fields of cracked structures were investigated by using various optical methods, among them moiré [5–10] and speckle techniques [11–14]. Holographic interferometry was also applied to study fracture problems [15–23]. To calculate stress intensity factors K using experimentally measured displacement fields in the neighbourhood of crack tips, various approximate approaches were proposed based on asymptotic theoretical dependencies [5, 6, 10, 15–17, 23] and on regression analysis of discrete experimental data [24]. A combined numerical–experimental technique was developed where the displacement components distributions along the contour surrounding the crack region is used as initial boundary conditions for two-dimensional fracture problems which are solved by finite or boundary elements codes [11].

The experimental evaluation of numerical methods is of considerable importance. Comparison of the displacement distributions in the crack borders, measured by holographic interferometry and speckle techniques and corresponding finite elements code results, was carried out for a cylindrical shell with an axial semi-circular surface flaw [20].

In this chapter a newly developed method to determine the stress intensity factor K_I for the shell structures with surface cracks is discussed. This approach is based on the line spring model which was further developed by the holographic measurements of the crack borders displacements and slopes. A computer simulation and combined experimental–numerical solution were carried out to verify the derived approximate equations.

Displacements fields are presented for cylindrical thin-walled shells with through-cracks of different location and orientation under tension and torsion. The stress intensity factor for a through-crack is determined by using the weight function, which is obtained experimentally.

6.1 Determination of stress intensity factor K_I for shells with surface flaws

A relatively simple and accurate technique to calculate K_I for shell structures with surface flaws has been developed based on the line spring model. This technique was first proposed by Rice and Levy [25] and further developed by Erdogan and Delale [26, 27], and Cheng and Finnie [28, 29]. By implementing this approach, the finite element numerical K_I solutions can be obtained for various typical shell structures containing semi-elliptical surface cracks, as shown by Parks and White [30]. These results provide the same accuracy as most distinctive three-dimensional numerical solutions for the similar fracture problems. An experimental evaluation of the line spring model was carried out by Ezzat [31] and King [32].

The line spring model is suitable from the point of view of employment of experimentally measured displacement fields, because the unknown parameters in this approach are the displacements and rotations of the crack border. Indeed, an approximate technique to calculate K_I along the surface crack front may be developed by combining the hypothesis of the mechanical line spring model and the boundary conditions in the form of crack border displacements

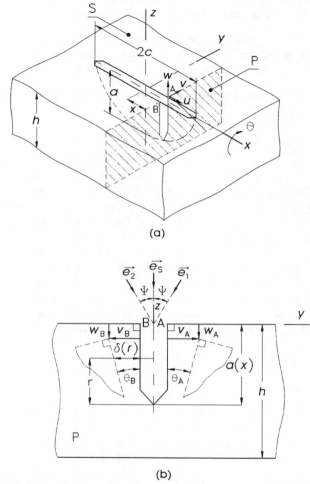

Figure 6.1. Shell element with surface flaw (a) and its cross section with holographic interferometer optical arrangement (b)

on a shell surface obtained by holographic interferometry. This type of approach would be useful for any real shell structure with a surface flaw located anywhere on the outer object surface which is accessible for viewing.

Let us consider the basics of the line spring model. A cracked shell element having a wall thickness h, a surface semi-elliptical crack with a border length $2c$, and maximum flow penetration a, is shown in Figure 6.1(a). The local Cartesian coordinate system (x, y, z) is introduced so that the z-axis is aligned with the surface normal and the x-axis is parallel to the crack borders. Usually, notations of the displacement vector components are:

u —tangential to the shell surface and crack borders
v —tangential to the shell surface and normal to the crack borders
w —normal to the shell surface.

We define the averaged crack mouth open displacement (CMOD) δ_m as:

$$\delta_m = \frac{|v_A| + |v_B|}{2} \tag{6.1}$$

where v_A and v_B are the displacement components at points A and B located on the opposite crack borders on the shell surface (Figure 6.1(b)).

Any shell fragment with a surface crack can be represented by a set of plane elastic spring elements. These elements are similar to strips with a width h which is equal to the shell wall thickness. There is an edge crack of varying depth $a(x)$ in each spring element as shown in Figure 6.1(b). The basic hypothesis of the line spring model consists of the assumption that the stress intensity factor K_I at any point of the partially through-crack front is equal to the corresponding value of K_I for a spring element under plane strain state.

To express analytically the stress intensity factor through the experimental displacement data, an asymptotic dependence for displacement distribution near the crack tip is used. In particular, the crack opening displacement δ for the edge crack is expressed as

$$\delta(r) = \frac{K_I}{H} \sqrt{\frac{r}{2\pi}} \tag{6.2}$$

where r is the distance from the crack tip (see Figure 6.1(b)) and $H = E/(1 - \mu^2)$ is the generalized elasticity modulus for the plane strain state. In previous research, experimental approaches for K_I determination based on the asymptotic expression (6.2) were applied to transparent specimens with surface cracks. For opaque objects, displacements can be measured by experimental methods only in the outer surface, i.e. at a distance from the surface crack front. To account for additional influences on the shell structure's overall mechanical behaviour, the asymptotic expression (6.2) should be modified. In this way, a reasonably good approximation of the crack opening displacement δ for an edge crack in the strip may be described by a truncated power series [33]:

$$\delta(r) = \sum_{n=1}^{N} A_n \left(\frac{r}{a}\right)^{n/2} \tag{6.3}$$

where A_n $(n = 1, 2, 3, \ldots)$ are unknown coefficients.

From asymptotic dependence (6.2) and by way of a limited analysis of function (6.3), the stress intensity factor K_I is expressed:

$$K_I = \frac{H}{4}\sqrt{\frac{2\pi}{a}}A_1 \tag{6.4}$$

Accounting for the equivalence between the plane strip elements with the edge cracks and corresponding cross-sections of shell structure with a surface flaw, the boundary conditions to calculate the unknown coefficients A_n can be formulated. For the simplest case, if CMOD is available from the experimental data only, then the first condition will be

$$\delta_m = \sum_{n=1}^{N} A_n \tag{6.5}$$

where δ_m is the crack mouth opening displacement on the shell surface corresponding to the cross-section with coordinate x, shown in Figure 6.1(b).

The next condition is formulated for another quantity, which can be obtained experimentally in the object surface, namely crack borders slope θ shown in Figure 6.1(b):

$$\theta = \frac{\partial \delta}{\partial r}(r = a) = \frac{1}{a}\sum_{n=1}^{N}\frac{n}{2}A_n. \tag{6.6}$$

The average crack border slope can be estimated in the same way as expression (6.1):

$$\theta = \frac{|\theta_A| + |\theta_B|}{2}$$

where θ_A, θ_b are the slopes of opposite crack borders in any section P (see Figure 6.1(b)).

When analysing the stress–strain state in the corner points A and B of the crack separation surfaces, we can observe that shearing strain $\gamma_{yz} = 0$. This leads to the additional boundary condition derived in references [34, 35]:

$$\frac{\partial^2 \delta}{\partial r^2}(r = a) = \sum_{n=1}^{N}\frac{n}{2}\left(\frac{n}{2} - 1\right)A_n = 0. \tag{6.7}$$

Also, this results in the expression to determine the crack separation surfaces rotation angle θ:

$$\theta = \frac{\partial w}{\partial y} \tag{6.8}$$

which is more accurate for shallow shells. Derivative $\partial w/\partial y$ can be determined by a finite difference method for three uniformly spaced nodes:

$$\frac{\partial w}{\partial y} = \frac{1}{2\Delta y}(-3w_0 + 4w_1 - w_2). \tag{6.9}$$

An alternative approach consists of an experimental measurement of the outer shell surface normal slopes θ along crack borders by using optical differentiation techniques (see Chapter 4). An approximate expression to estimate the stress intensity factor K_I follows from equation (6.4), where unknown coefficient A_1 is determined by boundary conditions (6.5)–(6.7).

If the power series (6.3) is truncated with a single first member ($N = 1$), then

$$K_I = \frac{H}{4}\sqrt{\frac{2\pi}{a}}\,\delta_m. \tag{6.10}$$

Note here that only one experimentally obtained parameter δ_m appears in the last equation.

For a two-member function (6.3) ($N = 2$) the estimation equation with two experimental quantities (δ_m, θ) is derived:

$$K_I = \frac{H}{8}\sqrt{\frac{2\pi}{a}}(3\delta_m - 2a\theta). \tag{6.11}$$

Thus, for a three-member truncated series (6.3) ($N = 3$) we obtain

$$K_I = \frac{H}{32}\sqrt{\frac{2\pi}{a}}(15\delta_m - 12a\theta). \tag{6.12}$$

The same experimentally obtained quantities (δ_m, θ) appear in equation (6.12) which differs by the constant coefficient values from the previous equation (6.11).

Surface flaws in pressurized shells are usually deformed in normal stretch mode (or mode I). Thus, the problem to be solved takes advantage of the symmetry and only two displacement components of the crack borders v and w have to be used for SIF determination.

The reflection hologram technique is most effective in analysing the displacement fields in the neighbourhood of surface flaws.

A step-by-step procedure of how to record and interpret reflection holograms was discussed in Sections 2.4 and 2.5. Consequently, we can use a holographic interferometer with sensitivity matrix K (2.70) to obtain the two above-mentioned displacement components in previously known directions. The corresponding optical set-up for any cross-section of the cracked shell structure is illustrated in Figure 6.1(b).

Combining expressions (2.32) and (2.70), one can derive an expressions to interpret two fringe patterns corresponding to symmetrical viewing directions $\vec{e}_1$ and $\vec{e}_2$:

$$v = \frac{\lambda(n_1 - n_2)}{2 \sin\psi} \tag{6.13}$$

$$w = \frac{\lambda(n_1 + n_2)}{2(1 + \cos\psi)} \tag{6.14}$$

where v and w are the tangent and normal displacement components; λ is the light wavelength; ψ is the viewing angle; and n_1 and n_2 are the absolute fringe numbers.

Relative crack border displacements have to be used to calculate K_I for surface flow problems only, thus it is not necessary to find the zero-order fringe. For this case, the fringe pattern interpretation procedure is modified in the following way.

Displacement components for any surface point under consideration are determined with reference to one fixed point R near the crack borders. Using equations (6.1) and (6.13) an averaged crack mouth opening displacement δ_m may be expressed in the form

$$\delta_m(x) = \frac{\lambda}{4 \sin\psi} \left[|n_{1R}^A - n_{2R}^A| + |n_{1R}^B - n_{2R}^B| \right] \tag{6.15}$$

where n_{1R}^A, n_{2R}^A, n_{1R}^B, n_{2R}^B are the absolute fringe orders in points A and B counted from an arbitrary reference point R on the fringe pattern for $\vec{e}_1$ and $\vec{e}_2$ viewing directions.

The crack mouth opening displacement δ_m is the most convenient experimentally measured fracture parameter capable of verifying numerical codes because of its weak dependence on finite element model discretization. Also, CMOD is an integral characteristic of local strain concentration resulting from both the mechanical behaviour of the whole shell structure and the presence of surface flaw. Therefore, it is the most convenient criterion to prove the adequacy of multiparametric models describing the ductile fracture process [36].

To verify the accuracy of approximate expressions (6.10)–(6.12) a finite element method investigation of a strip with an edge crack and a plate with a semi-elliptical surface crack, loaded by tension and bending, was carried out. In this case the numerically obtained displacement components near a crack

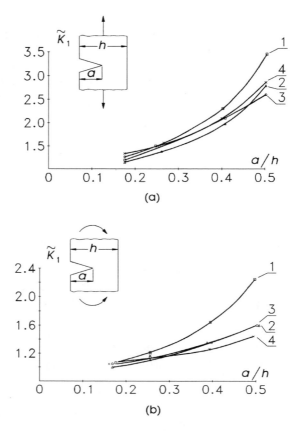

Figure 6.2. Dependencies of normalized stress intensity factor against reference deepness of the edge crack in a strip under tension (a) and bending (b)

are used instead of real data of the holographic experiment in order to reduce the body of specimen testing.

Numerical simulations were made for a set of strips (see Figure 6.1(b)) which have edge notches with reference lengths $a/h = 1/6$; $1/4$; $2/5$; $1/2$. Figure 6.2 illustrates the normalized stress intensity factors

$$\tilde{K}_I = \frac{K_I}{\sigma\sqrt{\pi a}}$$

where σ is the maximum nominal stresses in the strip under tension or bending loads. Curves 1, 2 and 3 correspond to results calculated by expressions

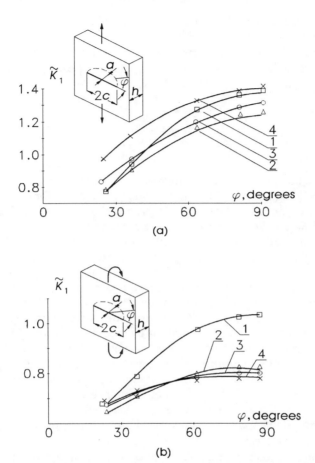

Figure 6.3. Distributions of normalized stress intensity factor along the front of semi-elliptic crack in a plate under tension (a) and bending (b)

(6.10)–(6.12). Curve 4 shows the results obtained by the accurate asymptotic equation for two nodes located near the crack tip [37]:

$$K_{\mathrm{I}} = \frac{H}{4}\sqrt{\frac{2\pi}{L}}\left(4\delta\left(\frac{L}{4}\right) - \delta(L)\right) \qquad (6.16)$$

where L is the node distance from the crack tip.

A comparison of these results shows that equations (6.11) and (6.12) provide a reasonable accuracy, but the simplified estimation K_{I} by equation (6.10), based only on the crack mouth opening displacement δ_m, is acceptable for small depth cracks ($a/h = 1/6$; $1/4$) and overestimates the SIF for edge cracks with the ratio $a/h > 1/2$.

Also, numerical tests were carried out for the set of plates with surface semi-elliptical flaws under tension and bending loads. For example, the normalized stress intensity factor

$$\tilde{K}_{\mathrm{I}} = \frac{K_{\mathrm{I}}}{\sigma\sqrt{\pi a/Q}} \tag{6.17}$$

(where σ are maximum nominal stresses in a plate under tension or bending and $Q = 1 + 1.464(a/c)^{1.65}$ is an elliptic integral) distributions along the non-through-crack front in a plate of thickness h with surface semi-elliptical crack, having a semi-axis $a/c = 1/6$; $a/h = 3/8$ are shown in Figure 6.3.

For tension loading, equations (6.10), (6.11) and (6.12) provide reasonable accuracy as illustrated by curves 1, 2 and 3, according to Figure 6.3(a). Curve 4 results from accurate numerical computations by the asymptotic formula (6.16). In this way, for the plate bending problem, the distributions $\tilde{K}_{\mathrm{I}}$ along the crack front are illustrated in Figure 6.3(b). Curves 2 and 3, obtained by using expressions (6.11) and (6.12), are correlated with curve 4, which corresponds to an accurate computation by expression (6.16). In addition, the simplified expression (6.10) provides an overestimation of $\tilde{K}_{\mathrm{I}}$, which reached 30% at the point of maximum crack penetration (see curve 1 in Figure 6.3(b)).

Crack flaws existing in real shell structures, as a rule, have an irregular shape. However, available solutions were obtained for some elementary crack front shapes, such as semi-circular or semi-elliptic. First, a schematic representation of the real flaws must be chosen, which provides a reasonable accuracy. The simplest way is to include a plane flaw into a rectangular fragment and to inscribe a semi-ellipse in it. The largest overestimation that can be obtained for such a schematic representation is not more than 20% for maximum K_{I} [38]. The main conclusions from the discussed approximate approach verification by numerical simulation are the following:

- Expressions (6.11) and (6.12) provide accurate K_{I} estimations for plane strips with edge cracks as for three-dimensional surface cracks in a plate under typical loading conditions.
- The simplified expression (6.10) can be used for a fast K_{I} estimation for shallow cracks ($a/h < 1/2$), taking into account the overestimations of the SIF which are limited to the range 30–50% for both membrane and bending loadings.
- Maximum K_{I} values are not strongly influenced by the surface crack front shape. An overestimation of 20% may occur.

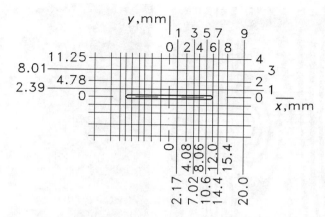

Figure 6.4. Coordinate grid on the shell surface around flaw borders

Previous experience of spring model solutions for typical surface flaws in shells demonstrates that more accurate K_I estimations may be obtained for central cross-sections where the crack front has deeply penetrated the shell wall, but less accurate estimations for crack front segments intersected with the outer shell surface. Also, the proposed approach is most reliable for sufficiently long surface flaws, satisfying the condition $2c > h$.

A combined application of holographic interferometry and the numerical finite element code was made to verify the approach by using a cylindrical shell with an axial semi-elliptical flaw under internal pressure. The shell specimen was made of aluminium alloy with an outer surface diameter $D = 90$ mm and a wall thickness $h = 13$ mm. An artificial axial flaw on the outer shell surface was produced in the cylinder centre by an electro-discharge machine with a thin foil electrode and having a width of 0.3 mm. The flaw shape was semi-ellipse with a semi-axes: $c = 12$ mm half-length and $a = 6.6$ mm maximum penetration depth, so the reference flaw dimensions were approximately $a/c = 1/2$ and $a/h = 1/2$. An internal pressure loading was introduced in such a way as to exclude an axial tension in the tested shell.

An orthogonal coordinate grid on the outer specimen surface was painted around the crack borders. Nodal points coincided with the finite element mesh nodes of the same shell. The development of this map graticule is shown in Figure 6.4.

Double-expose reflection holograms were recorded under an internal pressure increment $q = 6.0$ MPa. Typical holographic interferograms, obtained from two symmetrical viewing directions $\vec{e}_1$ and $\vec{e}_2$ (shown in Figure 6.1(b)) corresponding to viewing angle $\psi = 30°$ are illustrated in Figure 6.5.

In order to analyse the outer surface deformations around the flaw borders of the shell, the displacement components v and w, corresponding to the

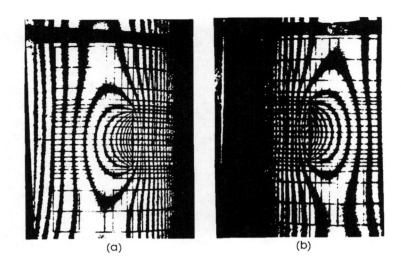

Figure 6.5. Holographic interferograms of shell with surface flaw corresponding to symmetric viewing directions $\vec{e}_1$ (a) and $\vec{e}_2$ (b)

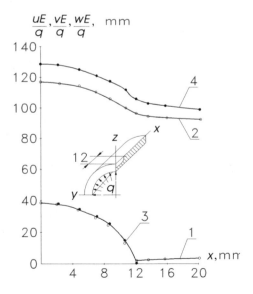

Figure 6.6. Distributions of normalized displacement components along the border of axial surface crack in a cylindrical shell

cylindrical coordinates expressed by equation (2.78), are determined in the grid nodes (see Figure 6.4). For example, normalized displacements components vE/q and wE/q ($E = 70\,000$ MPa is the elasticity modulus) distributions along the surface flaw borders are shown in Figure 6.6 by curves 1 and 2, respectively.

The numerical solution for an analogous cylindrical shell with an axial surface flaw was obtained by a software package for three-dimensional fracture mechanics analysis, which was adapted by special spring finite elements [37]. Corresponding normalized displacement components vE/q and wE/q are compared with the experimental data in Figure 6.6 (see curves 3 and 4, respectively).

An analysis of the experimental and numerical data shows that the crack borders opening displacements v accurately coincide. The normal displacement component w distributions (curves 2 and 4) differ in constant only. Apparently this is conditioned by differences between the finite element model and real specimen gross mechanics. An analogous stable correlation of the experimental and numerical displacement fields is observed along the rest of the nodal lines on the shell surface.

To calculate K_I values using the approximate expressions (6.11) and (6.12), the crack borders slopes θ should be determined. For this purpose the finite difference formula (6.9) for computing derivative $\partial w/\partial w$ in the crack border by three uniformly spaced nodes data belonging to $y0$, $y1$ and $y2$ grid sections (see Figure 6.4) was applied. The distribution of the normalized slope $\theta E/q$ along the crack border is illustrated in Figure 6.7. An alternative way to determine slopes θ may be based on the optical displacement differentiation techniques, described in Chapter 4.

A comparison of the normalized stress intensity factor analogous to (6.17)

$$\tilde{K}_I = \frac{K_I}{\dfrac{qR}{h}\sqrt{\dfrac{\pi a}{Q}}}$$

(where $Q = 1 + 1.464(a/c)^{1.65}$ is an elliptic integral) distributions along the surface crack front, obtained by experimental and numerical methods, is illustrated in Figure 6.8. Curve 1 results from equation (6.12), based on experimental dependencies for crack mouth opening displacements δ_m and crack border slopes θ. Curve 2 represents the numerical code results. The maximum experimental value of K_I at the flow central cross-section exceeds the corresponding numerical result by 12%. But from an engineering point of view, the experimental and numerical curves are reasonably correlated.

Thus, let us consider that the above-discussed experimental–analytical approach to determine SIF in shells with surface flaws is reasonably based. This technique may be effectively applied to the study of any real shell structure

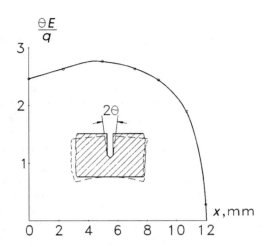

Figure 6.7. Distribution of normalized crack border slope on a shell surface

with surface flaws, located both in regular stress zones and in zones characterized by stress gradients and concentration.

6.2   Determination of displacement fields and fracture parameters for cylindrical shells with through-cracks

We shall now discuss the deformations of the thin walled cylindrical shells with through-cracks [1, 4]. Displacement fields for shell structures in the neighbour-hood of through-cracks, which can be determined by holographic interferometry with a high accuracy, are of considerable practical interest. Similar experimental data may be used for the development of various deformation fracture criteria, either as initial information for experimental–numerical methods to calculate fracture mechanics parameters or for verification of numerical codes. In addition, combined experimental–analytical approximate approaches may be developed to determine typical fracture mechanics parameters.

Experimental investigations were carried out for shell specimens analogous to those shown in Figure 2.27 with basic dimensions: outer surface radius $R = 30$ mm, cylindrical shell length $L = 100$ mm and wall thickness $h = 1.5$ mm (Figure 6.9). They were made of hardening aluminium alloy. Locations and orientation of through-cracks were chosen based on typical fracture accidents of the real shell structures.

Three specimens were prepared for the investigations:

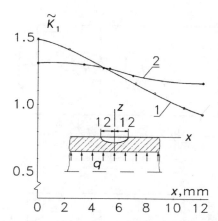

Figure 6.8. Normalized stress intensity factor distribution along the front of axial surface flaw

- Cylindrical shell no.1 with a circumferential through-crack, located in the central cross-section (Figure 6.9(a)).
- Cylindrical shell no. 2 with an inclined through-crack, $\alpha = 45°$ (Figure 6.9(b)).
- Cylindrical shell no. 3 with a circumferential through-crack located at distance $a = 2\,mm$ from the rigid shell end (Figure 6.9(c)).

Through-crack lengths are the same for all specimens and equal to $2l = 24\,mm$. The geometrical parameter

$$\lambda = \left[\sqrt{\frac{12(1 - \mu^2)}{Rh}} \right]^{1/2} \cdot l \qquad (6.18)$$

analogous to parameter β (5.91) for a cylindrical shell with a hole and characterizing a reference curvature of the shell fragment with a crack are presented in Table 6.1.

Through-cracks in shell specimens were made artificially by an electro-discharge technique using a thin-foil electrode. The slits produced had a small width almost equal to $0.3\,mm$.

The shell specimens with through-cracks were investigated under static loading by tensile force P and torsion moment M.

A three-dimensional displacements determination was carried out by the reflection hologram technique. An optimal hologram interferometer set-up with twelve viewing directions was used for quantitative hologram interpretation as discussed in Section 2.4.

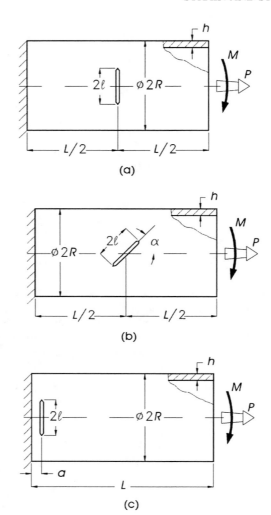

Figure 6.9. Cylindrical shell specimens with through cracks: no. 1—(a); no. 2—(b); no. 3—(c)

Table 6.1 Geometric parameters and nominal stresses for cylindrical shells with through cracks

Specimen, no.	1	2	3
λ	3.25	3.25	3.25
σ_0, MPa	3.6	5.4	6.8
τ_0, MPa	0.2	0.1	0.25

Typical interferograms of the outer surface of the shell in the neighbourhood of the through-cracks, obtained by the double-exposure technique under tension loading, are shown in Figure 6.10 for specimens 1, 2 and 3, respectively. Analogous interferograms for torsion loading are shown in Figure 6.11.

Nominal tensile stresses σ_0 (5.92) and nominal shearing stresses τ_0 (5.93), corresponding to the above-mentioned fringe patterns, are shown in Table 6.1.

A quantitative interpretation of holographic interferograms was carried out in the equidistant nodes along the upper (UB) and lower (LB) crack borders, shown in Figure 6.12. Discrete experimental distributions of three displacement components u, v, w with reference to the cylindrical coordinates (2.78) are shown by points in Figure 6.13 for tension loading and in Figure 6.14 for torsion. In both figures the dependencies designated as 1, 2 and 3 correspond to the upper through-crack border (UB in Figure 6.12), but the same ones designated as 4, 5 and 6 correspond to the lower crack border (LB in Figure 6.12).

Since the displacement component distributions along the through-crack borders are characterized by harmonic dependencies, their analytical approximation may be based on the truncated trigonometrical series (see Section 4.1). Using the even type or odd type periodical extension for the functions defined in the interval $-l < s < l$, it was proposed to describe the displacements components u, v, w in the crack borders by approximating the Fourier series of the form in reference [39], for example for displacement component u:

$$u = \sum_{k=0}^{N} a_k \cos\frac{k\pi s}{l} + b_k \sin\frac{k\pi s}{l} \qquad (6.19)$$

where l is the half-length of the through-crack. The algorithm for determination of the unknown coefficients a_k, b_k is similar to that in the technique used for the closed contour (see equation (4.11)). The computational expressions have the form

$$a_k = \frac{1}{l}\int_{-l}^{l} u\cos\frac{k\pi s}{l}\, ds$$

$$\qquad (6.20)$$

$$b_k = \frac{1}{l}\int_{-l}^{l} u\sin\frac{k\pi s}{l}\, ds$$

where u is an experimentally determined discrete displacement function. Integrals in expressions (6.20) should be computed by a numerical integration technique, for example using Simpson's rule. The curves in Figures 6.13 and 6.14 correspond to analytical approximations (6.19).

Some comments should be made subsequent to the analysis of the dependencies shown for the displacement fields in shells with through-cracks. In addition, tangent displacement components, for example axial u for a

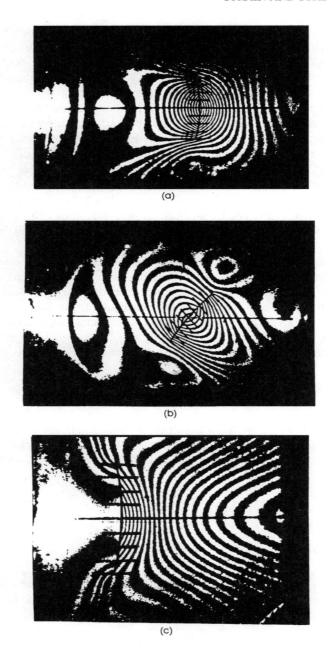

Figure 6.10. Holographic interferograms of cylindrical shells with through cracks under tension: specimen no. 1 (a), specimen no. 2 (b) and specimen no. 3 (c).

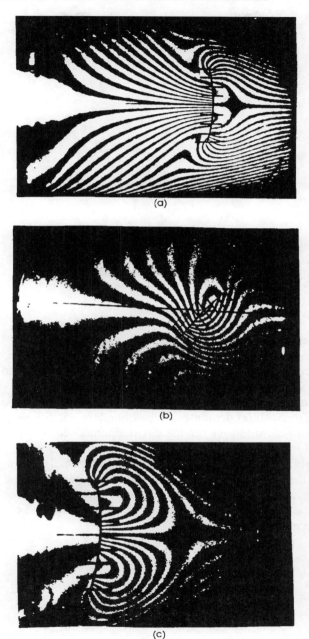

Figure 6.11. Holographic interferograms of cylindrical shells with through cracks under torsion: specimen no. 1 (a), specimen no. 2 (b) and specimen no. 3 (c)

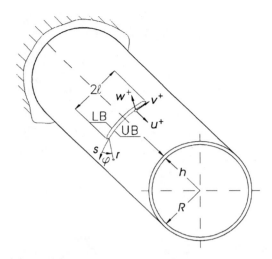

Figure 6.12. Notations and coordinate systems used for cylindrical shell with through crack

cylindrical shell under tension loading and circumferential v for torsion loading are dominating, and a significant deflection w appears in the cracks' edges. This experimentally proven fact showed the necessity to take into account the bending components of the stress–strain state for the overall mechanical analysis of shell structures with through-cracks. Also, Figure 6.13(c) and Figure 6.14(c) demonstrate the boundary conditions' influence on the displacement distributions for the circumferential crack located near the rigid shell end.

The next problem to discuss is how to use experimental information about displacement fields in a cracked shell to determine basic fracture mechanics parameters. In most cases, approaches to determine the stress intensity factor K for cracked bodies, based on experimental displacement field data, use asymptotic dependencies of the form

$$
\begin{bmatrix} u \\ v \\ w \end{bmatrix} = \frac{1}{\mu_0} \frac{\sqrt{r}}{2\pi} F \begin{bmatrix} K_I \\ K_{II} \\ K_{III} \end{bmatrix}
\tag{6.21}
$$

where $\mu_0 = E/2(1 + \mu)$ is shearing modulus; E is the elasticity modulus; and μ is Poisson's ratio. F is the matrix of the geometrical functions of the form [37]

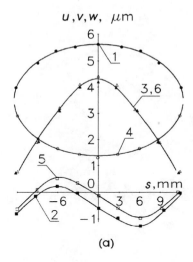

(a)

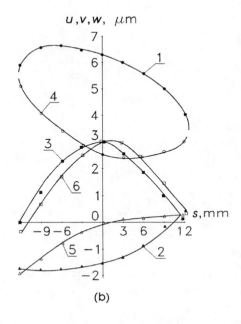

(b)

Figure 6.13. (a) and (b)

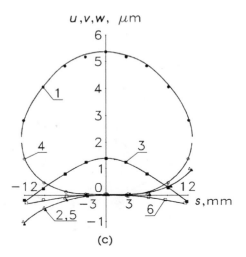

(c)

Figure 6.13. Distributions of three displacement components along the through crack borders in cylindrical shells under tension: specimen no. 1 (a), specimen no. 2 (b) and specimen no. 3 (c)

$$F = \begin{bmatrix} f_{11} & f_{12} & 0 \\ f_{21} & f_{22} & 0 \\ 0 & 0 & f_{33} \end{bmatrix}$$

with components

$$f_{11} = \cos\frac{\varphi}{2}\left(1 - 2\mu^* + \sin^2\frac{\varphi}{2}\right) \qquad f_{11} = \sin\frac{\varphi}{2}\left(2 - 2\mu^* + \cos^2\frac{\varphi}{2}\right)$$

$$f_{21} = \cos\frac{\varphi}{2}\left(2 - 2\mu^* - \cos^2\frac{\varphi}{2}\right) \qquad f_{22} = \cos\frac{\varphi}{2}\left(-1 + 2\mu^* + \sin^2\frac{\varphi}{2}\right)$$

$$f_{33} = \sin\frac{\varphi}{2},$$

where (r, φ) are the polar cordinates, centred in the crack tip (see Figure 6.12); μ^* is equal to μ for the plane strain state, but $\mu/(1+\mu)$ is for the plane stress state.

Thus, if displacement vector components u, v, w are available in any point near the crack tip, then the stress intensity factor may be expressed as:

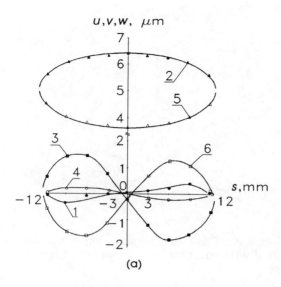

(a)

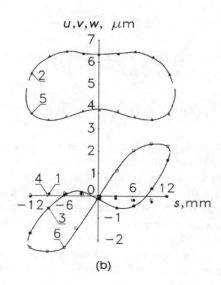

(b)

Figure 6.14. (a) and (b)

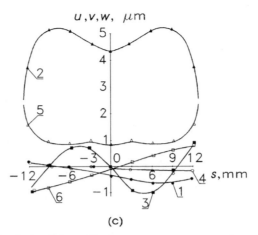

(c)

Figure 6.14. Distributions of three displacement components along the through crack borders in cylindrical shells under torsion: specimen no. 1 (a), specimen no. 2 (b) and specimen no. 3 (c)

$$K_I = \mu_0 \sqrt{\frac{2\pi}{r}} \frac{vf_{12} - uf_{22}}{f_{21}f_{12} - f_{11}f_{22}} \qquad K_{II} = \mu_0 \sqrt{\frac{2\pi}{r}} \frac{vf_{11} - uf_{21}}{f_{21}f_{12} - f_{11}f_{22}}$$

$$K_{III} = \mu_0 \sqrt{\frac{2\pi}{r}} \frac{w}{f_{33}}. \tag{6.22}$$

In particular, for the symmetrical crack border deformations, the SIF values may be calculated by the expressions:

$$K_I = \mu_0 \sqrt{\frac{2\pi}{r}} \frac{u}{f_{11}} \text{ or } K_I = \mu_0 \sqrt{\frac{2\pi}{r}} \frac{v}{f_{21}}. \tag{6.23}$$

The main disadvantage of applying the asymptotic equation (6.22) is the lower accuracy of the displacement measurements in the neighbourhood of the crack tip. Reliable experimental data may be obtained, as a rule, in the points located outside the zone where the asymptotic equations (6.22) are correct. In order to implement expressions (6.22) for experimental determination of SIF with a reasonable accuracy, special approximating functions containing non-singular terms should be derived [23].

An implementation of the displacement fields obtained by means of holographic interferometry for calculation of SIF values can be effectively carried out through the use of the approach based on an energy balance. Therefore, in a general way, if the crack has grown, the strain energy release rate G may be expressed as [37]

$$G = \frac{1-\mu}{2\mu_0} K_{\mathrm{I}}^2 + \frac{1-\mu}{2\mu_0} K_{\mathrm{II}}^2 + \frac{1}{2\mu_0} K_{\mathrm{III}}^2 = \frac{1}{2}\int_0^{l^+} \sigma_{ij} n_j^+ \frac{\partial u_i^+}{\partial l}\,\mathrm{d}s + \frac{1}{2}\int_0^{l^-} \sigma_{ij} n_j^- \frac{\partial u_i^-}{\partial l}\,\mathrm{d}s$$

(6.24)

where σ_{ij} are stress components in the crack line for the initial shell structure without a crack under designated loading conditions; $n_j^+(n_j^-)$ are normal unit vector projections on the upper (lower) crack border; $u_i^+(u_i^-)$ are displacement vector components in the upper (lower) crack border; $i, j = 1, 2, 3$.

Derivative $\partial u_i/\partial l$ is called the weight function [35]. By the weight function method, stress intensity factors for cracked shells under arbitrary loading conditions may be calculated quite simply. The relevant computational formula follows from equation (6.24):

$$K_{\mathrm{I}} = \int_0^{l^+} \sigma_{ij} n_j^+ t_i^+\,\mathrm{d}s + \int_0^{l^-} \sigma_{ij} n_j^- t_i^-\,\mathrm{d}s$$

(6.25)

where

$$t_i = \frac{H}{2 K_{\mathrm{I}}^*} \frac{\partial u_i}{\partial l} \qquad (i = 1, 2, 3)$$

is the normalized weight function; $H = E/(1 - \mu^2)$ is the generalized elasticity modulus; and K_{I}^* is the stress intensity factor for the reference loading conditions of the same shell structure with a through-crack.

Let us demonstrate how the crack driving force approach can be applied to determine stress intensity factor K_{I} by using as an example a cylindrical shell with a circumferential through-crack (specimen no.1, Figure 6.9(a)) under tension. From the generalized energy expression (6.24) for the illustrative case under consideration, a simplified formula follows:

$$K_{\mathrm{I}} = \left[\frac{\sigma_0 E}{2(1 - \mu^2)} \int_0^l \left(\frac{\partial u^+}{\partial l} - \frac{\partial u^-}{\partial l} \right)\mathrm{d}s \right]^{1/2}$$

(6.26)

where σ_0 is the nominal tension stress in the shell specimen and $\partial u^+/\partial l (\partial u^-/\partial l)$ are weight functions for axial displacement component in the upper (lower) circumferential crack border (see Figure 6.12).

The proposed approach should be based on the experimentally determined weight function. To solve this problem specimen no. 1, described above, was modified. Using an electrodischarge technique, the initial through-crack was lengthened by increment $\Delta l = 3$ mm in both directions. The finished crack dimension became $2(l + \Delta l) = 30$ mm. An experimental procedure to determine the displacement component distributions in the lengthened crack borders was repeated. In Figure 6.15 a comparison of the axial displacement component u distributions along the borders of the circumferential through-cracks of the initial (1, 2) and finished (3, 4) length is illustrated which corresponds to the

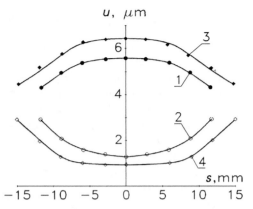

Figure 6.15. Crack opening displacements in a tensile cylindrical shell for initial and final length of circumferential crack for the upper (1, 3) and lower (2, 4) border

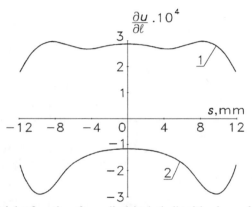

Figure 6.16. Weight function for cylindrical shell with circumferential through crack under tension for the upper (1) and lower (2) border

nominal tension stress $\sigma_0 = 3.6$ MPa. Using these experimental data, the weight function $\partial u / \partial l$ was approximately determined as shown in Figure 6.16.

After calculations using expression (6.26), where quantities $E = 70\,000$ MPa and $\sigma_0 = 3.6$ MPa were substituted, an experimental estimation of the stress intensity factors, corresponding to both through-crack tips, was obtained and $K_I^{ex} = 0.63$ MPa$\sqrt{m}$. To verify the result obtained from the reference source of numerical solutions [1], the required information was extracted:

$$K_I = \sigma_0 \sqrt{l}(F_m - F_b) \tag{6.27}$$

where F_m, F_b are normalized tabulated stress intensity factor membrane and bending components; σ_0 is the nominal tension stress in the shell specimen; and

l is the half-length of the circumferential through-crack. Using handbook data, corresponding to the curvature parameter $\lambda = 3.25$ (see equation (6.18)), and by calculations with formula (6.27), the reference value $K_I^{ref} = 0.53\, MPa\sqrt{m}$ was obtained. Thus, in spite of experimental determination of the derivative $\partial u / \partial l$ with a rough crack length increment, a reasonably accurate K_I estimation was obtained.

It seems that experimental determination of the weight functions for shells with through-cracks under reference loading conditions is very promising.

References

1. Muracami, Y. (ed.) (1987) Stress intensity factors handbook. In 2 Vol. Pergamon Press, Oxford.
2. Atluri, S. N. (ed.) (1986) *Computational Methods in the Mechanics of Fracture.* North-Holland, Amsterdam.
3. Raju, I. and Newman, I. (1979) Stress intensity factors for wide range of semielliptical surface cracks in finite-thickness plate. *Engng. Fract. Mech.* **11**, 817–829.
4. Panasyuk, V. V., Savruk, M. P. and Dacyshin, A. P. (1976) *Stress Distribution near the Cracks in Plates and Shells.* Nauk.Dumka., Kiev.
5. Chiang, F. and Wiiliams, R. C. (1985) Simultaneous generation of 3-D displacement contours of fracturing specimen using moiré. *Exp. Fract. Mech.* **22**, 731–735.
6. Chiang, F. and Harish, T. (1988) Three-dimensional crack tip deformation: an experimental study and comparison to HRR field. *Int. J. Fract.* **36**, 243–257.
7. Smith, C., Post, D., Hiatt, G. and Nicoletto, G. (1983) Displacement measurement around cracks in three-dimensional problems by hybrid experimental technique. *Exp. Mech.* **23**, 15–20.
8. Kenneth, M. and Liecht, M. (1985) Moiré of crack-opening interferometry in adhesive fracture mechanics. *Exp. Mech.* **25**, 61–67.
9. Kockelmann, H. and Kragelan, F. (1981) Application of moiré methods in experimental fracture mechanics. *Nucl. Engng Design* **67**, 181–190.
10. Smith, C., Post, D. and Epstein, J. (1981) Algorithms and restrictions in the application of optical methods to the stress intensity factor determination. *Theor. Appl. Fract. Mech.* **2**, 81–89.
11. Balas, J. and Drzik, M. (1985) Determination of stress intensity factors by optical methods. *Mechanika Teoretyczna, Stosovana* **23**, 136–142.
12. Chiang, F. and Asundi, A. (1982) Objective white-light speckles and their application to the stress intensity factor determination. *Opt. Letters* **7**, 378–379.
13. Chiang, F. and Haresh, T. (1990) Surface and interior stress intensity factor measurement by a random speckle method. *Int. J. Fracture* **43**, 185–194.
14. Kaufmann, G. and Lopergolo, A. (1983) Evaluation of finite element calculations in a cracked cylinder under internal pressure by speckle photography. *J. Appl. Mech. Trans. ASME* **50**, 896–899.
15. Ovchinnikov, A. V., Safarov, Yu. V. and Gorlinsky, R. N. (1983) Stress intensity factor determination by holographic interferometry. *Fiz. Khim. Mekh. Mater.* N2, 59–63.
16. Akimkin, S. A., Balalov, V. V., Kaydalov, V. B., Nikishkov, G. P., Pichkov, S. N. and Shchepinov, V. P. (1992) Stress intensity factors determination on models made of lower-modulus materials by experimental-analytical approach. In *Experimental Investigations of Stresses in Structures*, pp. 97–104, Nauka, Moscow.

17. Tyrin, V. P. (1990) Holographic interferometry application to determine stress intensity factor. *Zavod. Labor.* **56**, 155–158.
18. Dudderar, T. and Doerries, E. (1976) A study of effective crack length using holographic interferometry. *Exp. Mech.* **16**, 300–304.
19. Dudderar, T. and Gorman, H. (1973) The determination of mode I stress intensity factors by holographic interferometry. *Exp. Mech.* **13**, 174–180.
20. Kaufmann, G., Lopergolo, A., Idelson, S., *et al* (1987) Evaluation of finite element calculations in part-circular crack by coherent optics techniques. *Exp. Mech.* **27**, 154–157.
21. Will, P., Totzauer, W. and Michel, B. (1988) Analysis of surface cracks by holography. *Theor. Appl. Fract. Mech.* **9**, 33–38.
22. Otsemin, A. A., Deniskin, S. A., Sitnikov, L. L., *et al* (1987) Stress intensity factors determination for inclined crack by holographic interferometry. *Zavod. Labor.* **53**, 66–68.
23. Dudderar, T. D. and O'Regan, R. (1971) Measurement of strain field near a crack tip in polymethylmethacrylate by holographic interferometry. *Exp. Mech.* **11**, 49–56
24. Barker, D., Sanford, R. and Chona, P. (1985) Determining K and related stress field parameters from displacement field. *Exp. Mech.* **25**, 155–158.
25. Rice, J. R. and Levy, N. (1972) The part-through surface crack in elastic plate. *J. Appl. Mech. Trans. ASME* **39**, 185–194.
26. Delale, F. and Erdogan, F. (1982) Application of the line spring model to a cylindrical shell containing a circumferential or an axial part-through crack. *J. Appl. Mech. Trans. ASME* **49**, 97–106.
27. Delale, F. and Erdogan, F. (1981) The line spring model for surface cracks in a Reissner plate. *Int. J. Engng Science* **19**, 1331–1340.
28. Cheng, W. and Finnie, I. (1985) On the prediction of stress intensity factors for axisymmetric cracks in thinwalled cylinders from plane strain solutions. *J. Engng Mater. Technol.* **107**, 224–234.
29. Cheng, W. and Finnie, I. (1986) Determination of stress intensity factors for partial penetration axial cracks in thinwalled cylinders. *J. Engng Mater. Technol.* **108**, 83–92.
30. Parks, D. and White, C. (1982) Elastic-plastic line-spring finite elements for surface-cracked plates and shells. *J. Press. Vessel Technol.* **104**, 287–292.
31. Ezzat, H. (1985) Experimental verification of simplified line-spring model. *Int. J. Fracture* **28**, 139–151.
32. King, R. B. (1983) Elastic-plastic analysis of flaws using a simplified line-spring model. *Engng Fract. Mech.* **18**, 217–231.
33. Fett, T., Mattcheck, C. and Munz, D. (1989) On the calculations of crack opening displacement from stress intesity factor. *Engng Fract. Mech.* **34**, 883–890.
34. Fett, T., Caspers, M. and Muns, D. (1990) Determination of approximate weight functions for straight through cracks. *Int. J. Fracture* **43**, 195–211.
35. Fett, T. (1988) The crack opening displacement field of semielliptical surface cracks in tension for weight functions applications. *Int. J. Fracture* **36**, 55–59.
36. Wilkowski, G. and Eiber, R. (1978) Review of fracture mechanics approaches to defining critical size girth weld discontinuities. *WRC Bull.* N239.
37. Morozov, E. M. and Nikishkov, G. P. (1980) *Finite Element Method in Mechanics of Fracture*. Nauka, Moscow.
38. Nikishkov, G. P., Ignatyuk, N. M., Beizerman, B. R. and Azerovich, A. A. (1987) Estimation of stress intensity factor determination error for cracks modelling. *Fiz. Khim. Mech. Mater.* **23**, 118–120.
39. Anpilov, A. V., Balalov, V. V., Morozov, E. M. and Shchepinov, V. P. (To be published). Holographic interferometry and weight function method application to determine stress intensity factors in shells with through cracks. *Zavod.Labor.*

7 DETERMINATION OF MATERIAL MECHANICAL PROPERTIES

The method of holographic interferometry provides us with the possibility to develop non-conventional techniques to determine material mechanical properties because of its capability for recording the displacement fields of the deformed object surface. Information on the whole-field displacement components distributions over the surface under study leads to an increase in the reliability of data obtained, since the actual conditions of specimen deformation can be taken into account.

A high sensitivity of holographic interferometric techniques with respect to the normal to the object surface displacement component leads to a wide implementation of these techniques for specimen testing under bending. Some non-conventional approaches to the determination of elasticity modulus, Poisson's ratio and material elasticity limit based on the method of compensation holographic interferometry are presented in this chapter.

An investigation of parameters describing a material microplasticity deformation is of great scientific and applied interest. The relations between microplastic strain and net stress (so-called microplasticity diagrams) are needed for the design of various high-precision elastic elements of different devices and to ensure the dimensional instability of specific structural elements used in high-precision mechanical engineering [1–3]. It should be noted that holographic interferometric techniques are capable of measuring microplastic strain values in the range from 10^{-4} to 10^{-6}.

This chapter describes the techniqes of microplasticity diagram construction based on both one-axis and two-axes bending of the specimen to be tested. The approach to the specimen testing under non-uniform bending moment distribution allows us to reduce the body of experimental investigations considerably. The residual strain measurements needed for microplasticity diagram plotting can be carried out by using a conventional approach to both fringe pattern interpretation and compensation holographic interferometry.

The last two sections of this chapter deal with the study of material elasto-plastic behaviour in irregular zones of a structure. The parameters describing the material mechanical properties in the so-called root of the notch of the specimen under testing are of great importance for correct life-time prediction in the low-cyclic fatigue deformation range [4].

7.1 Determination of the elastic constant of a material by means of bending testing

Specimen testing to determine the material mechanical properties under bending has some advantages compared to conventional methods of material testing. Among these advantages, the following should be pointed out:

- A high sensitivity with respect to strains.
- A high immunity of the measuring parameters (deflections, curvatures) with respect to random variations of temperature.
- A capability to ensure the required type of the specimen's deformation by an appropriate choice of the geometrical form of the specimen.

It should be noted, however, that bending is a geometrically non-linear type of deformation. The linear relations between displacements, strains and stresses are valid in the small deformation range only.

It is known that conventional interferometry has been effectively implemented for determination of Poisson's ratio. According to this approach a specimen made of an isotropic material and having the form of a beam with a rectangular cross-section is subjected to a pure anticlastic bending [4–8]. Specimens in this case must to have a mirror surface or be made of a transparent material [7]. The implementation of holographic interferometry can extend this approach in testing of specimens with optically rough surfaces [9–11].

A deflection field of a beam loaded with bending moment M that is constant over the beam length is described by equation (4.56). Let us locate the origin of the Cartesian coordinate system at the centre of the specimen and assume, without loss of generality, that $B = C = 0$. The validity of the latter condition can be ensured by means of an optical compensation of the slopes. If the direction of deflection coincides with the direction of illumination $\vec{e}_S$ and observation $\vec{e}_1$ (see Figure 2.19), the interference fringes, representing the lines of the deflection equal levels, will form a set of hyperbolae (see Figure 7.1), since for isotropic materials the condition $\ae_1\ae_2 < 0$ are valid (see Section 4.3). The angle α between the asymptotes to these hyperbolae, according to the theory of beam bending, is related to the value of Poisson's ratio μ in the following way:

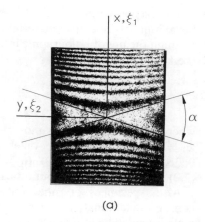

(a)

Figure 7.1. Fringe patterns corresponding to the four-point pure bending of a rectangular plate obtained for beryllium (a) and aluminum (b)

(b)

$$\mu = \frac{|\ae_2|}{|\ae_1|} = \mathrm{tg}^2\alpha2. \tag{7.1}$$

When the value of the angle α is measured on the conventional interferogram of the beam deflection (see Figure 7.1), the accuracy of Poisson's ratio μ to within 5 per cent can be achieved, if the conditions of the deforming specimen correspond to the analytical model adopted [10–11].

As follows from the analytical solution, formula (7.1) is valid when the ratio of the specimen length l to its width b is not less than five. If the length of the specimen l is limited, the value of b must be too small in order to satisfy the above-mentioned condition. In such a case, decrease of the specimen width that is a basis for a reliable measurement of the curvature $\ae_2$ leads to an increase in the error made in the determination of Poisson's ratio. Moreover, the decrease of the specimen length results in an influence of the boundary conditions on the measurement accuracy of the technique.

The elasticity modulus E is the other important parameter describing elastic behaviuor of isotropic materials. The value of E can also be determined through the use of holographic interferograms shown in Figure 7.1 by means of the known relation

$$E = \frac{M}{æ_1 I} \tag{7.2}$$

where I is the inertial moment of the beam cross-section. The value $æ_1 = \partial^2 w / \partial \xi_1^2$ should be determined by numerical differentiation of the deflection function $w(\xi_1, 0)$. In order to obtain the value of $æ_1$ from equation (7.2) the deflection distribution can be approximated with a square parabola.

Also, specimens of other forms can be implemented for material elastic constant determination under bending. A specimen having a form of a rhombic plate is loaded by four concentrated forces as shown in Figure 7.2. Such a plate is deformed under conditions of two-axial pure bending with reduced moments [12]

$$M_1 = -\frac{P}{2\mathrm{tg}(\gamma/2)},$$

$$M_2 = \frac{P\mathrm{tg}(\gamma/2)}{2}$$

where γ is the lesser angle in the rhomb apex. If the principal curvatures $æ_1$ and $æ_2$ of the deformed plate surface have been measured, the values of Poisson's ratio and elasticity modulus can be calculated in the following way:

$$\mu = \frac{M_2 æ_1 - M_1 æ_2}{M_2 æ_2 - M_1 æ_1},$$

$$E = \frac{12(1 - \mu^2)}{h^3} \frac{M_2 æ_2 - M_1 æ_1}{æ_2^2 - æ_1^2}$$

where h is the plate thickness.

Let us consider the case when the specimen to be tested is made of an anisotropic material. It is assumed that the directions of the anisotropy axes are known with the third anisotropy axis directed normal to the object surface. A determination of elastic constants in this case requires testing of two specimens, the longitudinal axes of which are oriented along the anisotropy axes. These axes will be denoted below by superscripts 0 and 90. The values of the principal curvatures being measured by means of holographic interfero-metry are connected with the so-called technical elastic constants through the following relations [13]:

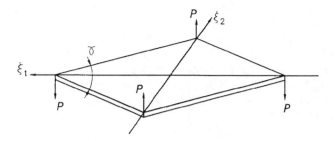

Figure 7.2. Load arrangement for the four-point pure bending of rhombic plate

$$E_1 = \left(\frac{M}{I}\right)^0 \frac{1}{\text{æ}_1^0}, \quad E_2 = \left(\frac{M}{I}\right)^{90} \cdot \frac{1}{\text{æ}_1^{90}}$$
$$\mu_{12} = -\frac{\text{æ}_2^0}{\text{æ}_1^0}, \quad \mu_{21} = -\frac{\text{æ}_2^{90}}{\text{æ}_1^{90}}. \tag{7.3}$$

Measurements of the principal curvatures of the deformed surface containing expressions (7.1)–(7.3) can be effectively carried out by means of compensation holographic interferometry (see Section 4.3) for both isotropic and anisotropic materials. Since the deformed surface of the beam made of isotropic material subjected to a pure bending is a surface of the second order and is described by relation (4.56) (where $\text{æ}_2 = -\mu\text{æ}_1$ for an anisotropic material $\text{æ}_2^0 = -\mu_{12}\text{æ}_1^0$) in which the variable parameters of additional artificial displacement field (4.50) become equation (4.51), the total displacement field w^* takes the form

$$w^* = A + C\xi_2 - \frac{\text{æ}_1(1+\mu)}{2}\xi_2^2 \tag{7.4}$$

Expression (7.4) does not contain the coordinate ξ_1. The holographic fringe pattern in this case consists of a set of uniformly spaced straight lines which are oriented along the ξ_1 axis. Therefore the appearance of such a fringe pattern demonstrates that the longitudinal deformation of the specimen surface has been compensated and the compensation condition (4.51) has been met:

$$\text{æ}_1 = -\tilde{D}_1.$$

An analogous situation is observed for the transversal deformation of the specimen surface when

$$\text{æ}_2 = -\tilde{D}_2.$$

In the latter case the fringe pattern corresponding to the total displacement field w^* consists of straight lines which are parallel to the ξ_2 axis.

The accuracy of determination of the values of the principal curvatures is mainly defined by accuracy in establishing the instants of compensation. These instants are characterized by a change in the type of interference fringe pattern from the set of hyperbolae to the set of ellipses. An absolute error $\Delta æ_i$ made in curvature determination can be estimated by recording two fringe patterns immediately before and after the compensation instants. Before and after compensation the interference fringes have to be evidently identified as the hyperbolic set $(\tilde{D}_i^-)$ and the elliptical set $(\tilde{D}_i^+)$, respectively. If such information is obtained, the principal curvatures and error made in their determination can be represented as

$$æ_i = -\frac{\tilde{D}_i^- + \tilde{D}_i^+}{2},$$

$$\Delta æ_i = \frac{|\tilde{D}_i^+ - \tilde{D}_i^-|}{2}.$$

Accuracy in determination of the principal curvature values can also be estimated in the following way. The deflection field of the surface of the beam under pure bending is given by

$$w = A + B\xi_1 + C\xi_2 + \frac{æ_1}{2}(\xi_1^2 - \mu\xi_2^2). \tag{7.5}$$

Let us assume that when the field (7.5) is superimposed with the additional field of form (4.50), the conditions of compensation of the linear terms in expression (7.5) are exactly satisfied but the compensation of the curvature $\partial^2 w/\partial \xi_1^2$ is carried out to within some error so that the interference fringes can be visually identified as a set of straight lines. In this case the total field w^* can be represented as

$$w^* = A^* + \frac{\Delta æ_1}{2}\xi_1^2 - \frac{æ_1(1+\mu)}{2}\xi_2^2 \tag{7.6}$$

where $A^* = A + \tilde{A}$ and $\Delta æ_1 = \tilde{D} + æ_1$. In order to derive expression (7.6), the obvious assumption $|\Delta æ_1| \ll æ_1(1 + \mu)$ was taken into account. An estimation for the relative error of curvature determination is obtained from the condition that the deflections of the object surface points 1 and 2 lying on the same interference fringe have to be equal (see Figure 7.3):

$$\delta æ_1 = \frac{\Delta æ_1}{æ_1} = \frac{(1+\mu)t[2(\xi_2)_{min} + t]}{L_1^2} \tag{7.7}$$

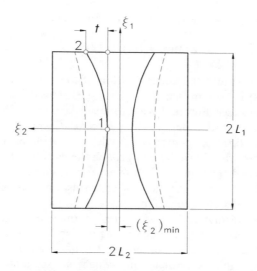

Figure 7.3. Schematic diagram of the interference fringe pattern configuration near the hyperbolic point of stationarity of deflection field

where t is the maximum deflection of the interference fringe at the distance L_1 from the ξ_1 axis.

It is evident that the maximum error will be made when the pair of symmetrical interference fringes nearest to the ξ_1 axis are used for determination of the principal curvature. In the limiting case when the above-mentioned fringes have a common point at the centre of the specimen surface $((\xi_2)_{\min} = 0)$, expression (7.7) takes the form

$$\delta\mathrm{æ}_1 = \frac{(1+\mu)t^2}{L_1^2}.$$

The expression for the error made in determination of the principal curvature $\delta\mathrm{æ}_2$ can be written by analogy:

$$\delta\mathrm{æ}_2 = \frac{(1+\mu)t^2}{L_2^2\mu}.$$

Assume now that the minimum value of fringe deflection t that can be resolved on the interferogram is equal to 1 mm, i.e. the maximum distance between two symmetrical fringes is 2 mm. Then, for typical magnitudes of geometrical parameters of the object surface area under investigation $L_1 = L_2 = 15$ mm and the principal curvature $\mathrm{æ}_1 = 5 \times 10^{-5}$ mm^{-1}, above-mentioned estimations gives for $\mu \simeq 0.3$ (see Figure 7.1(b)) $\delta\mu = \delta\mathit{æ}_1 + \delta\mathit{æ}_2 = 0.03,$

for $\mu \simeq 0.1$ (see Figure 7.1(a)) $\delta\mu = 0.06.$

It should be noted that the determination of the value of Poisson's ratio in the range 0.08–0.15 by measuring the angle between asymptotes to the interference fringes (see Figure 7.1(a)) in accordance with relation (7.1) may lead to an error of the order 30–50 per cent. Compensation holographic interferometry can be implemented for the determination of the elastic material constant in another way. In such an approach the pseudocurvature $\tilde{D}$ is kept as a constant value and the applied load is a variable parameter [14].

As an example of the application of compensation holographic interferometry to elastic constant measurement, let's consider a bending test of a laminated composite material. A specimen of thickness 4 mm and having a rectangular form was rigidly clamped along one short edge and loaded with a bending moment applied to the other short edge. Figure 7.4 illustrates three interferograms of this specimen on which subsequent instants of the compensation of the linear terms B and C of the field (4.56) (see Figure 7.4(a)), the principal curvature $\mathit{æ}_1^0$ (see Figure 7.4(b)) and $\mathit{æ}_2^0$ (see Figure 7.4(c)), are presented. The values of elastic constants obtained by means of substituting the compensation parameters (4.51) into expressions (7.3) are

$$E_1 = (5.6 \pm 0.2) \times 10^4 \quad \text{MPa},$$

$$\mu_{12} = 0.050 \pm 0.005.$$

These magnitudes are in good agreement with corresponding handbook data.

The capability of establishing the type of specimen deformation corresponding to the analytical or numerical model used for quantitative interpretation of the measurement results is one of the main advantages of holographic interferometry compared to conventional techniques of material testing. Information about the whole field displacement distribution in many cases directly reveals the incorrectness of specimen clamping, application of load, and so on. Compensation holographic interferometry is capable of fully indicating various defects of both the specimen under study and the loading device used.

The following illustration will make this clear. Interferograms presented in Figure 7.5 were obtained for the composite material similar to that described before Figure 7.4 in the same way. The inclinations of the interference fringes on these fringe patterns result from a mistake made in cutting the specimen from a sheet. In this case the axes of the specimen's anisotropy do not coincide with the direction of the bending moment and the direction perpendicular to it. Note that interferograms presented in Figure 7.5 show that compensation

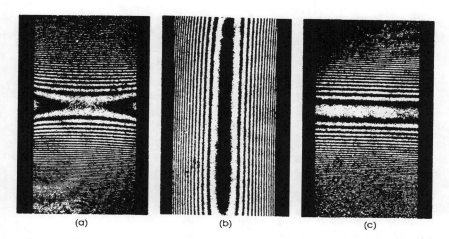

Figure 7.4. Instants of optical compensation of the slope (a) and two principal curvatures (b,c) of a composite plate under pure bending

holographic interferometry can be implemented for determination of the directions of the axes of anisotropy in laminated composite materials.

Let us now consider how the type of specimen clamping can influence the results obtained by compensation holographic interferometry. Three fringe patterns corresponding to bending of the same specimen that was used for recording the interferograms shown in Figure 7.5 are presented in Figure 7.6.

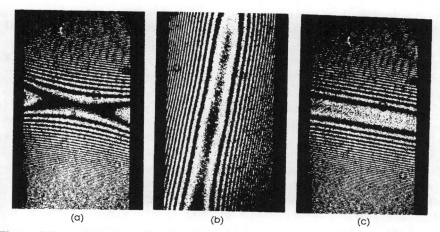

Figure 7.5. Instants of optical compensation of the slope (a) and two principal curvatures (b,c) of a composite plate under pure bending in the case of a noncoincidence of material anisotropy axes with the specimen's sides

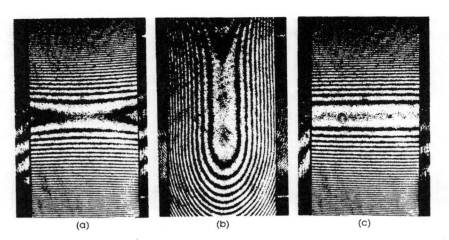

(a)   (b)   (c)

Figure 7.6. Instants of optical compensation of the slope (a) and two principal curvatures (b,c) of a composite plate under four-point pure bending

The obvious distinctions in the corresponding fringe patterns result from the different loading schemes implemented. In the last case the specimen is loaded in accordance with four-point pure bending so that its short edges cannot be subjected to torsion. Remember that the interferograms shown in Figure 7.5 correspond to the free motion of the specimen edge where the bending moment is applied. Fringe patterns presented in Figures 7.6(b) and (c) mainly consist of a set of straight lines parallel to the edges of the specimen. It should be noted, however, that the region of correct compensation is reduced compared to the interferograms presented in Figures 7.4 and 7.5 due to the influence of boundary conditions. The example clearly illustrates that the type of specimen clamping and loading exerts a major influence on the results of the determination of the elastic constants in materials.

Consider another example of application of the compensation technique to determination of the elastic constants of an anisotropic body. In references [15,16], by using some cubic crystals as example, it is shown that, depending on the crystallographic orientation of a specimen under testing, the coefficient of transversal deformation (an analogue of Poisson's ratio) can vary in a wider range compared to isotropic materials. For instance, for double-oblique crystal of nickel alloy of type MAR-M200, when the axis x_1 coincides with the direction $<011>$, the coefficient μ_{12} will have a great anisotropy. This anisotropy depends on the direction of the x_2-axis and the value of μ_{12} can vary from 0.65 to -0.06, i.e. it even becomes negative. When the specimen is cut in the direction corresponding to a negative value of the coefficient of transversal

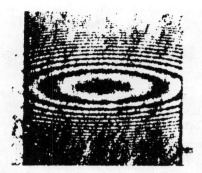

Figure 7.7. Interferogram of the four-point pure bending of a plate made of anisotropic monocrystal

deformation, all points of its surface after bending are the elliptical points $\mathfrak{x}_1 \cdot \mathfrak{x}_2 > 0$ (see Section 4.3). The interference fringes in this case represent a set of coaxial ellipses (see Figure 7.7). The value of the coefficient of transversal deformation in this case is given by

$$\mu_{12} = -\frac{a^2}{b^2} \tag{7.19}$$

where a and b are the dimensions of the axes of any ellipse in the directions x_1 and x_2, respectively. The magnitude of the μ_{12}-coefficient corresponding to the interferogram presented in Figure 7.7 is $\mu_{12} = -0.07$, that is, in good agreement with theoretical predictions. It should also be noted that the compensation technique of the elastic constant determination in this case completely coincides with the technique described above.

Determination of material elastic constants can be carried out on the basis of an analysis of resonance frequencies and vibration modes of the specimen under investigation [11,17]. The expression defining the resonance frequencies $v_{m,n}$ of the rectangular plate has the form

$$v_{m,n} = \frac{\pi}{2} \left(\frac{D}{\rho h} \left\{ \frac{A_m^4}{l^4} + \frac{A_n^4}{b^4} + \frac{2}{l^2 b^2} [\mu B_m B_n + (1 - \mu) C_m C_n] \right\}^{1/2} \right) \tag{7.8}$$

where ρ is the density of the plate material; h, l and b are the thickness, length and width of the plate, respectively; D is the cylindrical stiffness; A_i, B_i and C_i are the coefficients depending on the boundary conditions and the vibration mode of the specimen; and m and n are the integer numbers characterizing the vibration mode, which are equal to the number of nodal lines directed parallel to two plate sides having a common point. In the case of oscillation of a cantilever beam or plate $n = 0$, $A_n = B_n = C_n = 0$ and relation (7.8) can be simplified:

$$v_{m,0} = \frac{\pi}{2} \left(\frac{D}{\rho h} \frac{A_m^4}{l^4} \right)^{1/2}. \tag{7.9}$$

For bending-torsional vibration modes with a single nodal line along a length of beam $n = 1$, $A_n = B_n = 0$, $C_n = 12/\pi^2$. In this case expression (7.8) takes the following form:

$$v_{m,1} = \frac{\pi}{2} \left\{ \frac{D}{\rho h} \left[\frac{A_m^4}{l^4} + \frac{2\pi(1-m)C_m}{\pi^2 l^2 b^2} \right] \right\}^{1/2}. \tag{7.10}$$

The required relation for determination of the value of Poisson's ratio can be obtained by combining expressions (7.9) and (7.10):

$$\mu = 1 - \frac{\pi^2}{24} \frac{A_m^4}{C_m} \frac{b^2}{l^2} \left(\frac{v_{m,1}^2}{v_{m,0}^2} - 1 \right). \tag{7.11}$$

The main parameters which have to be determined for an implementation of expression (7.11) are the resonance frequencies $v_{m,1}$ and $v_{m,0}$. These frequencies can be effectively established by means of time-average holographic interferometry.

It was experimentally established that for $m > 3$ formula (7.11) ensures reliable results [17]. A spread in the values of Poisson's ratio thus obtained is not more than 5 per cent. A close coincidence was revealed between the data of the vibration mode analysis technique of Poisson's ratio determination and the corresponding results obtained by means of compensation holographic interferometry.

By way of example, holographic interferograms of an aluminium plane specimen ($133 \times 50 \times 2$ mm) corresponding to the bending vibration mode with a resonance frequency $v_{4,0} = 3062$ Hz and a bending-torsional vibration mode with resonance frequency $v_{4,1} = 4461$ Hz are shown in Figures 7.8(a) and (b) respectively. Substituting these resonance frequency values into formula (7.11) gives the value of the material Poisson's ratio $\mu = 0.33$, that is, in good agreement with handbook data.

7.2 Material microplasticity investigation by means of one-axis bending testing

The employment of holographic interferometry in testing specimens under bending provides us with the possibility to investigate material mechanical behaviour during microplastic deformation. The term 'microplastic

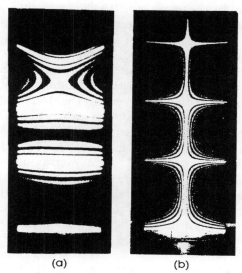

Figure 7.8 Fringe patterns corresponding to the bending (a) and bending-torsional (b) vibration resonance mode of a cantilever plate

deformation' in this section means that the value of the plastic strain ε_p is not more than 0.0001.

The output data of material testing under bending are the relations between the bending moment and the curvature of the specimen under study. Usually a transition to the standard one axis strain-stress diagram for the surface layer of specimen material should be made. Such a transition in the range of plastic deformation is complicated because of the non-linear character of dependencies between the strain and stress of the surface layer of the specimen and the applied bending moment. However, if the microplastic deformation range, where the condition $\varepsilon_p \ll \varepsilon_e$ (where ε_e is the maximum elastic strain) is valid, is of main interest, a strain calculation can be carried out using the relations of elasticity theory. As this takes place, the distribution of strains over the specimen cross-section in both the elastic and plastic deformation range are defined by Kirchhoff–Love's first hypothesis (see Section 5.1). In accordance with this assumption, the strain of the specimen surface layer is proportional to the curvature of the deformed specimen surface.

The pure bending of a plate or beam is the most wide spread type of material testing under bending. Today, the following designs of specimen loading have gained acceptance:

- A four-point bending [18].
- A cantilever bending by a concentrated moment [19].
- A quasi-pure bending of beam of equal bending strength [20].

Measurement of specimen sag over the fixed segment is the most frequently used to obtain the strain–stress relation by means of the conventional method of bending testing. The transition from a sag to a strain on the surface layer is connected with modelling the conditions of a specimen's clamping. Such modelling may lead to a considerable error in the strain calculation since the possible plastic strain of the specimen near its clamping is not taken into account. Techniques of material testing under bending based on the whole-field deflection measurement by holographic interferometry are free from similar disadvantages.

Holographic interferometry in material testing under bending can be applied for investigation of plastic strain accumulation. In order to describe this process quantitatively, a residual strain versus net stress diagram is constructed. The residual strains are obtained as a result of a step-by-step loading of the specimen and its subsequent unloading. Thus a process of a residual strain accumulation can be quantitatively described. Note that in this section, residual strain is defined as the strain maintained by the specimen after its loading and unloading.

The residual curvature of the deformed specimen's surface is derived from a double-exposure holographic interferogram where the first exposure was made for the initial unloaded state of the object surface and the second exposure after specimen loading and unloading [21]. To calculate the values of the residual curvature $æ_{1r}$, the discrete experimental distribution of deflection w is approximated with the square parabola of the following form:

$$w_r(\xi_1, 0) = A + B\xi_1 + \frac{æ_{1r}\xi_1^2}{2} \tag{7.12}$$

where the ξ_1 axis coincides with the longitudinal axis of the specimen having the form of a beam. When small strains resulting in the appearance of less than three fringes on the interferogram, or large strains resulting in the appearance of an unresolved number of fringes, are measured, holographic interferometry with finite width (carrier) fringes is used. An implementation of this technique allows us to choose a suitable number of fringes on the interferogram at each loading step by varying the linear terms of expression (7.12) [22].

The sequence of the experimental procedure is illustrated by means of the optical arrangement for reflection hologram recording shown in Figure 7.9. A cantilever beam having a segment of equal bending strength characterized by a linear change in the width is used. The beam is loaded with a concentrated force as shown in Figure 7.9. The stresses on the specimen face in this case are given by

$$\sigma_s = \frac{3P}{h^2 \mathrm{tg}\dfrac{\beta}{2}} \tag{7.13}$$

where the β-angle is shown in Figure 7.9, and h is the beam thickness.

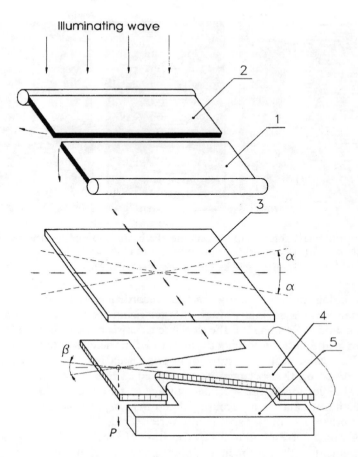

Figure 7.9. Schematic diagram of the set up for residual strain determination by means of reflection hologram recording

When the initial loading steps are investigated, the first exposure is made by illuminating the whole surface of the specimen. After specimen loading and unloading, the illumination direction of half of the specimen surface 4 and the unstrained and unmoved plate (the so-called 'witness' 5) mounted near the specimen in the same plane, is blocked with a mask 1, and the second exposure is made. Then mask 1 is removed, and mask 2 blocks the illumination direction of the other half of the specimen surface. The photoplate is rotated in the direction opposite to that of the specimen deflection and the third exposure is made.

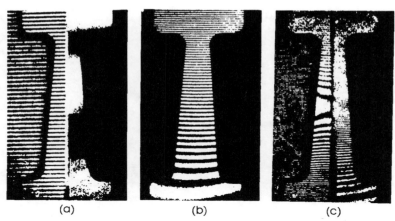

(a)   (b)   (c)

Figure 7.10. Interferograms of the bending of a beam obtained by means of introducing an additional field of initial (a), developed (c) residual deflection and without such an introduction at an intermediate deformation stage (b)

The holographic interferogram thus recorded contains on one half of the specimen surface a fringe pattern characterizing the residual strain caused by loading and unloading of the specimen between exposures. Such a fringe pattern is absent if the specimen was stressed in the elastic range. Starting from some load value, a small number of interference fringes (perhaps only one) can be observed on the specimen part involved. These fringes directly indicate a residual deflection of the specimen surface (see Figure 7.10(a)), but cannot be used as input data for an accurate determination of residual strains.

The other part of the specimen's surface image is covered with numerous fringes which correspond to a superposition of specimen deflection and mutual displacement of the specimen and photoplate. A quantitative interpretation of this superimposed fringe pattern ensures a reliable determination of the residual strain values. The uniformly spaced straight fringes on the surface of the witness 5 (see Figures 7.10(a) and (c)) correspond to the rotation of the photoplate between exposures.

After reaching a certain load level, the fringe spacing on the interferogram characterizing the residual deflection of the beam becomes rather high for a direct quantitative interpretation and accurate strain calculation. Starting from this instant, masks 1 and 2 are not used and the photoplate is not rotated between exposures. The whole specimen surface is covered with interference fringes corresponding to the residual deflection of the beam (see Figure 7.10(b)).

A further load leads to forming of unresolvable fringe patterns caused by a high level of residual deflection. In order to avoid this situation, the photoplate in this case is rotated between exposures in the direction of the specimen

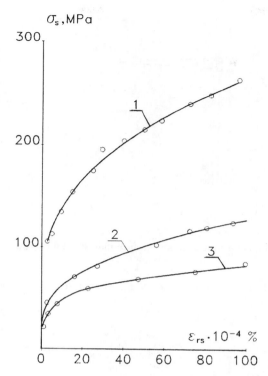

Figure 7.11. Residual strain versus net stress diagram obtained for an aluminum alloy (1), bronze (2) and M1 copper (3)

deflection. The consequence of inserting masks 1 and 2 remains the same as for recording of the interferogram shown in Figure 7.10(a). The fringe pattern corresponding to this case is presented in Figure 7.10(c). The point of stationarity corresponding to the total deflection field is clearly seen at the centre of the specimen. In this point the gradients of the actual and the artificial displacement field are mutually compensated.

The above-described technique is capable of measuring residual strains in the range of 10^{-4} per cent to 5×10^{-5} per cent. The total residual strain value ε_{rs} on the object surface accumulated after m loading and unloading steps can be determined by summing

$$\varepsilon_{rs} = \left| \frac{h}{2} \sum_{i=1}^{m} æ_{ri} \right|$$

where $æ_{ri}$ is the residual curvature increment after the ith loading step. It should be noted that the error of determination of ε_{rs} is conditioned by the

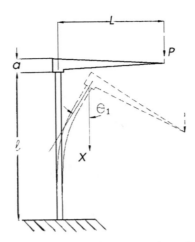

Figure 7.12. Loading arrangement for a cantilever specimen bending testing

errors made in each loading step. Therefore the number of loading steps should be as small as possible.

As an example of the application of the above-described technique, Figure 7.11 depicts the residual strain versus net stress diagrams of some materials. Material testing was carried out using the beam shown in Figure 7.9 of thickness $h = 4$ mm and length of the working segment $l = 50$ mm. The value of angle β was 9.5° (see Figure 7.9).

Bending a cantilever beam having the form of a thin rectangular plate of a constant cross-section loaded with a concentrated moment applied at the beam free edge is another widely used approach to material testing. In the simplest case such a moment can be stimulated by applying a concentrated force P acting parallel to the specimen's longitudinal axis on arm L as shown in Figure 7.12. The stress component resulting from the compressive effect of the force P can be neglected if the length of the beam L exceeds its thickness h by more than 20 times.

For small load values resulting in small beam deflections the stress σ_s on the specimen surface is given by

$$\sigma_s = \frac{6PL}{bh^2} \tag{7.14}$$

where b is the specimen width. With an increase in the load, the correct implementation of equation (7.14) is dramatically decreased. In such a case, the value of the angle of arm rotation and current curvature of the specimen surface have to be taken into account. The differential equation describing the deflection of the beam in this case has the form

$$\frac{d^2w}{dx^2} + \alpha^2 w = -\alpha^2 R \cos(\theta_1 - \varphi) \tag{7.15}$$

where $\alpha = 12P/Ebh^3$; $\mathrm{tg}\varphi = a/L$; $R = \sqrt{L^2 + a^2}$; a is defined in Figure 7.12. A solution of equation (7.15) can be represented in the following form:

$$w = R\cos(\theta_1 - \varphi)[\cos(\alpha x) + \mathrm{tg}(\alpha l) \cdot \sin(\alpha x) - 1]. \tag{7.16}$$

The value θ_1 in equation (7.16) is, in turn, a solution of the equation

$$\mathrm{tg}\theta_1 = \alpha R \cos(\theta_1 - \varphi)\mathrm{tg}(\alpha l).$$

The expression defining the value of the bending moment takes the form

$$M = -PR\cos(\theta_1 - \varphi) - Pw.$$

Let us now estimate the degree of non-uniformity of the stress distribution $\Delta\sigma$ along the beam length in the following way:

$$\Delta\sigma = \frac{\sigma(l) - \sigma(0)}{\sigma(l) + \sigma(0)} = \frac{M(l) - M(0)}{M(l) + M(0)} = \frac{1 - \cos(\alpha l)}{1 + \cos(\alpha l)}.$$

If the value of $\Delta\sigma$ is less that 0.05, we can consider that the specimen is subjected to a pure bending with the bending moment

$$\sigma_s = \frac{6M(0.5l)}{bh^2} = \frac{6PR\cos(\theta_1 - \varphi)\cos(0.5\alpha l)}{bh^2 \cos(\alpha l)}. \tag{7.17}$$

Measurement of the residual strains of a rectangular specimen can be performed by using the optical arrangement shown in Figure 7.9. Typical interferograms obtained as a result of testing invar alloy are presented in Figure 7.13 for initial (a) and developed (b) microplastic strain. The objects investigated were specimens of thickness $h = 3$ mm, width $b = 7$ mm and length $l = 50$ mm. The parameters of the load application shown in Figure 7.12 were: $a = 18$ mm, $L = 85$ mm. Figure 7.14 shows the results of invar alloy testing in the microplasticity range. Note that the stress values are derived from expression (7.17). The dependencies presented in Figure 7.14 plotted by using logarithmic coordinates show that the experimental results are well approximated with straight lines. This means that the relation between residual strains and stress for one-axial loading of the specimens investigated can be represented as a power function

$$\varepsilon_{rs} = A\sigma_s^n$$

where A and n are material constants.

The dependencies presented in Figure 7.14 can be reconstructed in the form of the microplastic strain ε_{ps} versus net stress σ_s diagram. To perform this, the

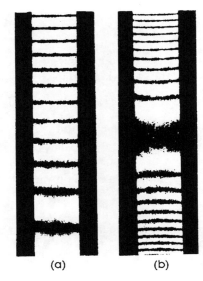

(a)   (b)

Figure 7.13. Interferograms of the bending of a beam made of invar alloy obtained without (a) and by means of (b) optical compensation of the central specimen cross-section slope

approach developed in reference [19] can be applied. The principle of this approach is as follows. After elastic unloading of the specimen strained outside the elastic range, the residual stresses σ_r which are brought about in the specimen material can be expressed as

$$\sigma_r = (\varepsilon_p - \varepsilon_r)E \tag{7.18}$$

where ε_p is the plastic strain; ε_r is the residual strain; and E is the material elasticity modulus.

Substituting equation (7.18) into an equilibrium equation

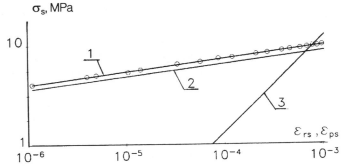

Figure 7.14. Residual (1) and microplastic (2) strain versus net stress diagram of the invar alloy. Line 3 indicates the microplastic deformation range

$$\int_{-0.5h}^{0.5h} \sigma_r z b \, dz = 0$$

yields

$$\int_0^{0.5h} \varepsilon_p z \, dz = \int_0^{0.5h} \varepsilon_r z \, dz \tag{7.19}$$

where the z-axis coincides with the normal to the object surface. Assume now that $\varepsilon_p = \varphi(\varepsilon)$ (where $\varepsilon = \varepsilon_e + \varepsilon_p$ is the total strain of the specimen resulting from loading). In accordance with the first Kirchhoff–Love hypothesis, the distribution of the total and residual strains in the normal to surface direction through the specimen thickness can be represented as

$$\varepsilon = \frac{2\varepsilon_s z}{h}, \quad \varepsilon_r = \frac{2\varepsilon_{rs} z}{h}. \tag{7.20}$$

By using expression (7.20), equation (7.19) can be transformed into the following form:

$$\int_0^{\varepsilon_s} \varphi(t) \, dt = \frac{\varepsilon_{rs} \varepsilon_s^2}{3}. \tag{7.21}$$

By differentiating equation (7.21) with respect to the variable ε_s we can obtain

$$\varepsilon_{ps} = \frac{1}{3} \frac{d\varepsilon_{rs}}{d\varepsilon_s} \varepsilon_s + \frac{2}{3} \varepsilon_{rs}. \tag{7.22}$$

Now we shall confine ourselves to consideration of the deformation range where the condition $\varepsilon_{es} \gg \varepsilon_{ps}$ is valid. The changing of the variable ε_s in equation (7.22) on the variable $\varepsilon_{es} = \sigma_s/E$ gives

$$\varepsilon_{ps} = \frac{1}{3} \frac{d\varepsilon_{rs}}{d\sigma_s} \sigma_s + \frac{2}{3} \varepsilon_{rs}. \tag{7.23}$$

Taking into account the linear character of the typical strain-stress diagram presented in Figure 7.14

$$lg\varepsilon_{rs} = lgA + nlg\sigma_s$$

expression (7.23) can be transformed into the following form:

$$\varepsilon_{ps} = \frac{A(n+2)}{3} \sigma_s^n = A_1 \sigma_s^n. \tag{7.24}$$

Therefore if the dependence between the residual strains and applied stresses for one-axial pure bending has the form

$$\varepsilon_{rs} = A\sigma_s^n$$

the microplastic strain versus net stress (or microplasticity) diagram plotted in the logarithmic coordinates can be obtained by means of a parallel rearrangement of the residual strain versus net stress diagram plotted in the same coordinate system by the value

$$lg\left(\frac{n+2}{3}\right).$$

The result of the transformation of the residual strain versus the net stress diagram (straight line 1) into a microplasticity diagram (straight line 2) is shown in Figure 7.14. The straight line 3 in the same figure depicts the dependence $lg(\bar{\varepsilon}_{es}) = lg(\sigma_s/10E)$ which reveals the region where the condition $\varepsilon_{es} \gg \varepsilon_{ps}$ is valid.

In conclusion to our discussion on testing invar alloy, one interesting effect should be pointed out. Interferograms of specimen bending are capable of revealing areas of material inhomogeneity on the specimen surface. These can be easily identified as areas where the interference fringes show an inflection (see Figure 7.15).

In the techniques of material testing in this section a constancy of bending moment over a part of the specimen being tested is ensured by proper specimen design and application of load. However, holographic interferometry is capable of a strain versus stress diagram plotting through the use of data obtained in a non-uniform stress field along specimen length. Let us consider a specimen having the form of beam of the length l loaded so that a distribution of bending moment along its longitudinal axis x can be represented by a smoothed known function $M(x)$. Under such a condition a distribution of the principal residual curvature of the specimen's deformed surface $\text{æ}_{1r} = \text{æ}_{1r}(x)$ represents a smoothed function. A correlation of these two dependencies allows us, after eliminating the variable x, to construct a dependence describing the deformation of the specimen in the coordinates 'residual curvature' versus 'bending moment'. Below we will call such a dependence a 'deformation diagram'. The main characteristic of such an approach is that the required deformation diagram can be obtained as a result of hologram recording corresponding to two or three loading steps. In contrast to the case of testing a single specimen loaded with a constant bending moment, this can be considered as simultaneous testing of the set of specimens of length dx each of which is loaded with a different bending moment. Note, however, that such an approach demands a constancy of material properties over the whole

Figure 7.15. Visualization of material inhomogeneity by means of bending testing

working part of the specimen under testing. Possible influence of the gradient of the stress field in the longitudinal direction $\partial\sigma_s/\partial x$ is also not taken into account.

Loading of the specimen under testing with a bending moment varying along the longitudinal axis of a beam can be performed as shown in Figure 7.16. In this case the load is transferred to the specimen through the arm subjected to the concentrated force P at the point with coordinate $x = a$. The distribution of the bending moment is represented by the linear function whose values in the edge points of specimen are equal to

$$M_1 = -Pa \quad \text{and} \quad M_2 = P(l - a).$$

Therefore the function $æ_{1r}(x)$ coincides exactly with the deformation diagram of the specimen material with the bending moment axis normalized by the factor

$$\frac{|M_2 - M_1|}{l}.$$

Different variants of the loading configuration can be implemented by varying the value and sign of parameter a (see Figure 7.16). The simplest variant is bending without an arm when $a = l$. In this case the moment distribution along the x-axis takes the form $0 \leqslant |M(x)| \leqslant Pl$. If after loading and unloading of the specimen the interferogram reveals an absence of residual deformation, the testing procedure should be repeated by using a greater value of the bending moment. Increasing the bending moment can be achieved by either increasing the value of force P or moving the point of the load application, keeping the force value as a constant when $a > l$.

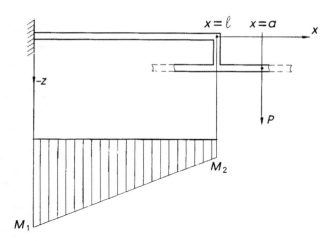

Figure 7.16. Schematic diagram of a specimen loading with a variable bending moment

When $a \geqslant l$ the value of the maximum deflection w corresponds to the least strained part of the specimen. If a relatively high level of plastic strain is reached in the specimen cross-sections near the clamping, a large fragment of the specimen is rotated as a rigid body. This hinders high-quality recording of the fringe pattern. Therefore, the load configuration where $a = 0$ is of great interest. Such an application of the load means that the maximum strain values will be reached in the specimens cross-sections of maximum deflection.

It should also be noted that increasing the load leads to the displacement at the point of load application in the direction of the x-axis due to arm rotation. This results in a false zero of the bending moment distribution. Therefore, the loading configuration where $a = l/2$ is of specific interest. Under such loading the specimen can be considered as though it consists of two segments which are equal in length and deformed in the same way (see Figure 7.17). The rotation angles of both specimen edges are equal to zero and, hence, the point of load application is not displaced with respect to the x-axis. In this case the function $æ_{1r}(x)$ has a symmetrical form with respect to the central cross-section of the specimen and, hence, the deformation diagram can be represented as

$$M(æ) = \frac{PD(æ)}{2} \tag{7.25}$$

where $D(æ)$ is the distance between cross-sections where the residual curvature function takes equal absolute values $æ$ which is used as tolerance placed on residual curvature. With this type of deformation diagram, the data obtained from the two above-mentioned segments of the specimen are averaged.

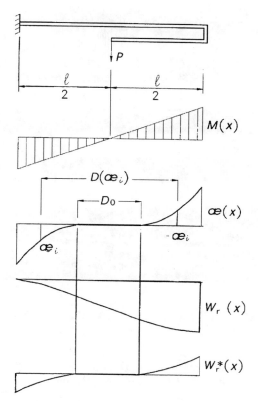

Figure 7.17. Symmetric loading arrangement for residual curvature determination in a nonuniform bending moment field

Figure 7.18(a) shows the interferogram of residual deflection of an aluminium specimen of length $L = 120$ mm, width $b = 10$ mm and thickness $h = 4$ mm. The concentrated force value applied according to the above-described symmetrical configuration is $P = 40$ N. After real-time optical compensation of the rotation angle $\theta_x(l/2)$ of the central cross-section of the specimen by means of superimposing the artificial displacement field $\tilde{w}_1 = -\theta_x(l/2)x$, the interferogram of the total field takes the form in Figure 7.18(b). The broad, bright fringe clearly seen in the central part of the specimen illustrates an elastic character of deformation of this specimen's segment. The two dark fringes surrounding the central bright fringe roughly indicate the demarcation lines between specimen fragments which underwent elastic and plastic deformation. The refinement of these lines can be achieved by superimposing the artificial deflection field of form $\tilde{w}_2 = \tilde{C}y$ that corresponds to a fictitious rotation of the specimen around its longitudinal axis.

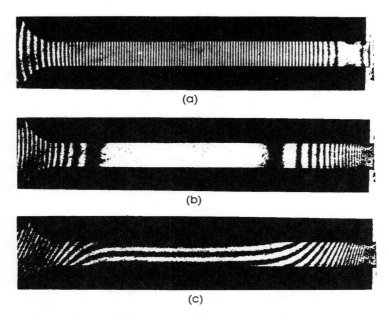

Figure 7.18. Interferograms of residual deflection of an aluminum specimen obtained
with infinite (a) and finite (b,c) width fringes

In accordance with equations (4.31) and (4.32) the inclination of the
interference fringes with respect to the x-axis is given by

$$\text{tg}\alpha = -\frac{(\theta_x^*)_2}{(\theta_y^*)_2} \tag{7.26}$$

where $(\theta_x^*)_2 = \partial(w_r^*)_2/\partial x$ and $(\theta_y^*)_2 = \partial(w_r^*)_2/\partial y$ are the partial derivatives of
the total displacement field recorded on the interferogram in Figure 7.18(c)
with respect to the Cartesian coordinates. Formula (7.26) demonstrates that
inside the segments of the specimen which are not plastically strained (where
the condition $(\theta_x^*) = 0$ is valid), the interference fringes must have straight lines.
The justification for this conclusion clearly follows from Figure 7.18(c). The
conditional lines between the interference fringes parallel to the greater
specimen sides and the inclined fringes in Figure 7.18(c) are the demarcation
lines between the segments of the specimen that have undergone elastic and
plastic deformation. The values of the bending moment in the cross-sections
coinciding with these lines can be used for precise determination of the value of
the so-called material elasticity limit σ_e. In accordance with equation (7.25) this
value can be expressed as

Figure 7.19. Interferogram of an aluminum specimen obtained with finite width fringes in the elastic deformation range

$$\sigma_e = \frac{3PD_0}{bh^2} \tag{7.27}$$

where b and h are the width and thickness of the specimen and D_0 is the distance between the two above-mentioned cross-sections. The value of the elasticity limit corresponding to the interferogram shown in Figure 7.18(c) is $\sigma_e = (36 \pm 2)\,\text{MPa}$.

In order to confirm once again the validity of the technique for visualization of plastically strained areas of the specimen surface, the double-exposure interferogram recorded according to the above-described compensation procedure in the elastic deformation range with concentrated force value $P = 1$ N is shown in Figure 7.19. In this case the whole surface of the specimen is elastically deformed between exposures and therefore the regions of the straight lines parallel to the longitudinal axis of the specimen are absent on the corresponding interference fringe pattern.

Figure 7.20(a) shows an interferogram of the same aluminium specimen obtained after its loading with force $P = 70$ N, unloading, and optical

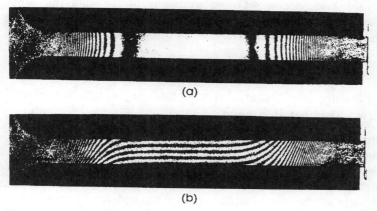

(a)

(b)

Figure 7.20. Visualization of the microplastic deformation zone on the surface of an aluminum specimen

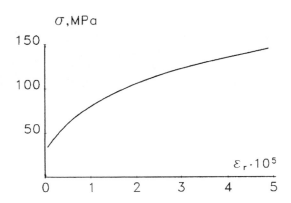

Figure 7.21. Residual strain versus net stress diagram for aluminum alloy

compensation of the rotation of the central segment of the specimen $\theta_x(l/2)$. The interference fringe pattern resulting from the superposition of the displacement field shown in Figure 7.20(a) and the artificial field $\tilde{w}_2 = \tilde{C}y$ are presented in Figure 7.20(b). Comparison of the interferograms shown in Figures 7.18(c) and 7.20(b) shows that in the latter case the area of segments of the specimens surface where plastic deformation occurred were enhanced due to a load increase. The value of the precise elasticity limit corresponding to the interferogram shown in Figure 7.20(b) is $\sigma_e = 40 \pm 3$ MPa, that is, in good agreement with the previous result.

In order to construct the deformation diagram the interferogram in Figure 7.20(a) was used. The second-order numerical differentiation of the deflection distribution (see Section 4.1) and averaging of the data thus obtained in accordance with equation (7.25) give us the dependence residual strains versus net stress shown in Figure 7.21.

7.3 Plotting of a material microplasticity diagram under two-axes bending conditions

Experimental investigations of the influence of the type of stress on the development of plastic and, in particular, microplastic strains are of great scientific interest. The efficiency of these investigations depends on the availability of an accurate and reliable technique capable of determining the components of the strain tensor. Below we will try to show that such a technique can be established upon the implementation of holographic interferometry.

An investigation of the process of microplasticity caused by the influence of the two-axial stress field can be effectively carried out by using a rhombic specimen, the load configuration of which is shown in Figure 7.2. For a square specimen when the value of the angle subtended by two neighbouring sides γ is equal to 90° the following relationship among the principal stress values is valid: $\sigma_1 = -\sigma_3$. This means that the square specimen is deformed under conditions of pure shearing. To an elastic approximation, the values of principal stress on the plate face are given by

$$\sigma_{1s} = -\sigma_{3s} = \frac{3P}{h^2} \qquad \sigma_{2s} = 0 \tag{7.28}$$

where P is the applied force value and h is the plate thickness.

The load configuration shown in Figure 7.2 is realized in the following way. Three corners of the specimen are simply supported and a concentrated force is applied at the fourth corner of the plate. For this reason, four shoulders with holes are made at the corners of the plate as shown in Figure 7.22. The shoulders at the centre of each side serve to mount various set-ups needed for hologram recording.

As in the case of microplasticity investigations under conditions of one-axial stress (see Section 7.2), the most effective approach to determination of microplastic strains consists of a step-by-step recording of the reflection

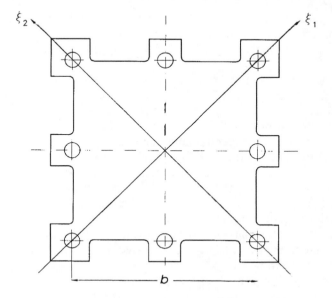

Figure 7.22. Drawing of a specimen for bending testing in two-axes stress field

Photoplate holder

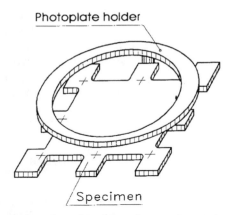

/ Specimen

Figure 7.23. Arrangement for photoplate mounting on the specimen

double-exposure holograms containing information on increments of residual strains when a load is increased. However, in this case the specimen is not rigidly fixed with respect to the elements of the optical arrangement and is capable of shifting as a rigid body after loading and unloading. Moreover, the local volumes of the specimen material can be crushed near the supports as a result of contact interaction when the load value reaches a relatively high level. This process also leads to undesired displacement of the specimen surface and recording medium. Therefore, in order to avoid the influence of all rigid body displacements on the resultant interference fringe pattern, the low mass plate holder is mounted on the specimen surface immediately as shown in Figure 7.23.

The deflection field of the deformed specimen surface represents the second-order surface of the form (4.56). The values of the principal curvatures of the deformed surface of the specimen are determined by approximation of the discrete deflection distributions along the diagonals of the square plate by square parabolae and sequential numerical differentiation.

In order to establish the viability of the set-up consisting of the specimen under investigation, the photoplate mounted on the specimen surface, and the unit of load application combined into a single aggregate, the set of multiple-exposure holograms were recorded by starting at different loading levels. A typical fringe pattern obtained for exposure number $N_0 = 5$ and load increment $P = 2\,\text{N}$, is shown in Figure 7.24. The presence of narrow bright fringes on the interferogram proves the correctness of the hologram recording procedure (see Section 2.1).

A typical interferogram characterizing an increment of residual strains of a specimen made of copper (the length of the side of the specimen $b = 50$ mm, and thickness $h = 3$ mm) is shown in Figure 7.25. The quantitative interpretation of

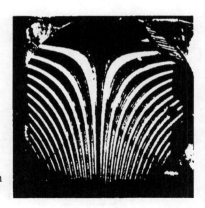

Figure 7.24. Typical multi-exposure interferogram of the specimen under four-point pure bending

the whole set of interferograms obtained for this specimen at different loading levels shows that to within an experimental error the values of the principal curvatures are equal to each other, $\ae_{1r} = \ae_{2r}$. This means that either the material has isotropic properties with respect to the initial plastic deformation or the axes of possible deformation anisotropy of the cold-rolled copper sheet are inclined with respect to the axes ξ_1 and ξ_2 (see Figure 7.22) by 45°.

The technique described above was implemented to establish the influence of the two-axial stress state on the development of microplastic strains in a specimen made of copper. Three sets of specimens of two types were manufactured and tested. The first type was a thin square plate shown in Figure 7.22. The second was prepared in the form of a beam having a segment of equal bending strength (see Figure 7.9). Before testing, the specimens were annealed in a vacuum for one hour. The annealing temperature was 300°C, 500°C and 800°C for the first, second and third set of specimens, respectively.

The results of the specimen tests allow us to construct the deformation diagram for each set of specimens in the form of dependence intensity of microplastic strains versus net stress intensity. It is assumed that the following relationships among the components of the plastic strain tensor are valid:

Figure 7.25. Typical interferogram of the residual deflection of a copper specimen under four-point pure bending

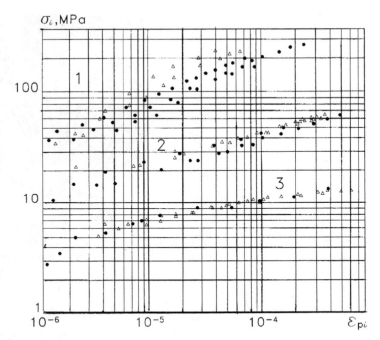

Figure 7.26. Microplastic strain intensity versus net stress intensity diagram of copper after annealing at temperatures 300C (1), 500C (2) and 800C (3) (dark and light dots correspond to one-axis and two-axes bending, respectively)

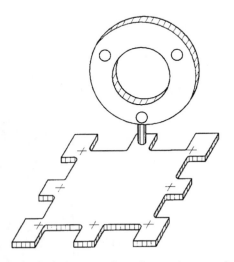

Figure 7.27. Kinematic device mounted on the specimen surface

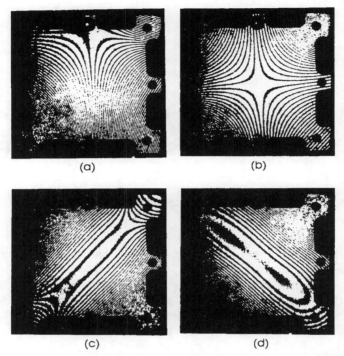

Figure 7.28. Typical interferograms of the residual deflection of plate under the four-point pure bending obtained by means of real-time technique (a) and corresponding to the different compensation instants (b,c,d)

$\varepsilon_{pr} = \varepsilon_{p3} = -0.5\varepsilon_{p1}$ in the case of one-axis stress and
$\varepsilon_{p1} = -\varepsilon_{p3}, \varepsilon_{p2} = 0$ in the case of two-axis stress.

The dependencies obtained are shown in Figure 7.26 by the set of dots. The coincidence of the values of the microplastic strains is shown only for the third set of specimens which underwent high-temperature annealing.

A numerical differentiation procedure needed to derive the values of the principal curvatures from double-exposure interferograms can be avoided if an optical compensation technique (see Section 4.3) is implemented for the same purpose [24]. The problem of eliminating possible rigid body displacements of the specimen with respect to the recording medium when the interferometer system shown in Figure 4.38 is used for optical compensation can be solved in the following way. For this purpose the lower unit of the precision kinematic device is mounted on one shoulder made in the centre of the specimen side (see Figure 7.27). This procedure allows us to spatially separate processes of the

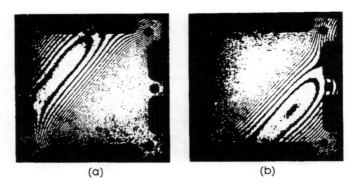

Figure 7.29. Vizualization of uniformly strained area above (a) and below (b) of one diagonal of the specimen under four-point bending

specimen loading and hologram recording. Note that residual strains only can be determined in this case.

The first step of the experimental procedure consists of recording the initial state of the object surface on a real-time hologram. Then the specimen is removed from the interferometer optical system and mounted in the loading device. After loading to the required level and then unloading, the specimen is accurately repositioned in the optical arrangement by using a kinematic device. A typical interferogram obtained by means of a real-time technique and characterizing the residual deflection of the specimen surface is shown in Figure 7.28(a). The form of this fringe pattern is analogous to that in Figure 7.25. Indeed, the plane containing the conditional point where a lower unit of the kinematic device is clamped on the specimen shoulder and tangential to the deformed specimen surface is the basic plane of interferogram recording as before.

The procedure for determining the values of the principal curvatures $æ_{1r}$ and $æ_{2r}$ consists of two steps of optical compensation. The first step involves the transversal displacement of the illuminating point source S (see Figure 4.38) while the point of stationarity of the total displacement field w^* will be observed at the centre of the specimen. A typical interferogram clearly illustrating this situation is given in Figure 7.28(b). The second step of optical compensation consists of varying the curvature of the illuminating wavefront by means of displacing the point source S along the optical axis of the interferometer (see Figure 4.38). Two sets of straight interference fringes oriented along each diagonal of the plate must result from the latter procedure. Interferograms corresponding to two compensation instants of the principal curvature along two diagonals of the plate are shown in Figures 7.28(c) and (d). The result of the quantitative determination of the principal curvatures values according to expression (4.57) are in good agreement with those obtained by the conventional approach containing the second-order numerical differentiation.

The holographic compensation technique allows us to visualize the uniformly strained region of the specimen surface simultaneously with determination of the principal curvature values. Interferograms shown in Figures 7.28(c) and (d) indicate that such a zone is placed inside the specimen surface at some distance from the points of support and load application. By varying the linear term of the artificial displacement field of form (4.50) the uniformity of the residual strain field can be inspected at any area of the specimen surface. Interferograms shown in Figure 7.29 clearly illustrate this.

7.4   Plotting of the static local strain diagram

The static local strain diagram represents the dependence between net stresses acting in the regular cross-sections of a structural element or specimen and the maximum local strains in the stress concentration region to be investigated obtained at the zero half cycle. Below we will also call this dependence the 'static local strain versus net stress diagram'. Similar dependencies constructed for different stress concentrators and different ranges of net stress value variations are the necessary connecting link for correct fatigue life-time prediction [4].

The static local strain versus net stress diagram obtained after testing a plane specimen with a central circular open hole under tension in the local strain range inherent in the low-cyclic-fatigue range is widely employed, for instance, in the analysis of fatigue behaviour and fatigue strength of various pin or rivet joints. Let us consider how this diagram can be obtained through the use of holographic interferometry.

The object investigated is a specimen made of aluminum alloy with a width $b = 60$ mm, length of 260 mm, and thickness $h = 6$ mm, at the centre of which a circular open hole with diameter $2r_0 = 12$ mm is located (see Figure 2.24). Note, as follows from the experimental results presented in Section 2.5, that for the relation between the specimen thickness and hole radius ($h = r_0$) the strain state of the hole edge vicinity is described with all three displacement vector components.

The basic principles of the experimental procedure based on reflection hologram recording by means of an overlay interferometer, and the approach to quantitative interpretation of the interference fringe patterns in order to, first, obtain the displacement component fields and, second, the circumferential strain distribution along the hole edge are described in Sections 2.5 and 4.1 in detail.

The local deformation diagram being constructed in the net stress range from $\sigma_0 = 0$ to $\sigma_0 = 220$ MPa. The tensile force causing the net stress occurrence in the regular cross-sections of the specimen is applied far from the hole in order to avoid the influence of the conditions of the specimen's clamping on the local deformation of the hole vicinity. The whole net stress range is divided into seven loading steps to ensure optimal fringe spacing near the hole at each interferogram to be recorded. The final points of each loading step are 40, 80, 115, 145, 175, 200 and 220 MPa.

A typical interferogram of the specimen surface obtained in the elasto-plastic deformation range for net stress increment from $\sigma_0 = 145$ MPa to $\sigma_0 = 175$ MPa is shown in Figure 7.30. Typical distributions of the in-plane displacement components u (1, 3) and v (2, 4) along the hole edge are presented in Figure 7.31(a). Curves 1, 2 and 3, 4 are plotted in the elastic and elasto-plastic deformation range, respectively, for the same net stress increment $\sigma_0 = 30$ MPa.

The extreme right point of the net stress range for displacement distributions obtained in the elasto-plastic deformation range $\sigma_0 = 175$ MPa corresponds to the value of the maximum tensile strain on the hole boundary $\varepsilon_x^A = 0.7 \times 10^{-2}$. The same loading range was used for recording the interferogram shown in Figure 7.30.

The three-dimensional character of the hole edge deformation is shown in Figure 7.31(b) with the distribution of the relative normal to the object surface displacement component $(w-w_\infty)$ (see expression (2.77)) obtained at different loading steps. Curve 1 corresponds to the elastic net stress increment $\sigma_0 = 30$ MPa, curves 2 and 3 are obtained for net stress increment from 145 to 175 MPa and from 200 to 220 MPa, respectively. As mentioned in Section 2.5, the dependence 5 in Figure 7.31(b) constructed in the elastic deformation range completely coincides with the results in both the analytical and numerical solution of the corresponding three-dimensional elasticity problem. The character of curves 6 and 7 obtained in the elasto-plastic deformation range clearly shows that the maximum decrease of the specimen thickness takes place at the point of the maximum tensile strain on the hole boundary. The dependence of this decrease on the net stress level is described by the essentially non-linear function, especially when the applied load is increased.

In order to plot the static local strain versus net stress diagram which is of main interest, the distributions of the circumferential strain ε_φ (5.87) along the hole edge have to be constructed for each loading step. The experimental distribution of the relative circumferential strain (stress) $\varepsilon_\varphi / \varepsilon_0$ (where $\varepsilon_0 = \sigma_0 / E$, E is the elasticity modulus of the material) along the hole boundary in the elastic deformation range constructed for $E = 71\,000$ MPa is shown by curve 1 in Figure 7.32. As was pointed out in Section 4.1, this distribution completely coincides

Figure 7.30. Typical interferogram of a plane specimen with hole under tension obtained in the elasto-plastic deformation range

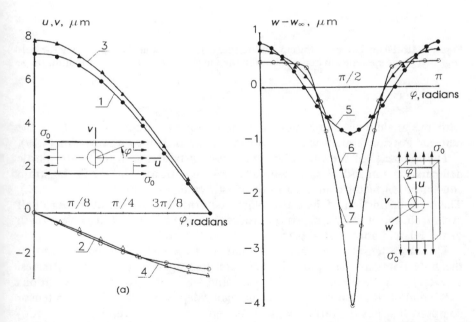

Figure 7.31. Typical distributions of the Cartesian in-plane (a) and normal (b) to the object surface displacement components along the circular hole edge in a plane specimen under tension

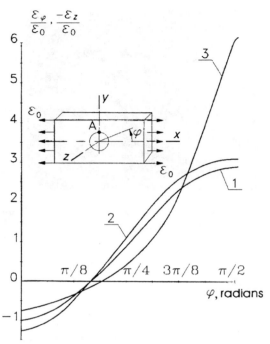

Figure 7.32. Distributions of the circumferential (1) and normal (2) strain along the hole edge in a plane specimen under tension obtained in the elasto-plastic deformation range

with the analogous distribution of relative normal strain $-\varepsilon_z/\mu\varepsilon_0$ (μ is the material Poisson's ratio) where ε_z is defined in accordance with equation (4.19).

The character of the circumferential strain distribution in the elasto-plastic deformation range does not essentially change with an increase in the strain concentration factor compared to the elastic case (see curve 2 in Figure 7.32). The main distinction of the elasto-plastic deformation process at the zero half cycle from the elastic deformation process consists of the considerable growth of the relative normal strains (see curve 3 in Figure 7.32).

The strain tensor components at point A of the maximum tensile strain at the hole boundary are plotted in Figure 7.33 against applied net stresses. Curves 1 and 2 are related to strains ε_x and $-\varepsilon_z$ and derived from equations (5.87) and (4.19), respectively. The analogous dependence for the strain tensor component ε_y has the form $\varepsilon_y = -\mu\varepsilon_x$ (where μ is the material Poisson's ratio) for the whole net stress range investigated and is not shown in Figure 7.33. The dashed line in the same figure indicates the conditional continuation of the linear segment of the ε_x-curve.

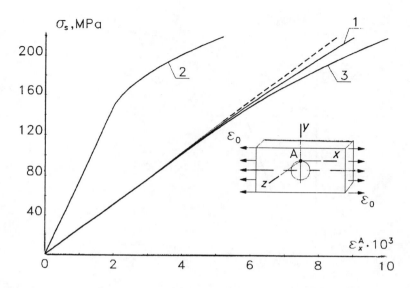

Figure 7.33. Local strain versus net stress diagram constructed at the root of the notch for maximum strain (1), normal strain (2) and strain intensity (3)

These dependencies denote that the transition of the specimen material near the hole into the stage of elasto-plastic deformation has an non-uniform character. Indeed, it can be clearly observed in Figure 7.33 that the linear relation $\varepsilon_z = -\mu(\varepsilon)\varepsilon_x$ is not valid when the value of the relative normal strain exceeds the magnitude $\varepsilon_z^0 = -2 \times 10^{-3}$. This phenomenon can be taken into account by calculating the strain intensity value [25]

$$\varepsilon_i = \frac{\sqrt{2}}{2(1+\mu)} \sqrt{(\varepsilon_x - \varepsilon_y)^2 + (\varepsilon_y - \varepsilon_z)^2 + (\varepsilon_z - \varepsilon_x)^2} \qquad (7.29)$$

where μ is Poisson's ratio of the material. When expression (7.29) is derived, it is taken into account that the strains ε_x, ε_y and ε_z at point A on the boundary of a load-free hole are the principal strains. The relation between the strain intensity ε_i and net stress level is depicted by curve 3 in Figure 7.33.

The required local strain diagram at point A on the hole edge can be constructed through the use of a standard one-axial strain versus stress diagram obtained for the same material and curves 1 or 3 in Figure 7.33. The corresponding dependence is shown in Figure 7.34. This graph allows us to compare the experimental result obtained with the data of known relations

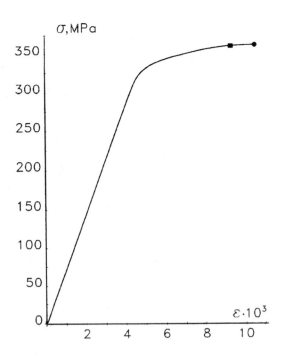

Figure 7.34. Local strain versus stress diagram obtained for the aluminum plane specimen with central circular open hole

for approximate calculation of local elasto-plastic strains at the zero half cycle. Neuber's formula gives the following value of the maximum strain on the hole edge for maximum net stress level inspected: $(\sigma_0^{max} = 220 \text{ MPa})$ $\varepsilon_x^{max} = 1.64 \times 10^{-2}$. Note that the maximum experimentally obtained magnitudes $\varepsilon_x^{max} = 0.93 \times 10^{-2}$ and $\varepsilon_i^{max} = 1.04 \times 10^{-2}$ (see Figure 7.34) are in good agreement with the Molsky–Glinka prediction $\varepsilon_x^{max} = 1.1 \times 10^{-2}$ [26]. The question concerning the influence of the specimen thickness or, in other words, the three-dimensionality of the surface strain state under plane stress conditions on the material transition into elasto-plastic deformation range will be discussed in Section 8.1, where other quantitative parameters of this process will also be presented.

7.5 Local plastic properties of the material surface layer in a high strain gradient region

As has been established in Section 7.4, the main characteristic of the process of material transition into the elasto-plastic deformation range in a high strain gradient region is the significant non-linearity of the function describing the relative normal strain distribution when the load is increased (see Figures 7.31(b), 7.32 and 7.33). Remember that the values of the ε_z-strain are determined by means of expression (4.19) is based on the assumption that the distribution of ε_z-strain in the normal to the surface direction through the specimen thickness is a constant function. The authenticity of this is confirmed by the numerical solution [27] and the results of a special holographic investigation [28] to within 5 per cent if the absolute value of the normal strain $|\varepsilon_z|$ does not exceed 0.4×10^{-2}.

The investigated specimen was made of rolled aluminium by cutting in such a way that the specimen face was parallel to the sheet face. In this case the value of the material yield limit in the normal to the surface direction through the specimen thickness may be less, according to handbook data, than the yield limit in the rolling direction coinciding with the direction of the specimen tension to within 3 or 5 per cent. Therefore, we have reason to believe that the behaviour of the ε_z-strain observed results from an anomalous decrease of the yield limit of the thin material surface layer [29]. It must be noted that such a decrease at point A and its vicinity occurs in the normal to the specimen surface direction only. This direction by data of the finite element method is characterized by the minimum value of the strain gradient compared to the strain gradient values in in-plane directions [27].

The above-described phenomenon is observed for two types of aluminium alloys investigated. Local strain versus net stress diagrams at point A on the hole edge characterizing the behaviour of the ε_z-strain at the zero half cycle of loading for two types of aluminium alloy are presented in Figure 7.33 by curve 2 (the first type) and in Figure 8.10(a) (the second type).

For the directions x (the axis of the specimen tension) and y in which the strains ε_x and ε_y near point A are characterized by essential values of gradient, the anomalous decrease in the yield limit is not observed. In fact, analysis of the static local strain versus net stress diagrams for the ε_x-strain presented in Figure 7.33 (the first type) and Figure 8.8(a) (the second type) gives the following conclusion. Taking into account the corresponding values of stress concentration factors on the specimen surface (see Table 8.1), it is established that the linear deformation process in the x-direction occurs when the value of the acting stresses belong to the linear segments of the corresponding material standard one-axial strain versus stress diagrams.

In consequence, we can safely assume that the parameters describing the initial plastic deformation of the material surface layer in the tension direction coincide with those for the internal layers of the specimen.

The observed anisotropy of the initial plastic properties of the material surface layer provides further experimental confirmation. The information on deformation kinetics contained in the above-mentioned local strain diagrams allows us to establish a quantitative link between the magnitudes of elasto-

Table 7.1 Relations between local strain values at the root of the notch

%	$\varepsilon_x^A 10^3$	$-\varepsilon_z^A 10^3$	$-e_{z/ex}$
0.02	0.6	0.23	0.38
0.05	0.74	0.28	0.38
0.01	0.92	0.35	0.38
0.15	1.04	0.41	0.39

plastic strains ε_x^A and $-\varepsilon_z^A$ which correspond to the equal values of tolerance placed on the residual strain values. Some of these values are listed in Table 7.1.

The data contained in the table show that at the initial stage of the elasto-plastic deformation the current values of strain ε_x^A and $-\varepsilon_z^A$ corresponding to the equal tolerance placed on the residual strain values R are related by the following dependence:

$$-(\varepsilon_z^A)_R = (0.38 \div 0.39)x(\varepsilon_x^z)_R \qquad (7.30)$$

Estimation (7.30) can be interpreted as the relationship between yield limits corresponding to different values of tolerance on the residual strain values in the directions z and x, i.e. in accordance with our assumption between the yield limit value of the surface layer of material and the analogous parameter for internal layers of material.

The data in reference [29] obtained for two types of steel and titanium alloy establish that the value of the yield limit of the material surface layer varies from 37 to 41 per cent of the analogous parameter value for the internal layers of the material. Therefore the quantitative data describing three-dimensional displacement distributions in a stress concentration region obtained by means of reflection hologram interferometry shows that in the high strain gradient zone an anomalous decrease in the material yield limit value on the object surface occurs in the direction characterized by a minimum strain gradient value. In the case when the plane stress conditions are valid, such a direction is the direction normal to the object surface.

Note, in conclusion, that anomalous elasto-plastic deformation in the normal to the surface direction is always observed on the specimen face at some vicinity of point A. The characteristic dimension of this vicinity at the zero half cycle of loading has an order of hole radius (see Figure 7.30).

References

1. Khenkin, M. L. and Lokshin, I. Kh. (1974) *Dimensional Stability of Metals and Alloys in Precise Machine and Apparatus Production.* Mashinostroenie, Moscow.

2. Kolesov, V. S. and Gicheva, S. D. (1980) About optical surface quality of laser mirrors. *Fiz. Khim. Metal Machin.* N6, 15–20.

3. Kolesov, V. S., Kostin, V. M., Lanin, A. G. *et al.* (1982) Metals surface layers deformation under beam heating. *Poverkhnost (physika, khimia, mechanika)* N12, 124–130.

4. Collins, J. A. (1981) *Failure of Materials in Mechanical Design. Analysis, Predicton, Prevention.* John Wiley, New York.

5. Timoshenko, S. P. and Goodier, J. N. (1970) *Theory of Elasticity.* McGraw-Hill, New York.

6. Cornu, M. A. (1969) Méthode optique pour l'étude de la surface extérieure des solides-elastiques. *Comp. Rend. Publ.* **69**, 333–341.

7. Sinha, N. K. (1977) Laser-target technique for determination of elastic constants of glass over a wide temperature range. *J. Mater. Sci.* **12**, 557–562.

8. Foster, C. G. H. (1976) Accurate measurement of Poisson's ratio in small samples. *Exp. Mech.* **16**, 311–315.

9. Yamaguchi, I. and Saito, H. (1969) Application of holographic interferometry to the measurement of Poisson's ratio. *Jap. J. Appl. Phys.* **8**, 768–771.

10. Jones R. and Bijl, D. A. (1974) Holographic interferometric study of the end effects associated with the four-point bending technique for measuring Poisson's ratio. *J. Phys. E.: Sci. Instrum.* **7**, 357–358.

11. Indisov, V. O., Osintsev, A. V., Shchepinov, V. P. and Yakovlev, V. V. (1984) Elastic constants determination by holographic interferometry. In *Strength Investigation of Materials and Structures of Nuclear Facilities*, pp. 51–55, Energoatomizdat, Moscow.

12. Marchant, M. J. and Snell, M. B. (1982) Determination of the flexural stiffness of thin plates from small deflection measurements using optical holography. *J. Strain Anal.* **17**, 53–61.

13. Lekhnitsky, S. G. (1977) *Elasticity Theory of Anisotropic Solid.* Nauka, Moscow.

14. Odintsev, I. N., Shchepinov, V. P. and Yakovlev, V. V. (1988) Material elastic constants measurement by holographic compensation technique. *Zhur. Tech. Fiz.* **58**, 108–113.

15. Krivko, A. I., Epishin, A. I., Svetlov, I. L. and Samoilov, A. I. (1988) Elastic properties of nickel alloy monocrystals. *Probl. Prochn.* N2, 68-75.

16. Svetlov, I. L., Epishin, A. I., Krivko, A. I., Samoilov, A. I., Odintsev, I. N. and Andreev, A. P. (1988) Poisson's ratio anisotropy for nickel alloy monocrystals. *Dokl. AN SSSR* **302**, 1372–1375.

17. Osintsev, A. V., Sivokhin, A. V., Shchepinov, V. P. and Yakovlev, V. V. (1985) Poisson's ratio determination by resonance technique using holographic interferometry. In *Strength of Materials and Structural Elements of Nuclear Reactors*, pp. 79–82, Energoizdat, Moscow.

18. Stepanov, V. V., Lanin A. G. and Kostin, V. M. (1980) A method to investigate small plastic deformation of materials under bending. *Zavod. Labor* **46**, 1143–1145.

19. Rakhshtadt, A. G., Zakharov, E. K. and Leshkovtsev, V. G. (1970) High-sensitivity method for determinaton of microplasticity deformation resistance characteristics for alloys under pure bending. *Zavod. Labor.* **36**, 980–983.

20. Ilyinsky, I. I., Shevelya, V. V. and Kruglik, A. P.(1983) A method to determine microyield parameters for sheet materials. *Probl. Prochn.* N2, 105–109.

21. Kostin, V. M., Odintsev, I. N., Stepanov, V. V., Shchepinov, V. P. and Yakovlev, V. V. (1988) Microplasticity diagrams construction under bending by using hologram interferometry. *Probl. Prochn.* N4, 111–114.

22. Odintsev, I. N., Shchepinov, V. P. and Yakovlev, V. V. (1987) Study of initial

plastic deformation under bending by holographic interferometry data. In *Analysis and Testing Materials and Structures Elements of Nuclear Equipment*, pp. 43–48, Energoatomizdat, Moscow.

23. Novikov, L. I., Odintsev, I. N., Shchepinov, V. P. and Yakovlev, V. V. (1985) Study of initial residual deformations of sheet materials under two-axes stress state using hologram interferometry. In *Strength of Materials and Structures Elements of Nuclear Reactors*, pp. 75–79, Energoatomizdat, Moscow.

24. Odintsev, I. N., Shchepinov, V. P. and Yakovlev, V. V. (1991) Materials testing under two-axes bending by compensation hologram interferometry. *Zavod. Labor.* **57**, 61–64.

25. Bezukhov, N. I. (1968) *The Principles of Elasticity, Plasticity and Creep Theory.* Vyssh. Shkola, Moscow.

26. Molsky, K. and Glinka, G. (1981) A method of elastic-plastic stress and strain calculation at a 'Notch root'. *Mater. Sci. Engng* **50**, 93–110.

27. Grishin, V. I. and Donchenko, V. Yu. (1983) Investigation of a stress concentration in thin and thick plates with a circular hole in the elastoplastic deformation range. *Proc. TsAGI* **14**, 85–90.

28. Begeev, T. K. and Gorodnichenko, V. I. (1989) Application of the holographic interferometry method for investigation of deformed state of internal cylindrical surface. *Zavod. Labor.* **55**, 93–95.

29. Prokopenko, A. V. and Torgov, V. N. (1986) Surface layer properties and fatigue limit of a metal.1: Relations between material yield limit and layer depth. *Probl. Prochn.* N4, 28–34.

8 INVESTIGATION OF ELASTO-PLASTIC LOCAL STRAIN HISTORY IN THE LOW-CYCLIC-FATIGUE RANGE

The process of gradual material damage accumulation conditioned by a variable stress influence leading to a change in the mechanical properties of the material (crack appearance and propagation to object fracture) is usually called fatigue. The variety of this process (known as low-cyclic-fatigue) is the main reason for crack occurrence and the rapid growth of such cracks thus resulting in the fracture of various structures [1–4].

Low-cyclic fatigue can be considered as a process characterized by a considerable level of local elasto-plastic strains arising at irregular zones of the structure. This process results in a decrease in the structure's life-time.

Joints of various design are widely used in the manufacture of machines, apparatus and structures. In many cases these joints are the primary sources of weakness in a structure, both from a static strength point of view and from a low-cyclic fatigue standpoint. Pin or rivet joints of thin sheets, especially in aircraft engineering, are the most characteristic examples of structures subjected to the influence of the low-cyclic-fatigue process. This can be explained by the fact that joints are usually the most loaded parts of a structure. It is well known that about 85 per cent of aircraft structure failures occur at joint elements [5].

Strain or stress concentration is the main reason for the occurrence of local elasto-plastic strains. This leads to a decrease in a structure's life-time in accordance with low-cyclic-fatigue laws. For example, a typical geometric stress concentrator for a pin or rivet joint is a circular through hole either open or filled with a cylindrical inclusion.

In a separate joint element the strain or stress concentration factor depends on both geometrical and force parameters. The geometrical parameters are:

thickness of sheets to be joined, design of joint, diameter of separate hole, and hole mesh configuration.

The parameters which should be considered as the force parameters are: the level of net stress acting in regular cross-sections of the structure, and the character of contact stress distribution between a hole edge and pin filling a hole. The latter parameter, which may be called a contact interaction character, depends on the fit type, the value of the force acting along the pin axis, materials of contact pair, and the contact surfaces roughness. Below, we shall consider the reasons which encourage this interest in the study of the local elasto-plastic deformation process of separate joint elements under cyclic loading by means of holographic interferometry.

Several approaches to the design of different types of joints (especially pin or rivet joints) based on static strength criteria have been developed with a high degree of reliability. The results of a local stress calculation under static loading coincide with experimental data in both the elastic and the elasto-plastic range [1,5]. However, these approaches cannot ensure the complete and detailed analysis of the reliability of the joint nor provide an accurate life-time prediction.

An optimal design of pin or rivet joints based on life-time criteria requires at least an analysis of the evolution of elasto-plastic deformation near the separate joint element under cyclic in-service loading, and must take into account the character of contact interaction [3,4].

A higher level of reliability of fatigue strength estimation and life-time prediction can be achieved by means of a quantitative description of the influence of the following factors:

- Determination of a change in stiffness of separate joint elements due to the process of local elasto-plastic deformation which results in a force redistribution.
- Estimation of local material adaptability parameters resulting in a change of the local material mechanical properties due to cyclic loading, and other factors which are related to features of the joints' manufacturing.

The above factors show that the problem of reliable life-time prediction of a pin or rivet joint and its optimal design from a fatigue strength standpoint cannot be solved without using data obtained by different experimental methods of deformable body mechanics. Moreover, we can affirm that experimental data describing the fields of local elasto-plastic strains in contact interaction regions and the quantitative description of the evolution of these strain fields during cyclic loading of different spectrum and amplitude are necessary conditions to create both numerical and analytical methods for the optimal design of the joint and its life-time prediction.

The need for various experimental investigations of local elasto-plastic strain fields near separate elements of pin or rivet joints to ensure a detailed description of the low-cyclic-fatigue process is based, for instance, on references [1–4]. Some examples of applications of different experimental techniques to the joint strength analysis are presented in references [6] and [7]. Some characteristic results of numerical and experimental studies, which clearly illustrate the set of main problems connected with analysis of the contact interaction influence on the fatigue life of joint elements, can be found in references [4–18].

An analysis of the above publications and many others revealed that there is an deficiency of experimental data on local elasto-plastic strain fields at the vicinity of separate joint elements subjected to both static and cyclic loading. The analysis also showed an absence of quantitative data describing processes of damage accumulation, fatigue crack appearance and propagation, with particular emphasis on contact interaction. The influence of different factors on the fatigue life of joint elements has been estimated through standard fatigue tests, often long-term and very expensive and usually providing only a small amount of information.

It should be noted that the well-known traditional methods of experimental mechanics, such as photoelasticity, strain gauges, moiré and others, can be used for partial solving of the above-mentioned problems only because of their particular features (use of modelling materials, presence of reference grating on the surface of interest, discrete character of measurement procedure, etc.).

We shall clarify the above statement in relation to moiré techniques. Various versions of moiré methods using laser or white-light illumination have been implemented for investigation of local strain fields under static loading [6, 7, 19–24]. The main disadvantage of these approaches applied to elasto-plastic local strain, especially under cyclic loading, is the presence of a reference grating attached to the object surface serving as an optical signal converter.

This grating is usually attached to the surface of the object to be investigated by an intermediate adhesive medium. The combination of grating and adhesive usually causes a rapid loss in the diffused reflection properties of the surface (needed to form the contrast fringe pattern) due to the influence of plastic strains. This loss occurs especially under cyclic loading in the low-cyclic-fatigue range. The application of moiré methods to the contact problem solution may be accompanied by further difficulties even under static loading. In many such cases only elastic problems can be solved [26].

As was shown in Chapters 1 and 2, the holographic interferometry method is free of such limitations because each fringe pattern is formed by means of superposition of light waves diffracted from the object surface with real microrelief. An optical signal converter, like a reference grating, whose properties can be changed during the entire experimental procedure, is not needed. This provides a unique possibility to investigate the elasto-plastic deformation

Table 8.1 The geometrical parameters of specimens and loading programs

Specimen No.	Thickness, h, (mm)	Hole radius r_0 (mm)	Testing Conditions	Type and Range of Loading	Stress Concentration factor on specimen surface, $k\sigma$
I	6	6	open hole	The cyclic loading in the range of net stresses $-120 \leq \sigma_0 \leq 230$ MPa	2.9
II	3	6	open hole	The single static tension in the range of net stresses σ_0 $0 \leq \sigma_0 \leq 244$ MPa	3.1
III	3	9	open hole	The single static tension in the range of net stresses σ_0 $0 \leq \sigma_0 \leq 232$ MPa	3.3
IV	6	6	hole filled with steel inclusion with interference $\beta^* = 0.75 \times 10^{-2}$	The cyclic loading in the range of stresses σ_0 $-120 \leq \sigma_0 \leq 230$ MPa	—
V	6	6	hole filled with steel inclusion with interference fit $\beta = 1.2 \times 10^{-2}$	The cyclic loading in the range of net stresses σ_0 $-120 \leq \sigma_0 \leq 230$ MPa	—
VI	6	6	hole filled with steel inclusion with push fit (no clearance or interference, $\beta = 0$)	The cyclic loading in the range of net stresses σ_0 $-120 \leq \sigma_0 \leq 230$ MPa	—

processes in irregular areas of structures under repeated and cyclic loading, taking into account the strain history and character of contact interaction.

A special holographic interferometry technique, described in Chapter 2, has been developed to obtain input data needed for determination of local elasto-plastic strains. This technique enables accurate measurements of 3-D or 2-D displacement fields near geometric stress/strain concentrators of any shape, line of contact interaction, and cracks on the body surface by means of reflection hologram interferometry.

Six plane specimens ($260 \times 60\,mm$) with different thicknesses made from a hardening aluminium alloy were used as objects for investigation. All specimens had a $2r_0$ diameter central hole. An analogous diagram of the specimens and coordinate system used is shown in Figure 2.24.

Three of the specimens had one open hole each. The holes in the last three specimens were filled with steel cylindrical inclusions (pins) with different types of fit. The specifications of the specimens, their geometrical parameters, elastic stress concentration factors for specimens with an open hole, uniform proportional interference for the last three specimens and testing programs used are listed in Table 8.1. Note that specimens IV, V and VI represent models of a separate element of pin or rivet joints.

All specimens were cut from a single 10 mm thick aluminium plate symmetrically in relation to its middle plane. The tension–compression direction coincided with the rolling one. A closed-loop servohydraulic computer-aided testing machine was used to perform the loading program automatically and accurately.

The experimental procedure for recording reflection holograms (both double-exposure and, in most cases, obtained by means of combined double-exposure and time-average technique to visualize the zero-motion fringe) in an opposite-beam arrangement and fringe patterns reconstruction completely conforms to the procedure described in Section 2.5.

A detailed description of the technique used for interpretation of quantitative fringe patterns to obtain the displacement component fields and then the corresponding strain and stress distributions is presented in Sections 2.3–2.5, 4.1.

Only one aspect of the experimental technique, which is of great importance for local elasto-plastic strain studies by means of holographic interferometry, should be considered here in more detail. This problem is connected with the need to divide the net stress range to be investigated into suitable parts to ensure recording of high-quality fringe patterns.

Step-by-step loading of the specimens and simultaneous recording of holographic interferograms must be performed to obtain the fringe pattern sets which describe the elasto-plastic deformation process in strain/stress concentration zones. This results from the high sensitivity of holographic interferometry techniques. For the reflection hologram approach involved, the sensitivity limit (the value of displacement component resulting in an appearance of one dark fringe) can be derived from relations (2.32) and (2.46). For out-of-plane component $d_1 \equiv w$ one can use the observation angle value $\psi = 0$, which leads to the evident result

$$w^L \geqslant \frac{\lambda}{4} = 0.16 \, \mu m. \tag{8.1}$$

The corresponding relation for in-plane components $d_2 \equiv u$ and $d_3 \equiv v$ has the following form:

$$u^L = v^L \geqslant \frac{\lambda}{4 \sin \psi}. \tag{8.2}$$

For a real practical value of observation angle $\psi = 50°$ the inequality in equation (8.2) gives the following result:

$$u^L = v^L \geqslant 0.2 \, \mu m.$$

The main advantage of the overlay interferometer with air clearance between the holographic emulsion and the object surface, which is shown in Figure 2.22, is its ability to retain the diffuse reflection properties of the investigated surface during the experimental procedure. It is a very important property because local elasto-plastic strain investigation usually requires the recording of about 10 fringe patterns under static loading and more than 100 fringe patterns under cyclic loading. These numbers result from estimations (8.2) and practical experience.

8.1 Influence of geometrical parameters on local surface elasto-plastic strains

The main topic of this section deals with the estimation of the specimen thickness and, of course, the elastic stress concentration factor on material transition in elasto-plastic deformation and the process of plastic strains development. The reason for this interest arises from the fact that displacement field measurements by optical methods, including holographic interferometry, can be performed on the surface (usually external) of non-transparent objects only.

The problem has two aspects. First, from a theoretical point of view, on a load-free surface of the object, loaded in accordance to plane stress conditions, the relation $\sigma_z = 0$ is valid (σ_z is the stress tensor component, acting along the normal surface direction—see equations (5.25), (5.83)–(5.85)). In fact, this condition is not fully valid for internal layers of material. The effect of the σ_z stress component influence on elastic strain and stress components distribution in the normal surface direction through the object thickness increases with respect to the growth of the plane specimen thickness.

Figure 8.1. Interferogram of specimen III subjected to a tension in the elasto-plastic deformation range

The second part of the problem is connected with the influence of the strain (or stress) tensor component gradients on material transition in the field of elasto-plastic deformation. As shown in Sections 7.4 and 7.5, the anisotropy of the material surface layer transition into the plastic stage is observed near the hole boundary for the first loading of the plane specimen under plane stress conditions. The significant decrease in the yield limit of the specimen material in the normal surface direction in comparison to the data of the standard one-axis strain-stress diagram is the main characteristic of this phenomenon. It was determined also that the value of the yield limit of the specimen material in directions lying in the specimen surface plane almost coincides with standard parameters from a one-axis strain–stress diagram, in spite of the presence of strain–stress gradients.

Thus for a correct interpretation of results of local elasto-plastic strain measurements under plane stress conditions by optical interferometric techniques it is necessary to make use of data from special test experiments for the most typical problems. One of them, and perhaps the most interesting problem, is the case of tension of a plate with a circular open hole.

Another objective of this part of the work is to show how the developed reflection hologram interferometry technique can be used for measurement of plastic strain values originating from the beginning of the elasto-plastic deformation process (approximately 0.5 per cent of strain for aluminium alloys) up to 1.8–2.0 per cent of strain. It should be noted that the latter values of plastic strain were obtained previously by moiré methods only.

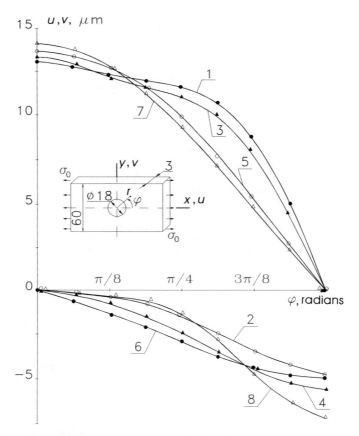

Figure 8.2. In-plane Cartesian displacement component distributions around the hole in specimen III.

To obtain experimental information needed for solving the above-mentioned problems, three plane aluminium specimens with different thicknesses and stress concentration factors on the specimen surface were tested (specimens I–III in Table 8.1). Some of the experimental information about the deformation study of these specimens was presented earlier in Chapters 2 and 4.

Typical fringe patterns, obtained for specimen I in the elastic deformation range, are shown in Figure 2.25. Corresponding displacement component distributions along the hole edge are presented in Figure 2.26.

A typical fringe pattern of specimen III in the elasto-plastic deformation range obtained for net stress increment σ_0 from 222 to 232 MPa is shown in Figure 8.1 This double-exposure interferogram illustrates the situation when all points of the specimen symmetry cross-section (see Figures 2.24 and 8.2) have

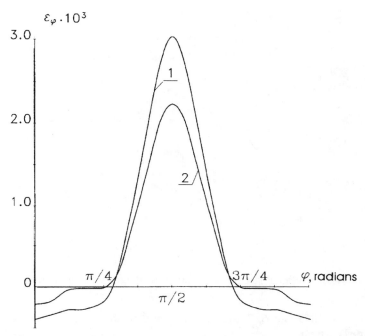

Figure 8.3. Distributions of the circumferential strain around the hole in specimen III

undergone plastic deformation. This case represents the upper loading limit (with respect to net stresses level) or deformation limit (with respect to strain values) in recording contrast holographic fringe patterns available for quantitative interpretation. The subsequent load increase is accompanied by an irreversible change in the surface microrelief over the area under study. As known from the theory of holographic interferometry, a change in the surface microrelief between two exposures results in decorrelation of the reconstructed light waves reflected from the object surface at different stages, causing a considerable reduction in the fringe contrast.

The most detailed analysis of displacement fields was carried out for specimen III. The displacement components were determined in the nodes of the mesh (see Figure 8.1) formed by means of the intersection of 9 circle lines $r = $ const ($r = 9, 11, 14, 16, 19, 21, 24, 27, 30$ mm) and 16 straight lines $\varphi = $ const ($\Delta\varphi = \pi/8$ radians).

Some typical distributions of in-plane Cartesian displacement components u, v along the circle lines $r = $ const are shown in Figure 4.10 for the elastic deformation range and in Figure 8.2 for the elasto-plastic deformation range. The latter figure corresponds to the net stress increment σ_0 from 208 to 222 MPA. In Figure 8.2 the curves with the uneven and even numbers depict the distributions of u and v displacement components, respectively, and curves

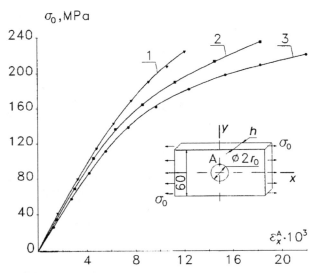

Figure 8.4. Maximum local strain versus net stress diagram for specimens I (1), II (2) and III (3)

1 and 2, 3 and 4, 5 and 6, 7 and 8 correspond to the circle lines $r = 9$, 11, 21, 30 mm, respectively.

Obtained discrete sets of experimental data (different shaped points in Figures 4.10 and 8.2) were approximated in the angular direction with the Fourier series to obtain the functional relations needed for following numerical differentiation, as described in Section 4.1.

Distributions of the circumferential strains ε_φ along the hole edge for specimens I and II and along the hole edge and 8 circles $r = $ const for specimen III were obtained by means of the Fourier approximation procedure at each loading step. For example, the distributions of elasto-plastic circumferential strains ε_φ along the hole edge (curve 1) and the circle line $r = 11$ mm (curve 2), obtained in accordance with relation (5.87) and corresponding to Figure 8.2, are shown in Figure 8.3.

The relations between maximum local strains ε_x^A in point A on the hole boundary and net stresses σ_0 are shown in Figure 8.4 for specimens I, II, III by curves 1, 2, and 3, respectively. The values ε_x^A are derived from ε_φ distributions along the hole edge (5.87) for $\varphi = 90°$.

The distributions of elasto-plastic strains ε_x along the most loaded symmetry cross-section $y = 0$ (see Figure 2.24) in specimen III in the elastic range and for different loading stages in the elasto-plastic range are presented in Figures 4.11 and 8.5, respectively. In the latter figure curves 1, 2, 3, 4 and 5 correspond to the net stress increments σ_0 from 142 to 168 MPa, from 168 to 190 MPa, from 190 to 208 MPa, from 208 to 222 MPa and from 222 to 232 MPa, respectively.

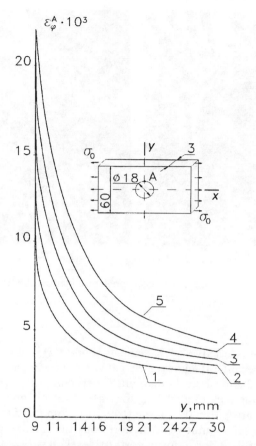

Figure 8.5. Distributions of the elasto-plastic strain in the tension direction along the central symmetry cross-section in specimen III for different loading steps

The relations in Figure 8.4 clearly show the influence of geometrical factors on the transition of material in stress/strain concentration zones in the plastic stage. The comparison of curves 1 and 2 clearly illustrate the influence of the specimen thickness on the process of local elasto-plastic deformation. It should be noted that the maximum plastic strain value in the direction of tension for specimen I is in good agreement with the Molsky–Glinka prediction [25, 26]. The decrease by double the specimen thickness results in: the maximum plastic strain value on the hole boundary for specimens II and III increasing to a value which is in better agreement with the Neuber formula [3] than with that of Molsky and Glinka.

Comparison of dependencies 2 and 3 in Figure 8.4 shows that (with respect to plastic strains development) the increase in the elastic stress concentration

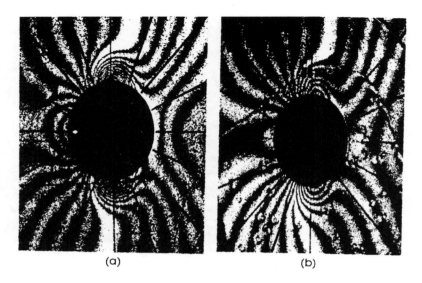

(a)   (b)

Figure 8.6. Typical interferograms of specimen I obtained at the 1st (a) and 3rd (b) loading cycles

factor by 6 per cent (see Table 8.1, specimens II and III) is almost equivalent to doubling of the specimen thickness (see Table 8.1, specimens I and II).

Comparison of the local strain diagram for specimen III in Figure 8.4, and the strain distribution in Figure 8.5, illustrates that all points of symmetry cross-section $y = 0$ are within the plastic strain range under maximum net stress level $\sigma_0 = 232$ MPa. It is important to note that the maximum plastic strain value in Figures 8.4 and 8.5 for net stress level $\sigma_0 = 232$ MPa is equal to 2.2×10^{-2}. This is the first time that this measurement range has been achieved by holographic interferometry. Earlier it was achievable only by means of moiré techniques [20, 21, 23, 24].

The sensitivity of the reflection hologram technique used for determination of in-plane displacement determination is approximately equal to $0.2\,\mu$m—see equation (8.2). This value corresponds to gratings with a frequency of not less than 5000 lines/mm. Therefore, it is clear that measurement accuracy and the capability of providing information with the developed holographic technique cannot be compared with data obtained by means of moiré gratings with frequencies less than 500 lines/mm.

Obtained results illustrating the influence of an elastic stress concentration factor and specimen thickness on the development of local elasto-plastic strains under plane stress conditions must be taken into account for analysis of local cyclic deformation processes presented in subsequent sections of this chapter.

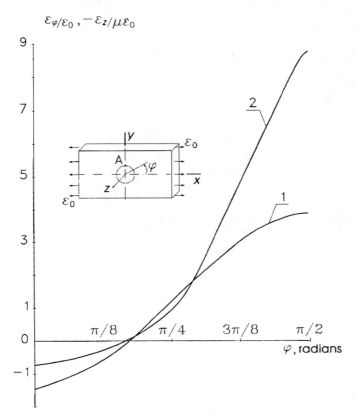

Figure 8.7. Typical distributions of the relative elasto-plastic circumferential strains along the edge of hole in specimen I at 3rd loading cycle

8.2 Tension and compression of a plate with an open hole

Local elasto-plastic deformation study of a plate with an open hole subjected to cyclic loading under plane stress conditions (tension–compression) is of both scientific and practical interest from different standpoints. Dependencies between local strains near an open hole and net stresses are required as foundations for the numerical models used for life-time analysis of pin and rivet joints. Investigation of the initial stage of the local cyclic deformation process is of great interest. The reason for this consists of the fact that the material mechanical properties in the deformation stage, described by a static one-axis strain versus stress diagram, convert to those corresponding to an analogous cyclic diagram. The quantitative description of this kinetic transition process and accurate determination of its range are very important for reliable life-time predictions of different structures [2, 3, 28].

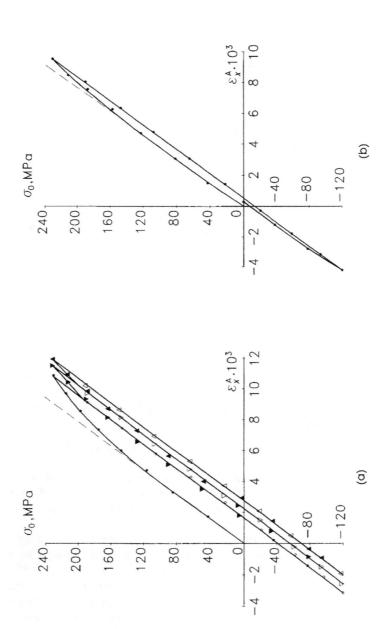

Figure 8.8 Maximum local strain versus net stress diagram for specimen I obtained at first three (a) and 14th (b) loading cycles

The quantitative description of the fatigue crack's appearance and propagation processes, which is necessary for both numerical modelling of the crack growth and formulation of fracture criteria, can be effectively established by means of local cyclic deformation kinetic study near an open hole [2, 3, 29].

This section describes the application of reflection hologram interferometry in the investigation of the elasto-plastic local strain history near an open hole before fatigue crack propagation. The stage of local cyclic deformation process which was studied is characterized by subsurface damage accumulation followed by the appearance of a short crack on the specimen surface.

The plane specimen used was specimen I whose parameters are presented in Table 8.1. The testing program consisted of cyclic soft loading of the specimen by tension–compression in the net stress range $-120 \leqslant \sigma_0 \leqslant 230$ MPa with a zero beginning point and a positive zero half cycle. The parameters of the loading cycle used are those most typical for aircraft structures with a considerable life-time.

The deformation process of an open hole neighbourhood in plane specimen I at the first three, 14th, 218th, 521st, 1017th and 1418th loading cycles was obtained by means of step-by-step recording of the reflection holographic interferograms. The experimental technique based on the use of overlay interferometer is described in Section 2.5.

First, it is necessary to consider the initial stage of the local cyclic deformation process. A typical interferogram of specimen I which was obtained at the compression part of the first loading cycle (net stress increment σ_0 from -62 MPa to -96 MPa) is shown in Figure 8.6(a). A typical interferogram, relevant to the maximum net stress at the third cycle (net stress increment σ_0 from 212 to 230 MPa) is shown in Figure 8.6(b).

The distribution of the relative circumferential strain $\varepsilon_\varphi/\varepsilon_0$ (where ε_φ is defined by expression (5.87), $\varepsilon_0 = \sigma_0/E$, $E = 74\,000$ MPa is the elasticity modulus of the specimen material) along the hole edge (corresponding to the interferogram in Figure 8.6(b)) is shown in Figure 8.7 by curve 1. The analogous distribution of the relative normal strain $-\varepsilon_z/\mu\varepsilon_0$ (where ε_z is defined by expressions (5.83)–(5.85); μ is the material Poisson's ratio) is presented in the same figure by curve 2.

The difference observed between curves 1 and 2 in Figure 8.7 proves that the anomalous behaviour of the ε_z-strain (discovered at the first loading cycle and discussed in detail in Sections 7.4 and 7.5) also takes place at the third cycle. This shows that the above-mentioned transformation of the material strain–stress diagram from static to cyclic does not finish at the third loading cycle.

Local strain in the tension–compression direction ε_x^A versus net stress σ_0 diagrams and analogous diagrams for normal strain $-\varepsilon_z^A$, plotted at the first three loading cycles for point A of maximum strain concentration on the hole edge, are shown in Figures 8.8(a) and 8.9(a) (the open and filled points indicate the load increasing and decreasing, respectively). Note that the values of the

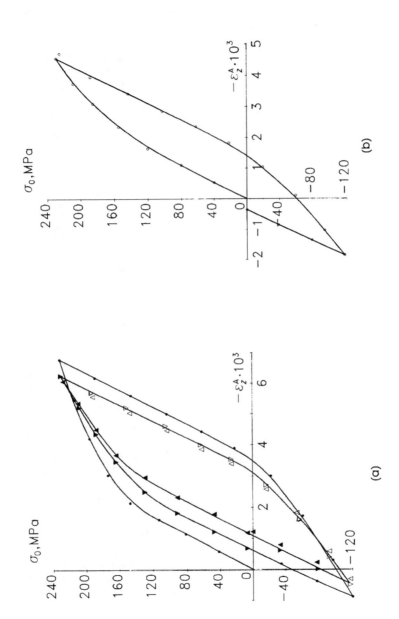

Figure 8.9. Normal local strain versus net stress diagram for specimen I obtained at first three (a) and 14th (b) loading cycles

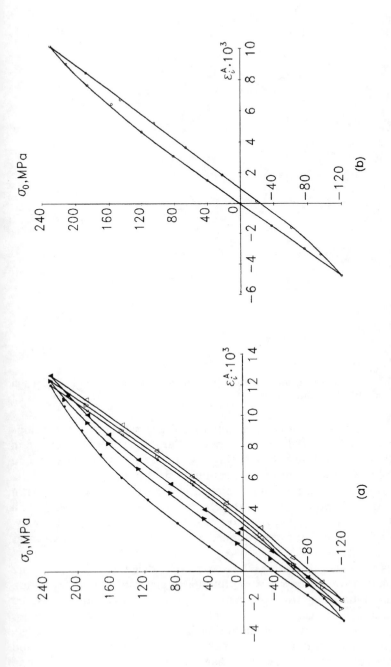

Figure 8.10. Maximum local strain intensity versus net stress diagram for specimen I obtained at first three (a) and 14th (b) loading cycles

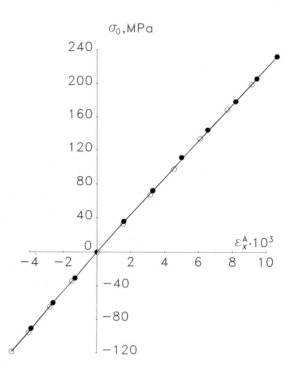

Figure 8.11. Maximum local strain versus net stress diagram for specimen I obtained at 218th, 521st and 1017th loading cycles

ε_x^A-strain are determined from expression (5.87) for $\varphi = \pi/2$. The same relations obtained for the ε_x^A-strain and the $-\varepsilon_z^A$-strain at the 14th cycle are presented in Figures 8.8(b) and 8.9(b), respectively (in these graphs the strain curve origin points coincide with the origin of the coordinate system used).

An analysis of experimental data presented in Figure 8.9 shows that, as in the case considered in Section 7.4, for the ε_z^A-strain the relation $\varepsilon_z^A = -\mu\varepsilon_x^A$ at all tension parts of cycles involved is valid up to the value $\varepsilon_z^A = -2 \times 10^{-3}$ only. Over this magnitude the above-mentioned relation has mainly a non-linear character.

The results presented in Sections 7.4, 7.5 and 8.1 demonstrate that a more accurate quantitative description of the local elasto-plastic deformation process in the initial stage, for specimen thickness $h = 6$ mm, can be ensured by using the strain intensity ε_i (7.29). The strain intensity versus net stress diagrams, plotted at the first three loading cycles for point A on the hole edge, are shown in Figure 8.10(a) (open and filled points depict loading and unloading, respectively). The analogous diagram for the 14th cycle is shown in

Figure 8.10(b). It should be noted that after each load reversal within one cycle for the following plot of the ε_i-curve it is necessary to remove the coordinate system origin to the corresponding loop peak and to use the opposite direction of the ε_i-axis. The need for this procedure derives from the fact that the strain intensity magnitude always has a positive meaning (see expression (7.29)).

The dependence between maximum local strain on the hole boundary ε_x^A and net stress σ_0 obtained for the 218th loading cycle is shown in Figure 8.11. This relation is a straight line and establishes that the normal strain versus the net stress diagram can be described by the following equation in the whole net stress range:

$$\varepsilon_z^A = -\mu \varepsilon_x^A.$$

At this point we can say that the anomalous behaviour of the normal local strain ε_z on the surface of the plane specimen subjected to cyclic tension–compression disappears up to the 218th loading cycle. Evidently, this case shows that the strain intensity curve coincides with the ε_x-curve.

It is interesting to consider the principal laws of material mechanical behaviour at point A of the maximum strain/stress concentration on the open hole edge (root of the notch) for the initial stage of the local cyclic deformation process which results from the local strain versus net stress diagram presented in Figures 8.8–8.11.

The above-mentioned relations for maximum plane strain and strain intensity show that the initial stage of the cyclic deformation process with an asymmetrical load amplitude is accompanied by a tensile residual strain accumulation. For the purpose of this section, residual strain is defined as the axial strain which would be maintained at the root of the notch if the external load was removed. The experimental data obtained allow us to estimate the minimum possible value of the residual strains accumulated from the first to 14th (or, the same, to the 218th) cycle as 0.8×10^{-2}. As shown in Section 8.1, the total value of residual strains in the middle plane of the specimen can be more than an analogous value on the specimen surface.

Comparison of the maximum strain value at the root of the notch $\varepsilon_x^{\max}$ corresponding to the first loading (at the zero half cycle) with data of the known approximations for local elasto-plastic estimation (see data in Section 7.4) shows its agreement with a good accuracy. An important note to remember is that the extrapolation of the data on elasto-plastic strains, obtained on a specimen surface, to a middle surface of a plane specimen should be carried out in accordance with the analysis presented in Section 8.1.

From the deformation kinetic standpoint, the anomalous behaviour of the normal strain at the initial stage of the local cyclic deformation process leads to hysteresis loop stabilization in the strain intensity versus net stress diagram at the 14th loading cycle (see Figure 8.10(b)). This means that an accumulation of the residual strains on the specimen surface ends. It should be noted, however, that an analogous process on the specimen middle plane may end some time later.

With regard to the problem of local elasto-plastic deformation process modelling, which is necessary, for instance, for numerical life-time predictions of different structures, it should be useful to obtain the dependencies between elasto-plastic strain concentration factors at the root of the notch for the maximum tensile load and the number of loading cycles N. These relations for the maximum strain in the direction of tension (curve 1), corresponding normal strain (curve 2) and strain intensity (curve 3), obtained without taking into account strains which were accumulated in the previous cycles, are shown in Figure 8.12. The graphs presented correspond to the maximum net stress value $\sigma_0 = 230$ MPa.

For the asymmetrical loading cycle involved, the stabilization of the strain concentration factors value (curve 1 in Figure 8.12) is reached almost after the first loading cycle. This is in good agreement with the known data on local adaptability of cyclic-hardening alloys [2].

The behaviour of curve 2 in Figure 8.12, which describes the normal strain concentration factor distribution, results from an anomalous decrease in the local material yield limit at the first loading cycles (see Sections 7.4 and 7.5). The coincidence of the two above-mentioned strain concentration factors at the 218th cycle apparently shows that the process of the material local adaptability to the conditions of the cyclic loading has ended. This can be considered as a criterion of a full transition of the material strain–stress curve from static to cyclic [3,4].

The change in the local material elasticity modulus is of importance especially for the future mechanical behaviour of the material under cyclic loading. This is connected with the dependencies presented in Figure 8.12. This elasticity modulus can be determined in accordance with equation (4.20).

An analysis of the linear parts of the local strain versus net stress diagrams at the first, second, third and 14th cycles provides the following value of the local elasticity modulus.

$$E = 74\,000 \pm 1000\,\text{MPa}. \tag{8.3}$$

The measurement of the straight-line inclination in Figure 8.11 leads to the following result:

$$E = 70\,000 \pm 1000\,\text{MPa}. \tag{8.4}$$

Evidently, this fact is partly connected with a constancy of the plane strain concentration factor after the first loading cycle and, in addition, with the non-linear character of the hysteresis loop parts corresponding to the tensile load at the first, second, third and 14th cycles. On the other hand, the local strain versus the net stress diagram at the 218th cycle is a straight line (see Figure 8.11). Due to an absence of a curved part that is inherent in the elasto-plastic

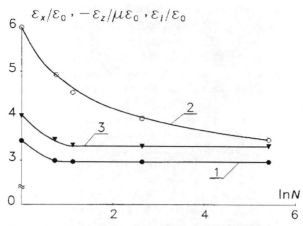

Figure 8.12. Dependencies between elasto-plastic strain concentration factors and the number of loading cycles for specimen I obtained on the stage of local adaptability of specimen material

local strain versus net stress diagram, the straight line in Figure 8.11 should have a smaller inclination in comparison with the inclination of the linear parts of the above-mentioned hysteresis loops to reach the maximum strain point $\varepsilon_x^{max} = 0.96 \times 10^{-2}$.

Therefore, taking into account the results of previous analysis, one can affirm that after the 218th loading cycle the open hole neighbourhood in the specimen considered is in the stable stage of the local cyclic deformation process. The mechanical behaviour of the specimen material in the region of the maximum strain at the hole edge during the above-mentioned deformation stage can be described by means of the linear elastic law with elastic modulus (8.4) changed in comparison to its initial value (8.3). The results of the following investigations also confirm this conclusion.

A typical interferogram of specimen I, obtained at the 521st loading cycle for net stress increment σ_0 from 38 to 76 MPa, is shown in Figure 8.13. The corresponding maximum local strain versus net stress diagram also has a linear character and almost completely coincides with the analogous dependence for the 218th cycle, presented in Figure 8.11. Almost the same result was obtained for the 1017th loading cycle, but with a small increase in the maximum strain value, corresponding in the positive net stress peak.

Analysis of the experimental data obtained for the 1418th loading cycle shows the transformation of a local deformation character near the root of the notch. The first evidence of this is a change in the fringe pattern behaviour in the range of the net stress positive values.

The typical interferograms of specimen I corresponding to the 1418th loading cycle obtained under tension from $\sigma_0 = 146$ MPa to 178 MPa and

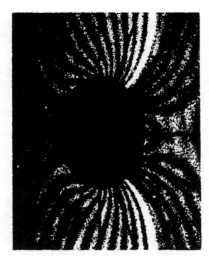

Figure 8.13. Typical interferogram of specimen I obtained at 525th loading cycle

unloading in the compression net stress range from $\sigma_0 = -120$ MPa to -90 MPa are shown in Figures 8.14(a) and (b), respectively.

For the following analysis one should reconsider the interferograms presented in Figures 8.6 and 8.13. The fringe patterns shown in Figures 8.6(a) and (b) are related to the stage of local adaptability of the material (see Figures 8.8(a) and 8.9(a)). The interferogram in Figure 8.13 is related to the stable stage of the local cyclic deformation process (see Figure 8.11). A common special feature of all the above-mentioned figures is that the interference fringes are either continuous between two specimen lateral sizes or end at the hole edge.

An opposite case occurs in Figure 8.14(a) where interruptions of interference fringes are clearly shown near the root of the notch (vertical diameter in the photograph). This is evidence of fatigue crack appearance on the specimen surface. The crack can be observed at the 1418th cycle for all load steps corresponding to the positive net stress range. When the net stress level reaches negative values, the fatigue crack is closed and does not result in interruptions of the interference fringes observed (see Figure 8.14b).

The qualitative data presented above, which provides the possibility to determine the number of loading cycles before crack appearance, also has a direct quantitative confirmation. The dependence between maximum local strain amplitude $|\Delta\varepsilon_x^A| = \varepsilon_x^{\max} + |\varepsilon_x^{\min}|$ at the hole boundary and the number of loading cycles is shown in Figure 8.15 by curve 1. It can be noted that the maximum local strain amplitude is calculated by using the strain intensity values (7.29) at the first three and 14th cycles, taking into account the analysis presented in Sections 7.4, 7.5 and 8.1 and the first part of this section.

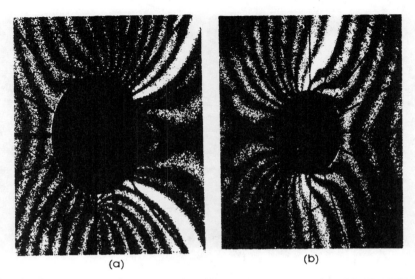

(a)   (b)

Figure 8.14. Typical interferogram of tension (a) and compression (b) of specimen I obtained at 1418th loading cycle

The maximum local strain amplitude versus the loading cycle number diagram for specimen I (curve 1 in Figure 8.15) consists of some characteristic parts. After strain amplitude stabilization at the initial stage of the local cyclic deformation process, which is characterized by a local mechanical properties adaptability of the material (cycles 2–14), an increase in the strain amplitude values, resulting from negative strain peaks growth, can be observed during the stable local deformation stage (cycles 218–1017). The appearance of a fatigue crack at the root of the notch is related to the large increase in the maximum strain amplitude value (the 1418th cycle).

It is quite clear that the relatively full set of the local strain versus net stress diagram, obtained for different stress concentration factors and loading cycle parameters, is an essential condition in order to formulate a reliable deformation criterion of a low-cyclic-fatigue life-time prediction for different types of structures.

A useful and interesting comparison of the data presented here can be made with the data of reference [30] obtained for an aluminium cyclic hardening alloy 2024-T351, the mechanical properties of which are in a very good agreement with the mechanical properties of the aluminium alloy used in our investigations. In reference [30] the objects investigated were plane specimens with complex-shaped cut-outs with root of notch radius $r_0 = 2.73$ mm.

The specimens underwent cyclic loading in the range of net stress and local strain which are comparable with analogous parameters described earlier in

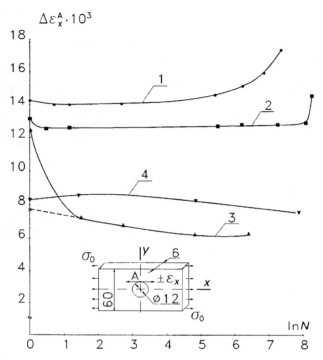

Figure 8.15. Maximum local strain amplitude versus the loading cycle number for specimens I (1), IV (3), V (4) and VI (5)

this section. Specimens used in reference [30] had the following values of stress concentration factors:

$$K_\sigma = 3.93 \text{ for static loading,}$$
$$K_f = 2.79 \text{ for cyclic loading.}$$

These factors are connected by the relation

$$K_f = q(K_t - 1) + 1$$

(8.5)

where q is the material sensitivity to a notch existence under cyclic loading.

Expression (8.3) for the above-mentioned parameters leads to the following value of q:

$$q = 0.61.$$

Otherwise, it is known that parameter q can be expressed in an alternative form [31]:

$$q = (1 + a/r_0)^{-1}$$

(8.6)

where r_0 is a notch radius and a is a material constant. Relation (8.4) can be used as a way of a-constant determination. The corresponding calculation for $q = 0.61$ and $r_0 = 2.73$ mm gives the following value of a:

$$a = 1.75 \text{ mm.}$$

It should be noted that the value of $K_f = \sqrt{K_0 K_\sigma}$ (K_σ and K_ε are the concentration factors of local stress and local strain, respectively) obtained by experimental data at all loading cycles investigated slightly exceeds the corresponding theoretical value of $K_f = 2.79$.

The value of a-constant can also be estimated from experimental data obtained by means of a modified Neuber's rule [3]:

$$\frac{(K_f \Delta \sigma_0)^2}{E} = \Delta \sigma \Delta \varepsilon \tag{8.7}$$

where E is the elasticity modulus of the material at an initial stage (before cyclic loading); $\Delta \sigma_0$ and $\Delta \sigma$ are amplitudes of net and local stress, respectively; and $\Delta \varepsilon$ is local strain amplitude.

The experimental data for positive net stress peaks: $\sigma_0 = 230$ MPa; $\Delta \varepsilon = 0.0098$ (this value is stable from the second to the 1017th cycle); $\Delta \sigma = 440$ MPa (this value corresponds to the value of $\Delta \varepsilon = 0.01$ on the cyclic strain–stress material diagram presented in reference [3]); and $E = 74\,000$ MPa can be used for calculation. In accordance with the above-mentioned parameters, expression (8.7) gives $K_f = 2.46$.

For specimen I $K_\sigma = 2.9$ (see Table 8.1) was used, $r_0 = 6$ mm, and in accordance with expressions (8.5) and (8.6) we obtain a-constant equal to 1.83 mm. This value is in good agreement with the a-constant value, corresponding to the data of reference [30].

Finally, we can conclude by mentioning that the well-known handbook by Peterson [31] contains the value of $a = 0.51$ for aluminium alloy.

8.3   Influence of radial interference and contact interaction along a filled hole edge on fatigue life-time

The study of joint element deformation kinetics, especially under cyclic loading, is of both scientific and practical interest, in particular because of the large number of pins and rivets found in engineering structures (especially aircraft structures). The reliability of these separate pins or rivets exerts its main influence on the structure's life-time [1,4,5].

Due to a deficiency of reliable experimental data describing a process of local elasto-plastic deformation in a plate with a hole both open and filled with a cylindrical inclusion (the latter holes are the models of separate pin or rivet joint element), it is very difficult to formulate a very effective approach to the prediction of a structure's life-time based on the deformation criterion of fatigue strength [3].

Thus, quantitative data on the characteristics of the local elasto-plastic deformation process of a plate with a filled hole under cyclic loading are needed for the engineering life-time prediction of joints of different structures, and as foundation or verification of the criteria of the appearance of fatigue cracks in contact interaction zones.

It is a known fact that the contact interaction of a hole edge and a pin filling this hole, and the preliminary plastic deformation of a hole vicinity resulting from radial interference, lead to an increase in the pin or rivet joint's life-time [5]. However, these facts are established as a result of long-term and very expensive standard fatigue testing which have not revealed the reasons for this phenomenon, especially in an elasto-plastic deformation range.

Data show a decrease in elastic circumferential stresses along the edge of a filled hole in a finite strip subjected to static axial tension in comparison with analogous loading of the same object with an open hole. These results were obtained for a plane specimen made of epoxy material with a hole filled by a pin without any clearance or interference by application of photoelasticity [8]. A numerical solution of the problem by means of a finite element method leads to the same result [15]. Some experimental data (obtained at the initial stage of cyclic loading of the plane specimen with a circular filled hole) which describe the effect of the contact interaction and preliminary local plastic deformation caused by an interference fit are presented in reference [32].

This section presents the results of a local elasto-plastic deformation kinetic study of two plane specimens with a filled hole, each representing the model of a separate pin or rivet joint element with high radial interference. To establish the amount of radial interference, the uniform proportional interference β can be defined as the ratio of excess of pin (inclusion) diameter over the hole diameter to the hole diameter. In our experiments which will be described below, two values of β are used: $\beta = 0.75 \times 10^{-2}$ and $\beta = 1.2 \times 10^{-2}$. Joints with these magnitudes of β are widely used in aircraft structures [5].

The specifications of the used specimens, their geometrical parameters, types of fit β, and the used testing program are listed in Table 8.1 (specimens IV and V). Note that again the geometrical parameters of plates with a hole presented in Table 8.1 coincide with those for the plane specimen shown in Figure 2.24.

To describe the deformation processes of a filled hole vicinity, sets of reflection holographic interferograms were recorded by means of a step-by-step procedure at the first, fourth, 14th, 115th, 518th and 1020th loading cycles for specimen IV, and at the first, fourth, 15th, 116th, 617th and 2618th loading cycles for specimen V.

Figures 8.16(a) and (b) present typical fringe patterns obtained for specimen IV at the same loading increment and almost equal load level at the first (σ_0 from 80 to 118 MPa) and 1020th (σ_0 from 76 to 114 MPa) cycles, respectively.

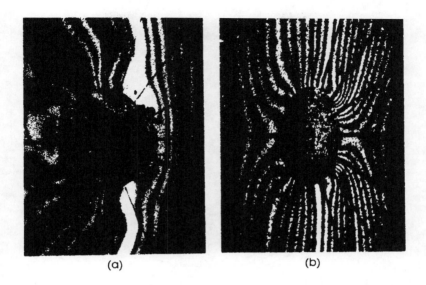

Figure 8.16. Typical interferograms of specimen IV obtained at 1st (a) and 1020th (b) loading cycles

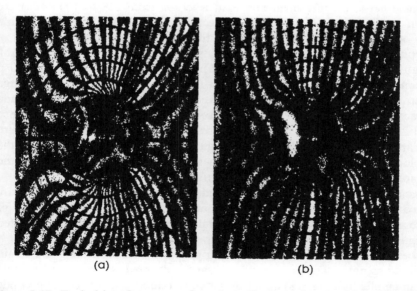

Figure 8.17. Typical interferograms of specimen V obtained at 1st (a) and 2618th (b) loading cycles

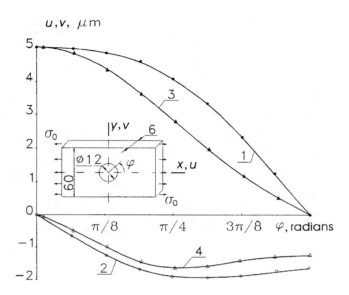

Figure 8.18. Distributions of the in-plane Cartesian displacement components along the line of contact interaction (hole edge) in specimen IV

The fringe patterns obtained for specimen V at the first and 2618th cycles at the same net stress increment and load level (σ_0 from 128 to 165 MPa) are shown in Figures 8.17(a) and (b), respectively, and illustrate other possible types of contact interaction between an aluminium plate and steel pin.

Typical distributions of Cartesian in-plane displacement components u and v along the line of contact interaction for interference fit with $\beta = 0.75 \times 10^{-2}$ (specimen IV) are shown in Figure 8.18. These relations correspond to fringe patterns presented in Figure 8.16. Analogous distributions of in-plane displacement components obtained for uniform proportional interference $\beta = 1.2 \times 10^{-2}$ (specimen V), which corresponds to fringe patterns in Figure 8.17, are shown in Figure 8.19. In both figures curves 1 and 2 correspond to the first loading cycle and curves with uneven and even numbers depict the displacement component u and v, respectively.

Circumferential strains ε_φ along a filled hole edge versus polar angle φ are plotted in Figure 8.20 for specimen IV and in Figure 8.21 for specimen V. These relations correspond to displacement component distributions presented in Figures 8.18 and 8.19, respectively.

To ensure a valid and correct comparison of the displacement and strain distributions presented above, it is necessary to estimate the values of the hole edge circumferential strain caused by radial interference only. The magnitude of the circumferential strain ε_φ^0 along a circular line of contact interaction with uniform radial interference β can be derived from the following expression [33]:

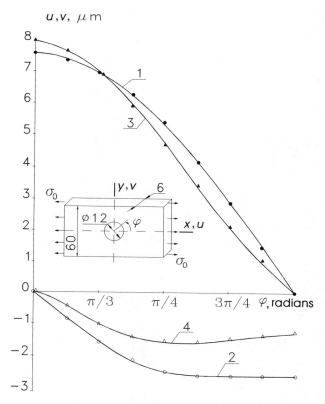

Figure 8.19. Distributions of the in-plane Cartesian displacement components along the line of contact interaction (hole edge) in specimen V

$$\varepsilon_\varphi^0 E = \frac{\beta}{(1+\mu)/E + (1+\mu_1)/E_1} \tag{8.8}$$

where $E = 74\,000$ MPa is an elasticity modulus of specimen; E_1 is an elasticity modulus of the pin; $E_1/E = 3$; $\mu = 0.34$ and $\mu_1 = 0.3$ are Poisson's ratio of materials of specimen and pin, respectively. It should be noted that equation (8.8) is valid in the elastic deformation range and in the initial stage of elasto-plastic deformation process only.

For the above-noted values of material parameters and $\beta = 0.75 \times 10^{-2}$ expression (8.8) provides the following meaning of uniform circumferential strain ε_φ^0 resulting from radial interference:

$$\varepsilon_\varphi^0 = 0.48 \cdot 10^{-2}. \tag{8.9}$$

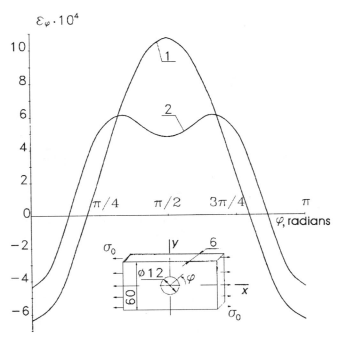

Figure 8.20. Distributions of the circumferential strain along the line of contact interaction (hole edge) in specimen IV obtained at 1st (1) and 1020th (2) loading cycles

That is in good agreement with the result obtained by application of the finite element technique to the solution of the corresponding contact problem [15].

By combining equation (8.9) and the one-axis strain versus the stress material diagram obtained (in a standard way) for a plane unnotched specimen, the following value of corresponding stress σ_φ^0 caused by radial interference only is obtained:

$$\sigma_\varphi^0 = 322\,\text{MPa} \simeq \sigma_{0.05}. \qquad (8.10)$$

Equation (8.10) shows that the value of preliminary uniform circumferential strain ε_φ^0 in specimen IV ($\beta = 0.75 \times 10^{-2}$) resulting from radial interference lies in the range of initial plastic deformation. The local strain versus net stress diagram for specimen I with an open circular hole obtained at zero half cycle, which for strain values $\varepsilon_x^A \geqslant 0.48 \times 10^{-2}$, has a non-linear character (see Figure 8.8(a)), also confirms this.

Moreover, estimations (8.9) and (8.10) obtained for a uniform proportional interference value $\beta = 0.75 \times 10^{-2}$ prove that when $\beta = 1.2 \times 10^{-2}$, the specimen material has undergone considerable plastic deformation in a hole neighbourhood.

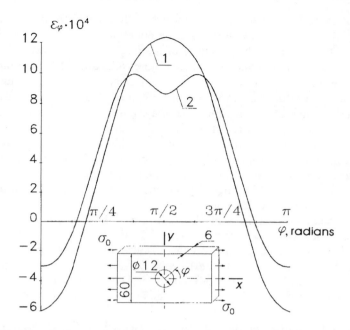

Figure 8.21. Distributions of the circumferential strain along the line of contact interaction (hole edge) in specimen V obtained at 1st (1) and 2618th (2) loading cycles

Now we shall return to an analysis on dependencies presented in Figures 8.18–8.21 which correspond to almost equal load levels and net stress increments.

First, it should be pointed out that the character of v displacement component distributions for the radial interference values involved is defined by a presence of a contact interaction at all points of a filled hole edge. This fact, which is valid for all net stress increments at all loading cycles being investigated for specimens IV and V, results from a comparison of the above-presented relations with distributions of a v-displacement component along an open hole edge in both the elastic and elasto-plastic range—see Figures 2.26, 4.10, 7.31(a) and 8.2.

On the other hand, the character of u-component distributions in all the above-mentioned figures does not allow us to draw a definite conclusion about the presence or absence of a contact interaction at a hole edge point of interest.

Comparison of the corresponding displacement component couples in Figures 8.18 and 8.19 obtained at the first loading cycles shows that the values of both in-plane displacement component in the case of $\beta = 1.2 \times 10^{-2}$ considerably exceed the corresponding values in the case of $\beta = 0.75 \times 10^{-2}$. This regularity can also be established at all investigated loading levels and is evidently connected with a different degree of plastic deformation of a hole vicinity.

Differences in the displacement component values result in differences in the circumferential strain values, although the character of ε_φ-strain distributions obtained at the corresponding couples of loading cycles is found to be almost the same for both values concerned—see Figures 8.20 and 8.21. In addition, the comparative analysis of the above couple of ε_φ-curves corresponding to the same values of uniform proportional interference β reveals one reason why radial interference increases the life-time of a separate pin or rivet joint element.

This reason consists of a redistribution of circumferential strains near point A which corresponds to the root of the notch in the case of an open hole with a growth in the number of loading cycles (see curve 2 in Figures 8.20 and 8.21). The observed redistribution leads to a decrease in material damage caused by the increasing number of loading cycles compared with the deformation of the same specimen with an open hole.

It should be noted that the analytical solution of the elasticity problem about axial tension of a plane specimen with a filled circular hole reveals the above-mentioned characteristic 'saddle' on the ε_φ-curve at some vicinity of point A ($\varphi = 90°$) for values of radial interference β which do not result in plastic deformation of a hole boundary [34]. However, this solution does not take into account friction forces and microplasticity along contacting surfaces due to roughness.

Relations which are analogous to those presented in Figures 8.18–8.21 were obtained at all loading steps and all loading cycles investigated for specimens IV and V in the net stress range from zero starting point to the maximum tensile peak $\sigma_0 = 230\,\text{MPa}$ then to the minimum compressive peak $\sigma_0 = -120\,\text{MPa}$ and then to the final zero point. The experimental data obtained allows us to construct relations between the circumferential strain ε_φ at any point of a hole edge and net stress σ_0 for all loading cycles investigated.

To compare the results obtained for specimens IV and V with those obtained for specimens I with an open hole (see Section 8.2), it is necessary to construct the above relations at point A on a hole edge (root of the notch, $\varphi = \pi/2$, $\varepsilon_{\varphi=90°} \equiv \varepsilon_x^A$). Note that with the growth of the number of loading cycles, the value ε_x^A becomes slightly less than the maximum value of ε_φ at a hole boundary—see Figures 8.20 and 8.21. It is evident that this difference has no influence on the life-time of the structure under investigation.

A local strain ε_x^A versus net stress σ_0 diagram obtained for specimen IV ($\beta = 0.75 \times 10^{-2}$) at different loading cycles plotted without taking into account the strain values accumulated in previous loading cycles are presented in Figure 8.22. The upper scale in Figure 8.22(a) corresponds to total strains resulting from a combined influence of preliminary interference and net stresses. Analogous dependencies obtained for specimen V are shown in Figure 8.23.

The above-mentioned local strain versus local stress diagrams allows us to plot dependencies between the local strain amplitude $\Delta\varepsilon_x = \varepsilon_x^{max} + |\varepsilon_x^{min}|$ and the number of loading cycles, which are presented in Figure 8.15 by curves 2

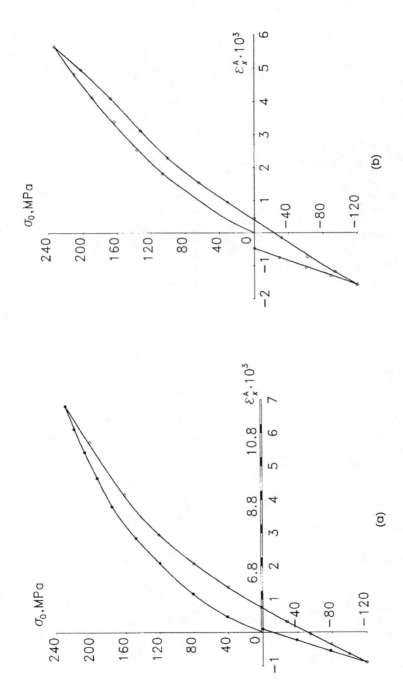

Figure 8.22. (a) and (b)

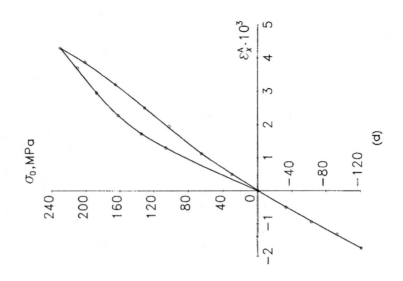

(d)

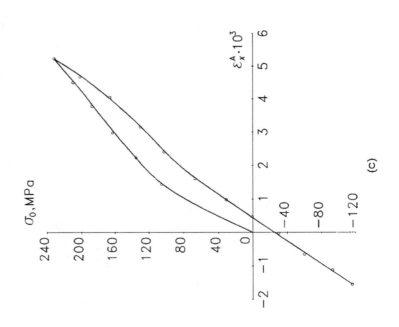

(c)

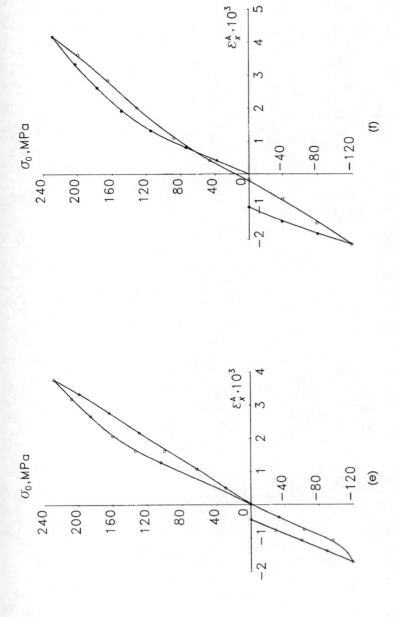

Figure 8.22. Local strain at the root of the notch versus net stress diagram for specimen IV obtained at 1st (a), 4th (b), 14th (c), 115th (d), 518th (e) and 1020th (f) loading cycles

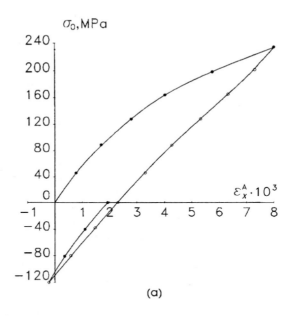

(a)

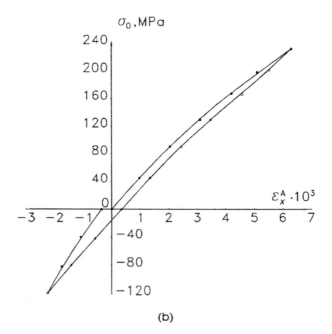

(b)

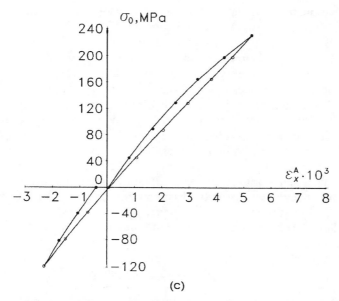

(c)

Figure 8.23. Local strain at the root of the notch versus net stress diagram for specimen V obtained at 1st (a), 4th (b) and 2618th (c) loading cycles

and 3 for specimens IV and V, respectively. The point marked Δ in Figure 8.15 illustrates the experimentally measured value of $\Delta\varepsilon_x$ for specimen IV at the first cycle only without the strain amplitude fraction caused by radial interference.

As a main result of an analysis on dependencies presented in Figures 8.22 and 8.23 it is necessary to note that the deformation trajectories have a non-linear character. Moreover, the processes of loading and unloading do not coincide with one another in most cycles investigated. This, together with the fact that mechanical properties of the material in contact interaction zones under cyclic loading are usually unknown, clearly shows the difficulties which must be overcome to establish a reliable analytical or numerical approach to quantitative determination of local strains under cyclic loading.

Relations 1, 2 and 3 in Figure 8.15 show that for the loading program involved the presence of radial interference, characterizing two values of the β concerned, causes a considerable decrease in the maximum local strain amplitude values in comparison with those obtained in the case of a specimen with an open hole. The difference between the maximum local strain amplitude values observed is evidently the main reason for the fatigue life growth of the pin and rivet joints with a growth (up to a certain limit) of a proportional uniform interference β which characterizes a fit type of separate joint element. This fact, which is well known as a result of numerous standard fatigue tests, has no reliable quantitative description, and the above-mentioned results are capable of clarifying the origin of the phenomenon.

One remarkable feature which results from the dependencies shown in Figures 8.20 and 8.21 and curves 2 and 3 in Figure 8.15 should also be noted. These dependencies directly prove that the optimal fit type, which ensures the maximum fatigue life can be established for a definite type of pin or rivet joint. Moreover, it is possible to design the optimal pin or rivet joint from a fatigue life standpoint by using the quantitative data describing the local cyclic deformation process of a separate joint element.

One other circumstance should be pointed out and must be taken into account for a correct comparison of curve 1 with curves 2 and 3 in Figure 8.15. It is evident that the material strain stage along the line of contact interaction cannot be completely described with circumferential strain ε_φ only. However, radial and shearing strains exert an influence on the values ε_φ and $\Delta\varepsilon_x$. This means that the two latter parameters can undoubtedly be used as a criterion for fatigue life analysis.

The problem of strain determination accuracy is an additional factor which defines a choice of the circumferential strain ε_φ as the basic parameter used for fatigue life prediction. As shown in Sections 4.1 and 4.2, the accuracy of circumferential strain determination along a closed hole boundary consider-ably exceeds the accuracy of radial strain determination at the end point of the segment.

In addition to the value of maximum local strain amplitude $\Delta\varepsilon_x$, a value of total local strain accumulated during the cyclic deformation process can be noted among the factors which influence crack appearance and, hence, fatigue life-time. As shown in Section 8.2, a strain accumulation process for specimen I with an open hole is localized during the first 14 cycles and leads to the accumulation of the tensile residual strains $\varepsilon_x^0 \geqslant 0.8 \times 10^{-2}$.

In specimens IV and V, except for the first loading cycle, each subsequent cycle is accompanied by an accumulation of insignificant compressive residual strains. This can be considered as a process which also delays the appearance of a fatigue crack.

The data presented in this section describing an image of the process of local cyclic elasto-plastic deformation and contact interaction influence may be effectively implemented for both fatigue life prediction of actual joints and for creating a reliable deformation criterion of fatigue fracture. A more detailed analysis of this problem will be the subject of the following section.

8.4   Fatigue crack initiation and propagation in a contact interaction zone

Elasto-plastic deformation around stress concentration zones exerts its influence mainly on the fatigue crack appearance moment and fatigue crack propagation stage in the low-cyclic-fatigue range [2-4]. Two problems should be noted.

The first problem is connected with the local elasto-plastic strain history under cyclic loading before the appearance of the crack. It is most important that we obtain a reliable estimation of this part of the structure's fatigue life period since most of the life-time occurs before the crack initiation and while the crack is still short. Some examples of investigations of this stage of the local cyclic deformation process using the holographic interferometry technique were given in Sections 8.2 and 8.3.

The second problem concerns local strain–stress description after the appearance of a fatigue crack. In order to assess the structural reliability of the cracked structures we must know both the stresses near the crack of a fixed length and the rate at which the crack grows under in-service fatigue loads. Both these factors depend on the strain–stress fields at the tip of the crack which are characterized either by the crack opening or stress intensity factors (SIF), or the J-integral. The relations between the parameters describing the strain–stress distributions near the crack and the number of loading cycles of different amplitudes are of main interest from the fatigue life point of view [3].

It was shown in Sections 8.2 and 8.3 that the relations between maximum stress amplitude on the hole boundary and the number of loading cycles can be used as a very effective approach to develop a reliable criterion of fatigue life-time prediction.

This section discusses the investigation of a local elasto-plastic deformation kinetic in the contact interaction zone from the zero half cycle up to fracture caused by the end of the low-cycle-fatigue process.

The object of the investigation was a plane aluminium specimen with a central circular hole, which is completely analogous to specimens IV and V listed in Table 8.1 with the exception of the fit type of the pin, filling a hole. In this case, the hole in the specimen was filled by a pin with a push fit (no clearance or interference, $\beta = 0$). See specimen VI in Table 8.1. The soft loading cycle parameters for specimen VI, completely coinciding with those for specimens III and IV, are presented in Table 8.1.

The choice of a fit type for a study of the local deformation kinetic of specimen VI up to fatigue fracture was influenced by the following reasons. The results of the local deformation process investigation of specimen I with an open hole in the low-cyclic-fatigue range (presented in Section 8.2) show the fatigue crack appearance just before the 1418th loading cycle. On the other hand, the data obtained in Section 8.3 demonstrate that the hole filling by a pin with a relative interference parameter $\beta > 0.75 \times 10^{-2}$ results in a considerable growth in the material's resistance to fatigue damage. In fact, there was no evidence of damage or fatigue cracks in specimens IV and V for over more than 2000 loading cycles. Therefore, the choice of the push fit was influenced by the need to maintain the conditions of the contact interaction along a hole edge, and at the same time to decrease the number of cycles before fracture.

As in other cases described in this Chapter, the local cyclic deformation process investigation was based on step-by-step loading of the specimen to be studied, and sequential recording of the reflection holographic interferogram sets. By means of this approach, the holographic interferograms of the filled hole neighbourhood in specimen VI were recorded for the first, second, 14th, 115th, 1316th, 2927th, and 3530th loading cycles for the complete net stress range, starting from zero to the maximum net stress value $\sigma_0 = 230$ MPa and then, through an intermediate zero point, to the minimum value $\sigma_0 = -120$ MPa and then to the zero final point for each loading cycle.

In addition, the interferogram sets were recorded on the 2900th, 3128th, 3931st, and 4133rd cycles for each zero to positive peak half cycle to obtain a detailed quantitative description of the fatigue crack development process.

Typical fringe patterns obtained for specimen VI on the first (σ_0 from 135 to 162 MPa) and 1316th (σ_0 from 143 to 174 MPa) cycles at almost equivalent net stress increments are shown in Figures 8.24(a) and (b), respectively. It should be noted that the interference fringes in these photographs do not have interruptions between the hole boundary and the lateral edges of the specimen. As shown earlier, such fringe behaviour is characteristic of the initial and stable stages of the local elasto-plastic deformation process (see Section 8.2). Apparently, accumulation of the material subsurface damages occurs during these stages only. The above-mentioned processes exert a very weak influence on the specimen material surface layers and cannot be discovered by means of holographic interferometry.

Typical distributions of the in-plane displacement components u (curves 1, 3) and v (curves 2, 4) in the Cartesian coordinate system along the line of contact interaction in specimen VI with push fit (no clearance or interference between the hole boundary and cylindrical steel inclusion) are presented in Figure 8.25. These relations were obtained on the first (curves 1, 2) and 1316th (curves 3, 4) cycles and correspond to the fringe patterns shown in Figure 8.24. Figure 8.26 demonstrates the circumferential strain distributions along the hole boundary, relevant to Figure 8.25.

Experimental results for specimen VI (push fit, $\beta = 0$) obtained on the first and 1316th loading cycles and presented in Figures 8.25 and 8.26 show that the in-plane displacement component u and v and circumferential strain ε_φ-distributions are similar to those obtained for specimens with an open hole (see Figures 7.31(a), 7.32, 8.2, 8.3 and 8.7). Some differences in maximum and minimun values of corresponding displacement components and circumferential strain in the specimens with an open hole and specimen VI are connected with the influence of contact interaction and depend on the loading level and cycle number.

It is interesting to compare the above-mentioned displacement and strain distributions with analogues obtained for specimens IV and V, in which the

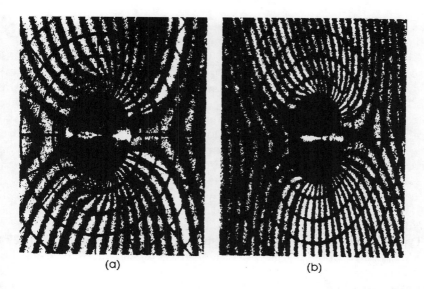

(a)   (b)

Figure 8.24. Typical interferograms of specimen VI obtained at 1st and 1316th loading cycles

contact interaction along the filled hole boundary is characterized by the magnitudes of relative interference $\beta = 0.75 \times 10^{-2}$ and $\beta = 1.2 \times 10^{-2}$.

Two main differences should be pointed out. The first consists of the absence of the relative minimum on the v-displacement component curves for specimen VI near the angular point $\varphi = \pi/2$. This minimum always presents itself on the v-displacement component distributions obtained for specimens IV and V (see Figures 8.18 and 8.19). Second, the comparison of circumferential strain ε_φ distributions shows (except for the difference in maximum and minimum peaks) an absence of the characteristic 'saddle' on the corresponding curves for specimen VI in the region of the angular point $\varphi = \pi/2$, which always appears on ε_φ-curves in specimens IV and V with the growth of the loading cycle number (see curves 2 in Figures 8.20 and 8.21). It is evident that these factors exert a direct influence on fatigue lifetime.

Holographic interferograms obtained under tension on the 2900th loading cycle clearly display the surface damage in the maximum net stress level used. To understand the influence of this factor on quantitative parameters of the local cyclic deformation process, interferogram sets were recorded on the 2927th and 3530th loading cycles for the complete net stress range $(-120\,\mathrm{MPa} \leqslant \sigma_0 \leqslant 230\,\mathrm{MPa})$.

Typical interferograms of specimen VI obtained on the 2927th and 3530th loading cycles for equal net stress increment σ_0 from 160 to 195 MPa are shown in Figures 8.27(a) and (b), respectively. The fringe interruptions can be

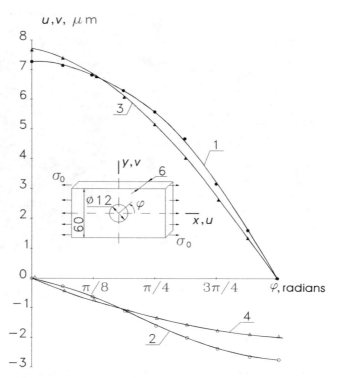

Figure 8.25. Distributions of the in-plane Cartesian displacement components along the hole edge in specimen VI obtained at 1st and 1316th loading cycles

observed in these interferograms near the point of maximum ε_φ values on the hole boundary (vertical diameter in the photographs). The interruptions are more clearly seen for the 3530th cycle. This clearly gives evidence of the short crack appearance on a specimen surface.

It is very interesting to find a quantitative description of the phenomenon observed. Distributions of the tangential displacement component u (curve 1) and v (curve 2) in the Cartesian coordinate system along the hole boundary in specimen VI obtained on the 2927th (circular dots) and 3530th (triangular dots) loading cycles are shown in Figure 8.28. These graphs correspond to interferograms presented in Figure 8.27(a) and (b). Corresponding circumferential strain ε_φ distributions are shown in Figure 8.29.

At first, even the relatively large-scale Figure 8.28 does not reveal the considerable difference between the corresponding displacement component for two loading cycles to be compared. Therefore, it is necessary to consider an analytical description of the experimental data, which were approximated by a Fourier series in the form of equation (4.16). An analysis of the validity of the

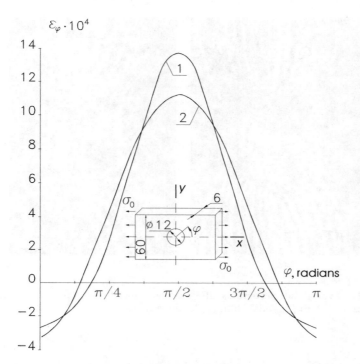

Figure 8.26. Distributions of the circumferential strain along the hole edge in specimen VI obtained at 1st (1) and 1316th (2) loading cycles

series coefficients (4.16), which has been carried out by means of the technique presented in Chapter 4.1, gives the following results for the u-displacement component:

$$A_1 = 8.51 \times 10^{-3} \text{ mm at the 2927th cycle,}$$
$$A_1 = 8.53 \times 10^{-3} \text{ mm at the 3530th cycle,}$$

and for the v-displacement component:

$$B_1 = 2.15 \times 10^{-3} \text{ mm, } B_3 = 0.23 \times 10^{-3} \text{ mm at the 2927th cycle,}$$
$$B_1 = 2.36 \times 10^{-3} \text{ mm, } B_3 = 0.20 \times 10^{-3} \text{ mm at the 3530th cycle.}$$

These data show why a notable difference between the strain distributions in Figure 8.29 results from relatively small differences between corresponding displacement component distributions shown in Figure 8.28.

Note that the above-mentioned example should be considered as a clear evidence of the analysis presented in Sections 2.4 and 4.1, and shows the

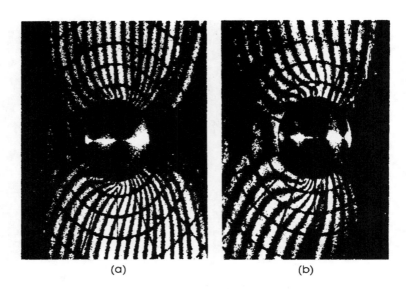

(a)   (b)

Figure 8.27. Typical interferograms of specimen VI obtained after a surface crack appearance at 2927th (a) and 3530th (b) loading cycles

accuracy of the displacement component which must be reached to ensure a reliable determination of strain/stress concentration.

Distribution of the circumferential strains along the hole boundary obtained for all load steps and all loading cycles allows us to obtain diagrams for net stress versus local strain. For maximum strains on the hole edge ε_x^A obtained for some loading cycles investigated, these diagrams are shown in Figure 8.30. The comparison of relations between maximum local strains on the hole edge ε_x^A and net stress σ_0 obtained at the zero half cycles for specimen I (open hole, Figure 8.8(a)) and specimen VI (Figure 8.30(a)), demonstrates a presence of almost equal linear parts in both graphs.

This provides us with the possibility to estimate the elastic stress (strain) concentration factor for the case of a hole in a plane specimen under tension filled by pin (inclusion) with push fit ($\beta = 0$). The corresponding magnitude can be calculated by means of expression (resulting from equation (4.20)):

$$K_\sigma = E\varepsilon_x^A/\sigma_0. \tag{8.11}$$

Equation (8.11) for $E = 74\,000$ MPa, $\sigma_0 = 105$ MPa and $\varepsilon_x^A = 3.9 \times 10^{-3}$ (see Figure 8.30(a)) leads to the following:

$$K_\sigma = 2.70. \tag{8.12}$$

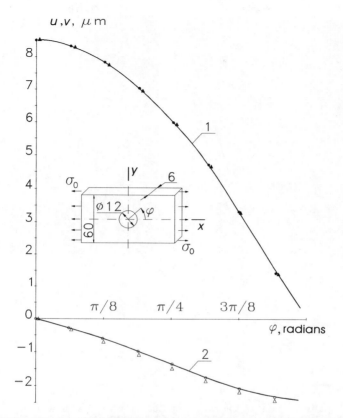

Figure 8.28. Distributions of the in-plane Cartesian displacement components along the hole edge in specimen VI obtained after a surface crack appearance at 2927th and 3530th loading cycles

Note that the value of the elastic stress concentration factor on the surface of specimen I with a circular open hole of the same diameter was calculated by FEM technique, $K_\sigma = 2.9$ (see Section 4.1).

The obtained value of K_σ (8.12) can be used for local elasticity modulus calculation in specimen VI in accordance with expression (4.20) for different loading cycles. These values are

$$E = 73\,000 \pm 1000 \text{ MPa}$$

for the 1316th loading cycle and

$$E = 68\,000 \pm 1000 \text{ MPa}$$

for the 2927th loading cycle.

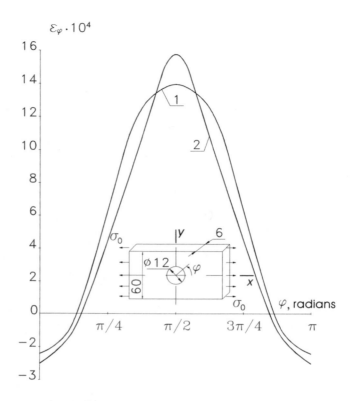

Figure 8.29. Distributions of the circumferential strain along the hole edge in specimen VI obtained after a surface crack appearance at 2927th (1) and 3530 (2) loading cycles

Two elements should be emphasized here to enable a better understanding of the above values. First, the local elasticity modulus magnitude for specimen I with an open hole is constant from the 218th up to the 1017th cycle and then smoothly decreases down to 70 000 MPa (see Section 8.2). This part of a local cyclic deformation process can be called a stable stage. Second, the above-mentioned presence of subsurface damages observed in specimen VI at the 2927th cycle should be taken into account.

It is evident that the difference of the local elastic modulus values in specimens I and VI is connected with the difference of maximum strain values in corresponding loading cycles during a stable stage of local cyclic deformation process. A more detailed analysis of this is given below.

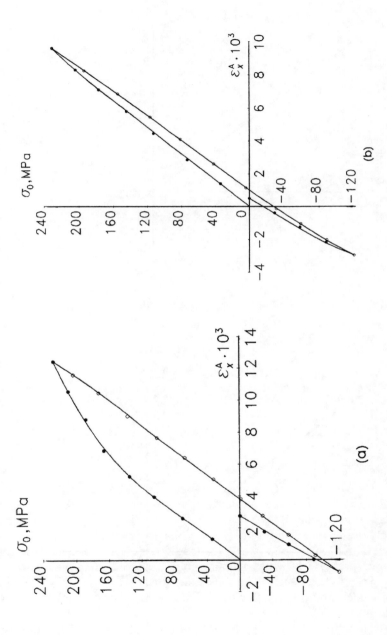

Figure 8.30. (a) and (b)

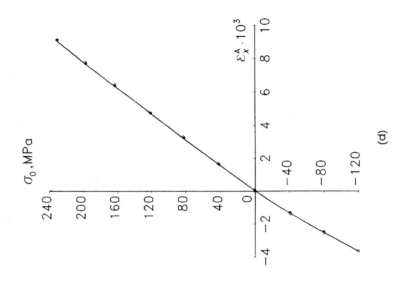

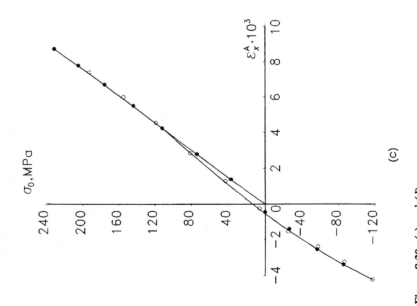

Figure 8.30. (c) and (d)

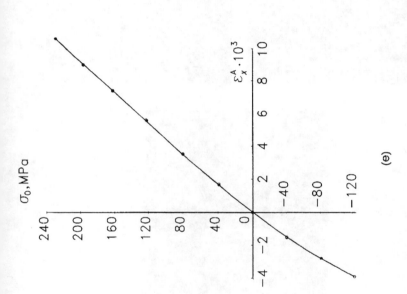

Figure 8.30. Maximum local strain versus net stress diagram for specimen VI obtained at 1st (a), 2nd (b), 1316th (c), 2927th (d) and 3530th (e) loading cycles

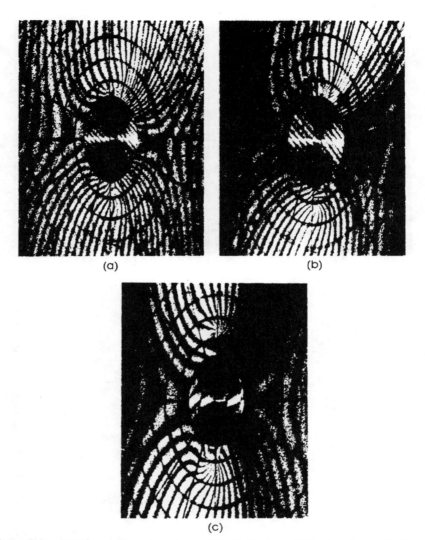

(a)

(b)

(c)

Figure 8.31. Interferograms of specimen VI obtained at 3931st loading cycle for net stress increment from 0 to 36 MPa (a), from 144 to 175 MPa (b) and from 206 to 230 MPa (c)

The quantitative description of a local cyclic deformation kinetic can be performed by means of an analysis of the local strain amplitude $(\Delta\varepsilon_x = \varepsilon_x^{max} + |\varepsilon_x^{min}|)$ versus the loading cycle number diagram. This diagram is based on relations presented in Figure 8.30 and shown in Figure 8.15 by curve 4.

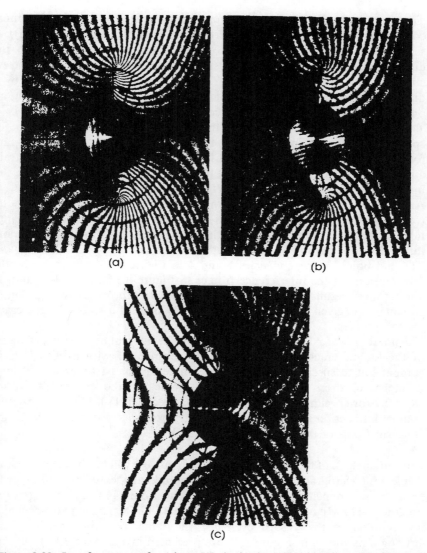

Figure 8.32. Interferograms of specimen VI obtained at 4133rd loading cycle for net stress increment from 0 to 36 MPa (a), from 144 to 175 MPa (b) and from 206 to 230 MPa (c)

A comparison of the above-mentioned diagrams for specimen I with an open hole (curve 1 in Figure 8.15) and specimen VI with a hole filled by a pin with push fit ($\beta = 0$, curve 4 in Figure 8.15), and also corresponding net stress versus local strain diagrams for the same specimens (Figures 8.10(a), (b); 8.11 and 8.30(a)–(e), reveals the following trends:

- The upper parts of the hysteresis loops during the first cycles coincide almost completely for both specimens concerned (see Figures 8.10(a) and 8.30(a)). A small difference in the inclination of loop linear parts in relation to the diagram's horizontal axis, which results from a difference of elastic stress concentration factors, is observed. This can apparently be explained by a different influence of a specimen thickness on the surface strains along an open hole edge and along a line of contact interaction (see Section 8.1).
- The local strain amplitude versus the loading cycle number diagram for specimen VI (curve 4 in Figure 8.15) has a stable part like the same diagram for specimen I (curve 1 in Figure 8.15). It should be noted that the strain amplitude value at the stable stage of the local cyclic deformation process are up to 10 per cent less in specimen VI than in specimen I.
- An increase of the maximum strain amplitude on open and filled holes (observed at the end of the diagrams) shows that subsurface defects turn into microcracks and emerge in the specimen surface. This process has a more fluent development in specimen I with an open hole.
- The end points of dependencies 1 and 4 in Figure 8.15, as proven by the interruptions of the interference fringes in Figure 8.14(a) and 8.27 and, in addition, the results of the subsequent deformation process investigation of specimen VI (given below), indicate the beginning of a stable fatigue crack growth process.
- The notable twofold (and more) increase of the interval before the beginning of the fatigue crack growth process in specimen VI with a filled hole in comparison to specimen I with an open hole can be partly explained by a difference of stress concentration factor values on the specimen's surface ($K_\sigma = 2.9$ for specimen III and $K_\sigma = 2.7$ for specimen IV). Undoubtedly, the contact interaction along the hole boundary exerts its main influence on the concerned part of the lifetime of specimen VI.

A detailed analysis of the fatigue crack growth process will be considered below.

Typical interferograms of specimen VI recorded at the 3931st loading cycle for net stress increment σ_0 from 0 to 36 MPa (a), from 144 to 175 MPa (b) and from 206 to 230 MPa (c) are shown in Figure 8.31. The fringe patterns which were recorded at the 4133rd cycle for the same load increments as the interferograms shown in Figure 8.31 are presented in Figures 8.32(a), (b) and (c), respectively. Both sets of photographs clearly show the fringe interruptions along the crack edges. The crack is so long that its length can be determined with a high accuracy by means of fringe pattern magnified photograph analysis. At the 3931st cycle the length L of the longest of the two cracks under observation on the surface of the specimen (top part of the photographs in Figure 8.31) is 3.4 mm. The analogous value for the 4230th cycle is 5.2 mm.

The crack opening value on the hole boundary Δ at the 3931st and at the 4130th loading cycles is so large that it is practically impossible to obtain a

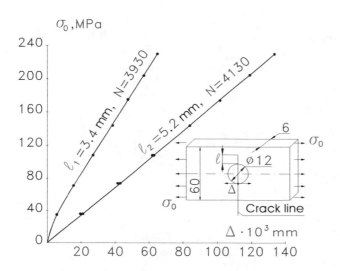

Figure 8.33. Dependencies between the crack mouth opening displacement and net stress level obtained at the stage of active fatigue crack propagation in specimen VI

corresponding circumferential strain distributions by means of Fourier series application. In this case the crack opening displacement can be determined by using the optimal two hologram interferometer with sensitivity matrix (2.70). A detailed description of the above-mentioned procedure is presented in Section 6.1. Note that an analogous approach can be implemented for residual stress determination (see Section 9.2).

Dependencies between crack opening displacement Δ on the hole edge (crack mouth opening displacement—CMOD) and net stress level at the 3931st and 4133rd cycles are shown in Figure 8.33. The main difference between these dependencies consists of the presence of non-linear parts at the beginning of the function, which describes CMOD for the 3931st cycle. CMOD versus the net stress level diagram for the 4130th cycle has a linear character. The latter factor is very important. The case is that the fatigue fracture of specimen VI occurred at the 4160th loading cycle and, therefore, we can be sure that the crack length at the 4160th cycle is practically a critical crack length. In other words, this crack is characterized by a critical CMOD and critical stress intensity factor (SIF) value.

The last parameter can be considered as a mechanical property of the material of the specimen involved. The corresponding meaning of critical SIF value for the material used lies in the range from 25 to 35 MPa $\sqrt{m}$.

An estimation of the SIF value in the case concerned can be based on relations (6.2) and (6.3). The corresponding distribution of the crack opening

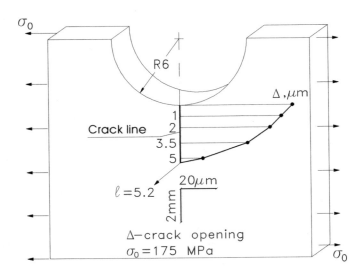

Figure 8.34. Opening displacement distribution along the crack length in specimen VI obtained at 4133rd loading cycle

displacement along the crack length obtained at the 4133rd loading cycle for net stress increment σ_0 from 0 to 175 MPa is shown in Figure 8.34. The approximation of this dependence with series (6.3) shows that only two first terms should be taken into account. Substituting the first of these coefficients into (6.4) for $H = E$ (where $E = 68\,000$ MPa is the elasticity modulus of the specimen's material after 2927 loading cycle) yields

$$K_{IC} = 26\,\text{MPa}\sqrt{m}.$$

Note that this magnitude is obtained taking into account a change of the material elasticity modulus due to cyclic loading.

In conclusion, it may be said that the quantitative relations obtained between displacement or strain fields and net stresses allow us to separate the following typical stage of the deformation process investigated: elasto-plastic deformation at the first and second cycles; the adaptability stage which is characterized by the change in the local material mechanical properties ($N = 14$–1316, where N is the loading cycle number); the stage of a subsurface defect initiation in a contact interaction zone ($N = 2000$–2700); surface crack appearance ($N = 2900$–3100); fatigue crack active propagation stage ($N = 3200$–4000); and the critical crack length stage ($N > 4000$, fracture cycle $N = 4160$).

References

1. Osgood, C. C. (1982) *Fatigue Design.* Pergamon Press, Oxford.
2. Makhutov, N. A. (1981) *Deformation Criteria of Fracture and Structure Elements Strength Analysis.* Mashinostroenie, Moscow.
3. Collins, J. A. (1981) *Failure of Materials in Mechanical Design. Analysis, Prediction, Prevention.* John Wiley, New York.
4. Vorobiev, A. Z., Olkin, B. I., Stebenev, V. N. and Rodchenko, T. S. (1990) *Fatigue Resistance of Structures Elements.* Mashinostroenie, Moscow.
5. Yarkovets, A. I., Sirotkin, O. S., Firsov, V. A. and Kiselev, N. M. (1987) *Manufacturing Technology for High-Fatigue-life Rivet and Bolt Joints in Aircraft Structures.* Mashinostroenie, Moscow.
6. Sukharev, I. P. (1987) *Experimental Methods of Deformations and Strength Study.* Mashinostroenie, Moscow.
7. Kobayashi, A. S. (ed.) (1987) *Handbook of Experimental Mechanics.* Prentice Hall, Englewood Cliffs, New Jersey.
8. Kozhevnikov, V. F. (1976) Stress state of strip in tension with filled hole. *Proc.TsAGI* 7, 90–98.
9. Roulands, R. E., Rahman, M. U., Wilkinson, T. L. and Chiang, Y. I. (1982) Single and multiple-bolted joints in orthotropic materials. *Composites* 13, 273–279.
10. Haer, M. W., Klang, E. C. and Cooper, D. E. (1987) The effects of pin elasticity, clearance and friction on the stresses in a pin-loaded orthotropic plate. *J. Compos. Mater.* 21, 190–206.
11. Ushakov, B. N. and Dunaev, V. V. (1981) Stress analysis in pin-joint with interference fit on model made of plastic. In *Plastic Utilization in Machinery Production*, 18, pp. 55–69, MVTU, Moscow.
12. Kozhevnikov, V. F. and Vidanov, N. A. (1986) Stress state of a plate, loaded by pin. *Probl. Prochn.* N10, 38–43.
13. Vidanov, N. A. (1989) Influence of some longitudinal joint characteristics on the stress concentration under one-axis loading. *Proc. TsAGI* 20, 76–82.
14. Mangalgiri, P. D., Ramamurthy, T. S., Dattaguru, B. and Rao, A. K. (1987) Elastic analysis of pin joints in plates under some combined loads. *Int. J. Mech. Sci.* 29, 577–585.
15. Grishin, V. I. and Begeev, T. K. (1988) Contact interaction study in aircraft structure elements. *Probl. Prochn.* N9, 84–87.
16. Yogeswaren, E. K. and Reddy, J. N. (1988) A study of contact stresses in pin-loaded orhotropic plates. *Comput. Structur.* 28, 1067–1077.
17. Zhilkin, V. A., Shevtsov, R. G., Kosenyuk, V. K. and Rakin, A. S. (1987) Stress investigation in pin-joints by holographic moiré method. *Probl. Prochn.* N12, 74–77.
18. Grishin, V. I., Begeev, T. K. and Litvinov, V. B. (1989) A study of stress–strain state of orthotropic material joints. *Mekh. Kompozit. Mater.* N4, 650–654.
19. Prigorovsky, N. I. (1983) *Methods and Measuring Devices for Strain and Stress Fields Determination:* Handbook. Mashinostroenie, Moscow.
20. Post, D. (1982) Developments in moiré interferometry. *Opt. Engineer.* 21, 458–467.
21. Sciammarella, C. A. (1982) The moiré method—a review. *Exp. Mech.* 22, 418–433.
22. Kang, B. S. and Kobayashi, A. S. (1988) J-resistance curves in aluminium SEN specimens using moiré interferometry. *Exp. Mech.* 28, 154–158.
23. Zhylkin, V. A. and Popov, A. M. Investigation of elasto-plastic strains. *Zavod. Labor.* 53, 65–68.
24. Post, D., Han, B. and Ifju, P. (1993) *High Sensitivity moiré. Experimental analysis for mechanics and materials.* Springer-Verlag, Berlin.

25. Zhylkin, V. A., Shevtsov, R. G., Koseniuk, V. K. and Rakin, A. S. (1987) Investigation of stress in key joints by holographic moiré technique. *Probl. Prochn.* N12, 74–77.
26. Gorodnichenko, V. I., Pisarev, V. S. and Shchepinov, V. P. (1990) Local strain diagram plotting under static loading by holographic interferometry data. *Zavod. Labor.* **56**, 59–64.
27. Molsky, K. and Glinka, G. (1981) A method of elastic-plastic stress and strain calculation at a 'notch root'. *Mater. Sci. Engng* **50**, 93–110.
28. Gorodnichenko, V. I. and Pisarev, V. S. (1990) Some laws of the initial stage of local cyclic deformation obtained by holographic interferometry method. *Fiz. Khim. Mekh. Mater.* N5, 106–113.
29. Gorodnichenko, V. I. and Pisarev, V. S. (1992) Local cyclic deformation of a plate with open hole by holographic interferometry data. *Zavod. Labor.* **58**, 52–55.
30. Guillot, M. W. and Sharpe, W. N. (1983) Local deformation investigation under cyclic loading. *Exp. Mech.* **23**, 354–360.
31. Peterson, R. E. (1974) *Charts and Relations useful in Making Strength Calculations for Machine Parts and Structural Elements*. John Wiley, New York.
32. Gorodnichenko, V. I., Lebed, E. V. and Pisarev, V. S. (1993) Initial stage of material local cyclic deformation in contact interaction zone near a hole by holographic interferometry data. *Probl. Proch.* N7, 90–95.
33. Birger, I. A. and Panovko, Ya. G., (eds) (1968) *Strength. Stability. Vibrations. Handbook.* Vol. 2. Mashinostroenie, Moscow.
34. Panasiuk, V. V. and Tepliy, M. I. (1975) *Some Contact Problems of the Elastic Theory*. Nauk. Dumka, Kiev.

9 RESIDUAL STRESS DETERMINATION BY THE HOLE-DRILLING TECHNIQUE

Residual stresses which appear during different stages of the manufacturing process of structural elements may exert a considerably negative influence on both the structure's static strength and fatigue life-time [1]. Therefore, the problem of determination of both the magnitude and sign of residual stresses in real structures is of great practical and scientific interest.

One of the most current and widely used methods of residual stress determination is the so-called hole-drilling technique [2–5]. According to this approach, a through or partially-through hole has to be drilled at the surface point of interest. After drilling, the hole vicinity is deformed due to unloading of material caused by a residual stress energy release. Measurement of these deformations in several directions allows us to determine the directions of the principal residual stresses (strains) and then to calculate their values by means of various analytical or numerical models.

The implementation of holographic interferometry for measurement of strains in the hole vicinity is capable of ensuring the highest sensitivity of the hole-drilling technique with respect to the residual stress determination. This is attributed to the fact that the determination of the displacement components, needed for calculation of the local strains at the hole vicinity, and further evaluation of the residual stresses can be performed along the hole boundary immediately. Moreover, the high sensitivity of the holographic interferometric techniques allows us to obtain reliable and accurate quantitative results when holes of small diameter (1–3 mm) are drilled. This means that a combination of the hole-drilling technique and holographic interferometry provides minimal possible destruction of the object surface in the region of interest which is necessary to detect residual stresses. Moreover, for large thick-walled structural elements the above-mentioned approach can be considered as a method of non-destructive testing when the geometrical parameters of the hole to be drilled do not exceed the damage dimensions that are permissible during the structure's service.

Another important feature of the approach consists of the fact that a holographic interferogram containing information on the displacement components fields near the hole boundary allows us to determine beyond doubt the directions of the principal strains (stresses) [6]. This feature is capable of simplifying the residual stress calculation and increasing the accuracy of the final results to a notable extent.

Any technique of residual stress determination based on hole-drilling implementation requiring determination of local deformation parameters only. Reflection hologram interferometry is the most suitable way for local strain investigation. As established in Chapters 2, 5, 6 and 8, a combination of the optimal optical set-up for a fringe pattern reconstruction with the technique of quantitative interferogram interpretation based on absolute fringe order counting is the most reliable and universal approach to determination of two- and three-dimensional local displacement fields.

To date, determination of the magnitude and sign of residual stresses by combining hole-drilling and the holographic interferometry method is mainly based upon analysis of the distribution along the hole boundary of the normal to the object surface displacement component w. The off-axis optical arrangement (see Figure 1.1) [6,7] as well as reflection hologram recording (see Figure 1.3(d)) [8–10] are used for this purpose.

However the approach mentioned above in many cases cannot be considered as an optimal method of residual stress analysis. In fact, determination of the tangential displacement components u and v can lead to more accurate and reliable evaluation of residual stresses for many practical cases [11,12]. The possibility of residual stress determination through the use of the tangential displacement components is particularly discussed in references [10,13,14]. However, a detailed metrological justification of similar approaches and descriptions of its practical implementation are absent in the current literature.

This chapter contains an unconventional description of the technique of residual stress determination based on the measurement of the tangential displacement components at four points of the hole edge by means of reflection hologram recording. An analysis of errors made in the residual strain and stress determination is presented. A special approach allowing us to define the absolute sign of the hole diameter change in two mutually orthogonal directions of the principal strains is also discussed. This means that the correct values of two principal residual stresses can be derived by interpreting two pairs of interferograms obtained for two pairs of symmetrical observations lying in two mutually perpendicular planes. The models and relations used for residual stress calculation are also presented. The accuracy of the approach involved is established through the use of special test experiments.

9.1   Basic relations between hole edge displacement components, principal strains and residual stresses to be determined

The unique feature of the approach to determination of residual stresses is that interference fringe patterns containing quantitative information on displacements at points of the hole boundary in the directions of the principal strains can be obtained by means of reflection hologram recording. According to the classical approach, the basic assumption required for quantitative interpretation of the displacement fields (resulting from hole drilling) in terms of residual stresses consists of the fact that a small hole is made in the two-dimensional stress field which is uniform independently of each two principal stress directions on the object surface. This means that the possible influence of stress gradients within the nearest hole vicinity is not taken into account.

We shall consider a small area of the surface to be investigated which is under plane stress conditions. Components σ_1 and σ_2 shall denote the stress field that satisfy the above-mentioned conditions (see Figure 9.1). A hole with diameter $2r_0$ is made at some point on the surface area under consideration. The centre of the hole is a conventional point where the residual stresses must be determined. The Cartesian coordinate system (x_1, x_2) with axes coinciding with the directions of the principal strains (stresses) ε_1 and ε_2, respectively, is also shown in Figure 9.1. These axes can be easily determined as the axes of symmetry of the interference fringe pattern.

After hole drilling and subsequent strain and stress redistribution, a one-axis stress state will exist along the loading-free boundary of the hole. In the case of a circular hole, the corresponding strain and stress components are given by

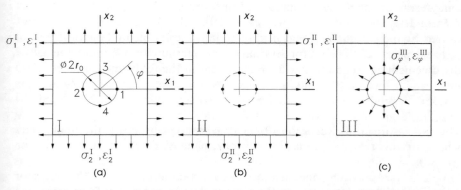

Figure 9.1. The model of residual stresses relaxation near the hole drilled in a uniform stress field

equations (5.87) and (5.49), respectively, which we rewrite here for the particular case involved:

$$\varepsilon_\varphi = \frac{1}{r_0}\left[\frac{\partial v}{\partial \varphi}\cos\varphi - \frac{\partial u}{\partial \varphi}\sin\varphi\right], \qquad \sigma_\varepsilon = E\varepsilon_\varphi \qquad (9.1)$$

where r_0 is the hole radius; u is the polar angle counting from the coordinate axis x_1 (see Figure 9.1); u and v are the displacement components of the point belonging to the hole edge in the directions x_1 and x_2 respectively; E is the elasticity modulus; and ε_φ and σ_φ are the circumferential strain and stress, respectively. It should be noted that expression (9.1) describes the strain and stress relaxation resulting from a combined influence of the σ_1 and σ_2 stress components.

The Roman numerals in Figure 9.1 conventionally denote different states of the object surface, each of which has to be quantitatively described in terms of principal strains and stresses to calculate the residual stresses.

State III (see Figure 9.1(c)) corresponds to the residual stress release after hole drilling and can be represented as a difference between the states I (see Figure 9.1(a)) and II (see Figure 9.1(b)). The parameters of state III have to be derived from the interference fringe pattern and those of states I and II should be described analytically in the general form proceeding from the elasticity theory relations.

State I corresponds to a two-axes loading with tension or compression of the element of the material volume containing the partially through (in the case of a thick-walled structure) or through (in the case of a thin-walled structure) hole. The components of this stress field are the principal residual stresses σ_1 and σ_2 to be determined. The relations between strains and stresses on the object surface in the case considered represent solution of the corresponding one-axis elasticity problem related to stress concentration and can be analytically or numerically represented in form (9.1).

State II is the initial deformed state of the object surface part of interest caused by the two-dimensional field of residual stress before hole drilling.

The general relationship for residual stress determination through the use of the hole-drilling technique can be represented in the following form:

$$I - II = III. \qquad (9.2)$$

The explicit form of expression (9.2) can be written by using any parameters which are inherent to the elasticity theory, namely, displacements, strains or stresses.

Within the approach concerned, we can most conveniently use the values of circumferential strains ε_φ having the most evident mechanical interpretation, at two characteristic points of the hole boundary shown in Figure 9.1(a):

$\varphi = 90°$ (point 3) and $\varphi = 0$ (point 1).

The following designations of these strains are introduced for further consideration:

$$\text{at point 1 } \varepsilon_{\varphi=0} \equiv \varepsilon_2^m$$
$$\text{at point 3 } \varepsilon_{\varphi=90°} \equiv \varepsilon_1^m \qquad m = \text{I, II, III.}$$

In order to determine the values of principal residual stresses σ_1 and σ_2, the strain values ε_1^I, ε_2^I, ε_1^{II} and ε_2^{II} have to be represented by the unknown values σ_1 and σ_2, and the strain values ε_1^{III} and ε_2^{II} should be derived through the experimentally obtained values of the tangential displacement components u and v in the corresponding directions. Subsequently, all these parameters must be absorbed into equation (9.2).

Let's consider in detail the relations between the parameters contained in equation (9.2) for all three above-mentioned states of the object surface.

STATE I—In this case the relationships between unknown values σ_1 and σ_2, on the one hand, and principal strains ε_1^I and ε_2^I, on the other, should be based on an analytical or numerical model describing an elastic deformation of the selected part of the object with a hole. In a general case, this model represents the solution of the corresponding elasticity problem concerning one-axis loading either a thin plate with through hole or a three-dimensional body with partially-through hole, the depth of which is usually equal to the hole diameter. A choice of the correct model of stress concentration factor determination is apparently the most complex step for all destructive methods of residual stress determination.

A more detailed analysis of the difficulties and peculiarities connected with the choice of a specific model for local stress determination lies beyond the scope of this book. It is important to us only that the output data of each such model can be represented in the form of a relative circumferential strain distribution along the hole edge. A typical form of such a distribution is shown in Figure 9.2. The values which are necessary for residual stress calculation are the magnitudes of the strain concentration factors α_1 and α_2 at points 1 and 3, respectively:

$$\alpha_1 = \frac{\tilde{\varepsilon}_{\varphi=90°}}{\varepsilon_1}, \qquad \alpha_2 = -\frac{\tilde{\varepsilon}_{\varphi=0}}{\varepsilon_1} \tag{9.3}$$

where E is the elasticity modulus of material; $\tilde{\varepsilon}_{\varphi=90°}$ and $\tilde{\varepsilon}_{\varphi=0}$ are the circumferential strains obtained from the solution of the corresponding one-axis elastic problem at points 1 and 3 lying on the hole edge (see Figure 9.1(a)). Note that generally when the value $\tilde{\varepsilon}_{\varphi=90°}$ is greater than zero, the magnitude

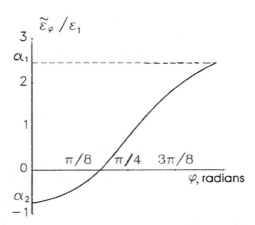

Figure 9.2. Typical distribution of the relative circumferential strain along the hole edge

of $\tilde{\varepsilon}_{\varphi=0}$ may be greater than, equal to, or less than zero. Figure 9.2 illustrates the most widespread practical situation when, for example, for one-axis tension, $\tilde{\varepsilon}_{\varphi=90°} > 0$ and $\tilde{\varepsilon}_{\varphi=0} < 0$.

In order easily to interpret the results of the following consideration we assume that the object surface near the hole can be referred to as the quasi-plane surface. This means that the directions of the principal strains ε_1 and ε_2 can be considered almost interchangeable with respect to the solution of the one-axis stress concentration problem. In such a case the corresponding relations for the principal strains ε_1 and ε_2 take the form

$$\varepsilon_1^{\mathrm{I}} = \alpha_1 \frac{\sigma_1}{E} - \alpha_2 \frac{\sigma_2}{E},$$
$$\varepsilon_2^{\mathrm{I}} = \alpha_1 \frac{\sigma_2}{E} - \alpha_2 \frac{\sigma_1}{E} \tag{9.4}$$

where α_1 and α_2 are the stress concentration factors from expression (9.3).

STATE II—The relations between the initial residual strains and stresses at points 1 and 3 of the surface area under consideration before drilling have the following form resulting from Hooke's law (5.13):

$$\varepsilon_1^{\mathrm{II}} = \frac{\sigma_1}{E} - \frac{\mu\sigma_2}{E},$$
$$\varepsilon_2^{\mathrm{II}} = \frac{\sigma_2}{E} - \frac{\mu\sigma_1}{E} \tag{9.5}$$

where μ is the Poisson's ratio of the material.

STATE III—To complete equation (9.1) with the right-hand side, we need to find how the strains ε_1^{III} and ε_2^{III} at points 1 and 3, respectively, at the hole boundary, caused by residual stress release, are related to the tangential displacement components u and v at some points on the hole edge derived from holographic interferometry data.

This problem can be overcome in the following way. It is known, for instance, that the Kirsch solution for one-axis tension of an infinite thin plate with a circular through hole gives the following form of circumferential stress distribution along the hole boundary [15]:

$$\sigma_\varphi = (1 - 2\cos\varphi)\sigma_1 \qquad (9.6)$$

where σ_1 denotes the net stress. The distribution of the circumferential stress along the circular hole edge in thin plate subjected to one-axis pure bending has an analogous form. It can be easily shown that the distributions of tangential displacement components u and v along a circular hole boundary corresponding to the stress distribution of form (9.6) have the following form:

$$u = A_1 \cos\varphi, \quad v = -B_1 \sin\varphi \qquad (9.7)$$

where $A_1, B_1 > 0$ in the case of one-axis tension and the polar angle φ is calculated from the hole diameter coinciding with the positive direction of x_1. Moreover, the spectrum with the same composition as expression (9.7) describes distributions of the tangential displacement components u and v along a circular hole boundary in a finite-width strip subjected to tension or compression under plane stress conditions in the elastic range. The experimental data presented in Sections 2.5, 4.2, 8.1 and 8.2 clearly confirm this. The same is also valid in the case of deformation of an object with a partially through circular hole. One of the most characteristic examples of the numerical solution of the latter problem will be given below. Substituting expression (9.7) into (9.1) yields

$$\varepsilon_\varphi^{III} = \frac{1}{r_0}[A_1 \sin^2\varphi - B_1 \cos^2\varphi]. \qquad (9.8)$$

Relation (9.8) directly gives the strain values ε_1^{III} which ε_2^{III} are necessary for solving equation (9.2):

$$\varepsilon_1^{III} = \frac{A_1}{r_0}, \quad \varepsilon_2^{III} = \frac{-B_1}{r_0}. \qquad (9.9)$$

Relations (9.9) can be reduced to a most convenient form for practical use. Relations (9.7) illustrate that the magnitudes $2A_1 = \Delta u$ and $-2B_1 = \Delta v$ can be defined as absolute values of the hole diameter change in the corresponding directions of principal strains ε_1 and ε_2. Using the introduced designations, relations (9.9) can be rewritten in the following form:

$$\varepsilon_1^{III} = \frac{\Delta u}{2r_0}, \ \varepsilon_2^{III} = \frac{\Delta v}{2r_0} \tag{9.10}$$

where Δu and Δv are the absolute values of change in the hole diameters passing through points 1, 2 and 3, 4 respectively (see Figure 9.1(a)).

Expressions (9.10) prove that the strain at the point of intersection of the hole boundary and the ith ($i = 1, 2$) direction of the principal stresses can be represented as the relative change of the hole diameter in the same direction. This is of great metrological importance in view of the accuracy of residual stress determination, since the values of strains ε_1^{III} and ε_2^{III} can be derived from holographically measured values Δu and Δv in the simplest manner without a numerical differentiation procedure.

Now we are capable of constructing the equation system from two equations of the form (9.2) written for points 1 and 3 at the hole edge which is sufficient to evaluate two unknown residual stress components σ_1 and σ_2:

$$\begin{cases} \varepsilon_1^{I} - \varepsilon_1^{II} = \varepsilon_1^{III} \\ \varepsilon_2^{I} - \varepsilon_2^{II} = \varepsilon_2^{III} \end{cases}. \tag{9.11}$$

The explicit form of this equation system is revealed by substituting expressions (9.4), (9.5) and (9.8) into (9.11):

$$\frac{1}{E}(\alpha_1\sigma_1 - \alpha_2\sigma_2 - \sigma_1 + \mu\sigma_2) = \frac{\Delta u}{2r_0},$$
$$\frac{1}{E}(\alpha_1\sigma_2 - \alpha_2\sigma_1 - \sigma_2 + \mu\sigma_1) = \frac{\Delta v}{2r_0}. \tag{9.12}$$

Solving the equation system (9.12) gives us the required relations for calculation of the residual stresses by means of parameters obtained by holographic interferometry:

$$\sigma_1 = \frac{E}{2r_0}\left[\frac{(\alpha_1 - 1)\Delta u + (\alpha_2 - \mu)\Delta v}{(\alpha_1 - 1)^2 - (\alpha_2 - \mu)^2}\right],$$
$$\sigma_2 = \frac{E}{2r_0}\left[\frac{(\alpha_1 - 1)\Delta v + (\alpha_2 - \mu)\Delta u}{(\alpha_1 - 1)^2 - (\alpha_2 - \mu)^2}\right]. \tag{9.13}$$

Note that in order to use relations (9.13) for the correct evaluation of the residual stresses, not only do the absolute values of the relative diameter change Δu and Δv in two directions of the principal strains, but its physical signs have to be established. As is well known (see Section 2.1) the method of holographic interferometry allows us to determine the displacement components to within a sign. This could reduce holographic interferometry application to mainly residual stresses analysis. To overcome this problem some additional information must be derived from holographic interferograms. The unconventional approach which provides us with the possibility to establish the physical signs of diameter changes Δu and Δv through the use of the same fringe patterns which serve for its absolute values determination will be presented in Section 9.2.

The representation of formulae (9.13) in the explicit form for two characteristic models widely used for the determination of the strains ε_1^I and ε_2^I would be appropriate in the conclusion of this section.

(1) A wide thin plate with a circular through hole subjected to a one-axis tension (compression). In this case the values of stress concentration factors (9.3) can be obtained from Kirsch's solution (9.6):

$$\alpha_1 = 3, \ \alpha_2 = 1.$$

Substituting these values into expression (9.13) yields

$$\sigma_1 = \frac{E}{2r_0} \left[\frac{2\Delta u + (1 - \mu)\Delta v}{(3 - \mu)(1 + \mu)} \right],$$

$$\sigma_2 = \frac{E}{2r_0} \left[\frac{2\Delta v + (1 - \mu)\Delta u}{(3 - \mu)(1 + \mu)} \right].$$

(9.14)

It should be noted that expressions (9.14) in this particular case coincide with the corresponding relations obtained in reference [11].

(2) An infinite thick plate with a circular partially through hole, the depth of which is equal to its diameter, subjected to a tension (compression). Values (9.3) were derived from the finite element solution of the corresponding elastic problem by means of the software package FITING [16]:

$$\alpha_1 = 2.60, \ \alpha_2 = 0.67.$$

In this case expressions (9.13) take the form

$$\sigma_1 = \frac{E}{2r_0} \left[\frac{1.6\Delta u + (0.67 - \mu)\Delta v}{(2.27 - \mu)(0.93 + \mu)} \right],$$

$$\sigma_2 = \frac{E}{2r_0} \left[\frac{1.6\Delta v + (0.67 - \mu)\Delta u}{(2.27 - \mu)(0.93 + \mu)} \right].$$

(9.15)

9.2 Interpretation of holographic interferograms recorded near the hole drilled in a uniform stress field

The recording process of the reflection hologram located immediately before the fragment of the quasi-plane object surface, at the centre of which the hole is drilled between exposures, is shown in Figure 9.3. The position of the hole edge with respect to the Cartesian coordinate system whose axes lying in the object surface coincide with the directions of the principal residual stresses can also be observed in this figure. The object surface area under study is illuminated with a plane wavefront (unit vector $\vec{e}_s$) directed along the normal to the surface (x_3-axis) at the hole centre. As shown in Section 2.4, this type of recording allows us to select the optimal parameters of the interferometer optical system (angle ψ between the unit observation vectors $\vec{e}_j$ ($j = 1, 2, 3, 4$) and vector $\vec{e}_s$) when the reflection hologram is being reconstructed.

The main feature of this hologram recording procedure is that the hologram has to be removed from its initial position after the first exposure and then

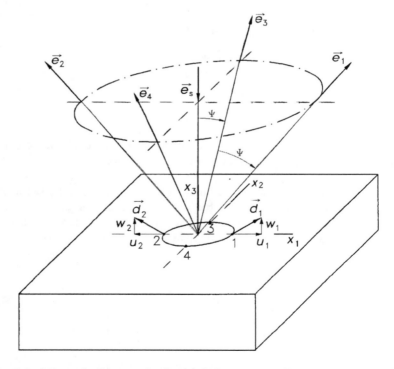

Figure 9.3. Schematic diagram of reflection hologram recording near the hole and fringe pattern reconstruction

returned to the same position with an accuracy to within less than half of the laser wavelength. This problem can be effectively solved by mounting a photoplate in the special kinematic device [17].

As mentioned in Section 2.4, a suitable choice of the number of viewing directions and its spatial configuration in order to reduce the influence of the experimental errors on the results of the holographic interferometric measurements depends on the number and directions of the displacement components to be determined. As follows from relations (9.13), two mutually orthogonal tangential displacement components in the definite directions have to be used as input data for a residual stress calculation. Remember that these directions coincide with the symmetry axes of the interference fringe pattern near the hole and can be easily established after hologram reconstruction.

The above-mentioned facts allow us to implement the optical set-up corresponding to the sensitivity matrix (2.70) to obtain all the needed information for residual stress evaluation. The spatial configuration of two pairs of viewing directions each of which allows us to determine the normal to the object surface displacement component w and one tangential displacement component u (or v) is shown in Figure 9.3.

To illustrate the procedure of relative diameter change determination in the direction of principal strain x_1 and x_2 we refer again to Figure 9.3. For example, we shall consider the diameter change in direction x_1 that passes through points 1 and 2 belonging to the hole edge (see Figures 9.1(a) and 9.3). Let's introduce the double-index system for the absolute fringe orders designation:

$$N_j^m$$

where $m = 1, 2, 3, 4$ is the number of the point at the hole edge and $j = 1, 2$ is the number of the viewing direction.

Note that for a hole made in a uniform stress field the tangential displacement component normal to the hole boundary of two points 1 and 2 lying on the same diameter u_1 and u_2 have the same value but an opposite sign. Figure 9.3 illustrates the case of an increase in the hole diameter in the direction x_1. The linear equation system (2.32) with the optimal sensitivity matrix (2.70) written for the point 1 at the hole boundary takes the following form:

$$\begin{bmatrix} \sin \psi & 1 + \cos \psi \\ -\sin \psi & 1 + \cos \psi \end{bmatrix} \begin{bmatrix} u_1 \\ w_1 \end{bmatrix} = \lambda \begin{bmatrix} N_1^1 \\ N_2^1 \end{bmatrix}. \tag{9.16}$$

The analogous expression for point 2 is given by

$$\begin{bmatrix} \sin \psi & 1 + \cos \psi \\ -\sin \psi & 1 + \cos \psi \end{bmatrix} \begin{bmatrix} -u_2 \\ w_1 \end{bmatrix} = \lambda \begin{bmatrix} N_1^2 \\ N_2^2 \end{bmatrix}. \tag{9.17}$$

Solving of the systems (9.16) and (9.17) results in, respectively,

$$u_1 = \frac{\lambda}{2\sin\psi}\,(N_1^1 - N_2^1),\ u_2 = \frac{\lambda}{2\sin\psi}\,(N_2^2 - N_1^2). \tag{9.18}$$

The error made in the tangential displacement component determination in the case concerned is estimated by expression (2.73). Residual stress analysis by means of this approach shows that a reliable quantitative determination of the tangential displacement components along the edge of the hole drilled in the stress field is possible when the viewing angle ψ is not less than 40°.

Let us return to the determination of the required diameter change in the x_1-direction Δu that is expressed as the sum of the first and second equations (9.18):

$$\Delta u = u_1 + u_2 = \frac{\lambda}{2\sin\psi}\,[(N_1^1 - N_1^2) + (N_2^2 - N_2^1)] \tag{9.19}$$

It is very important to point out that each of the two pairs of terms enclosed in brackets in the right-hand side of equation (9.19) can be defined as the difference of the absolute fringe orders counted on the single fringe pattern between points 1 and 2. This reveals the main advantage of the approach which consists of the possibility of quantitative interpretation of the fringe patterns obtained in terms of the absolute fringe orders without determination of the zero-order fringe. Solving the latter problem when an interference fringe pattern in the hole vicinity results from the residual stress redistribution caused by hole drilling between exposures is quite difficult. Naturally, equation (9.19) can be rewritten in terms of the diameter changing Δv in the direction x_2.

In order to establish the accuracy of the technique developed, a special test experiment was performed. The object investigated was a regular thin plate subjected to a load in the two-axes plane stress field (see the right-hand side of Figure 5.6). The specimen was loaded so that the uniform stress field with known parameters existed on the surface area of interest.

The interferograms obtained as a result of drilling a hole with a radius $r = 1.1$mm in the rectangular thin plate under plane stress conditions with principal stresses $\sigma_1 = -\sigma_2$ (so-called pure shearing) are presented in Figure 9.4. The symmetry of the fringe patterns corresponding to the displacement components of points lying in the same hole diameter should be noted as confirmation of the efficiency of the hologram recording and reconstruction process.

The quantitative interpretation of the interferogram presented in terms of absolute fringe orders provides the following input data needed for the determination of the tangential displacement component u at points 1 and 2 of the hole edge:

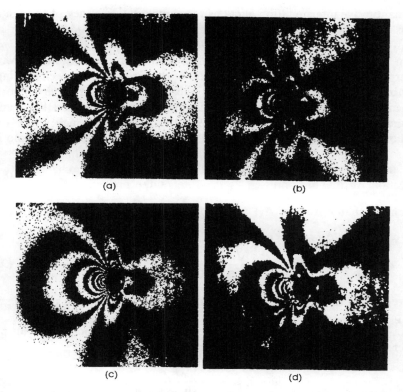

Figure 9.4. Interferograms for determination of the in-plane displacement components u (a, b) and v (c, d) near the hole drilled in a thin plate subjected to pure shearing

$$N_1^1 - N_2^1 = \Delta N_{12}^1 = 7.5 \text{ (see Figure 9.4(a))},$$
$$N_1^2 - N_2^2 = \Delta N_{12}^2 = 8.2 \text{ (see Figure 9.4(b))}.$$

Analogous data for the displacement component v at points 3 and 4 of the hole edge are:

$$\Delta N_{34}^1 = 8.5 \text{ (see Figure 9.4(c))},$$
$$\Delta N_{34}^2 = 8.5 \text{ (see Figure 9.4(d))}.$$

Substituting these values into expression (9.19) for $\sin \psi = 0.81$ yields

$$|\Delta u| = 6.1 \ \mu\text{m}, \qquad |\Delta v| = 6.6 \ \mu\text{m}. \tag{9.20}$$

In the first instance, the fringe patterns presented and the final result (9.20) of its quantitative interpretation allow us to estimate the accuracy of the residual stress calculation according to the expressions (9.13) (in the particular case involved (9.14)).

An analysis of the almost perfect interference fringe patterns shown in Figure 9.4 demonstrates that the value Δn of the error made in each reading of the absolute fringe order may be about half of a fringe. This value defines the real magnitude of the error in the residual stress determination that can be obtained by using expressions (2.73), (9.13) and (9.19). Inequality (2.73) and relation (9.19) give us the absolute error of the diameter change determination:

$$|\delta(\Delta u)| = 2|\Delta d_2| \leqslant \frac{\sqrt{2}\lambda}{\sin\psi}. \tag{9.21}$$

Assuming that $\delta(\Delta u) = \delta(\Delta v)$ and substituting expression (9.21) into (9.13) yields

$$|\delta(\sigma_1)| \leqslant \frac{E\sqrt{2}\lambda}{2r_0\sin\psi}\left[\frac{1}{(\alpha_1 - 1) - (\alpha_2 - \mu)}\right] \tag{9.22}$$

where $\delta(\sigma_1)$ is the absolute error made in determination of each residual stress component.

Let us estimate the value of error (9.22) for a through hole in a thin plate under plane stress conditions ($\alpha_1 = 3$, $\alpha_2 = 1$):

$$|\delta(\sigma_1)| \leqslant \frac{\lambda E}{\sqrt{2}r_0\sin\psi(1+\mu)}. \tag{9.23}$$

For aluminium alloys ($\mu = 0.32$, $E = 72\,000\,\text{MPa}$ and actual experimental conditions ($2r_0 = 3\,\text{mm}$, $\psi = 54°$, $\sin\psi = 0.81$, $\lambda = 0.633\,\mu\text{m}$) inequality (9.23) gives

$$|\delta(\sigma_1)| \lesssim 20\,\text{MPa}.$$

This estimation reveals the minimum possible value of error in the residual stress determination in aluminium alloys by means of a combination of the holographic interferometry and hole-drilling techniques.

The residual stress calculation in accordance with expressions (9.13) requires, apart from the absolute values, the identification of physical signs of the hole diameter change (increase or decrease) in two directions of the principal strains. A conventional quantitative interpretation of double-exposure holographic interferograms, in particular shown in Figure 9.4, is capable of finding the displacement components values to within a sign. Therefore, to identify the physical direction of each displacement component, additional information should be used.

A detailed analysis of interference fringe patterns obtained near the hole drilled in the stress field between two exposures, which are recorded as a result of a special test experiment where the directions of each tangential displacement component are known *a priori*, illustrates that these fringe patterns contain information on the sign of the hole diameter change in the directions of the principal strains (stresses). In order to demonstrate this we should return to Figure 9.4 in which four almost identical fringe patterns are presented. As is well known from elasticity theory, the pairs of interferograms shown in Figures 9.4(a) and (b) and 9.4(c) and (d) correspond to equal but opposite in sign stress components $\sigma_1 = -\sigma_2$. In the case being considered, subsequent to a definite load application, it is known that the fringe patterns shown in Figures 9.4(a) and (b) and 9.4(c) and (d) correspond to tension and compression, respectively.

The evident distinctions of fringe trajectories in the near vicinity of the hole edge, which have a most explicit form on the hole boundary segment containing the greater fringe number, are clearly observed in the different pairs of interferograms. These distinctions are revealed at the location of the point of intersection of two tangentials drawn to the same interference fringe at the point of its intersection with the hole edge. In the case of compression (decreasing of the corresponding hole diameter) the point of intersection of the above-mentioned tangentials lies inside the hole (see Figures 9.4(c) and (d)). In the case of tension (increasing of the hole diameter) this point is located outside the hole (see Figures 9.4(a) and (b)).

The physical reason of the phenomenon involved consists apparently of the different character of contact interaction between a drill and hole edge in the regions of tension and compression of the material when the hole is being made in the material fragment undergoing stress.

9.3 Some examples of stress determination by the hole-drilling technique

A series of characteristic examples of stress determination is considered in this section. These describe an implementation of the approach discussed in the two previous sections to different classes of structures. The main objective of all the test experiments which are presented below is to establish the accuracy that can be reached in a quantitative determination of residual stresses.

A THIN PLATE UNDER PLANE STRESS CONDITIONS

Consider now the final part of the investigation described in Section 9.2 that consists of a quantitative determination of stresses σ_1 and σ_2. The values of absolute change of the hole diameter in the directions of the principal strains are given by expression (9.20). As was established in Section 9.2, the fringe patterns presented in Figure 9.4 allow us to determine that $\Delta u > 0$ and $\Delta v < 0$. Substituting the values (9.20) with its physical signs into expression (9.14) gives the following results:

$$\sigma_1 = 75\,\text{MPa}, \quad \sigma_2 = -87\,\text{MPa}.$$

These magnitudes are in good agreement with the stress values resulting from known conditions of the plate loading:

$$\sigma_1 = -\sigma_2 = 70\,\text{MPa}.$$

ONE-AXIAL TENSION OF A THIN-WALLED CIRCULAR CYLINDRICAL SHELL

The objects investigated were cylindrical aluminium (elasticity modulus $E = 72\,000\,\text{MPa}$, Poisson's ratio $\mu = 0.32$) shells of different length $L = 100\,\text{mm}$ and $L = 180\,\text{mm}$. Other geometrical parameters of shells are similar to those presented in the description in Figure 2.27.

Interferograms which illustrate the redistribution of the displacement components caused by the through hole drilling in the thin-walled cylindrical shell of length $L = 100\,\text{mm}$ and subjected to a tension are shown in Figure 9.5. The almost perfect symmetry of the corresponding pairs of interference fringe patterns should be again noted.

The results of quantitative interpretation of these interferograms for the displacement component u (the x_1-axis is directed along the shell generatrix) have the following values:

$$\Delta N^1_{12} = 9.3 \text{ (see Figure 9.5(a))},$$
$$\Delta N^2_{12} = 8.1 \text{ (see Figure 9.5(b))}.$$

The analogous magnitudes for the displacement component v are:

$$\Delta N^1_{34} = 4.3 \text{ (see Figure 9.5(c))},$$
$$\Delta N^2_{34} = 4.3 \text{ (see Figure 9.5(d))}.$$

Calculations in accordance with relation (9.19) in this case gives ($\sin \psi = 0.87$)

$$\Delta u = 6.33\,\mu\text{m}, \quad \Delta v = -3.13\,\mu\text{m}. \tag{9.24}$$

The signs of the hole diameter changing Δu and Δv are identified proceeding from the rule established in Section 9.2. Note, however, that this sign can be easily derived from the known direction of the tensile force.

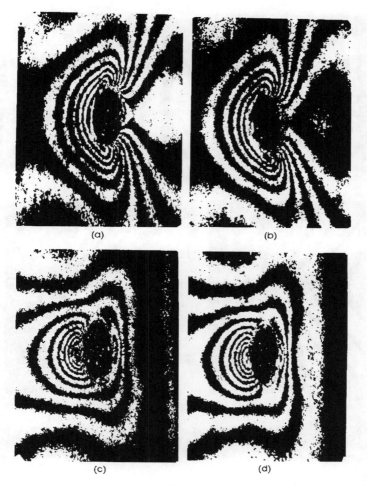

Figure 9.5. Interferograms for determination of the in-plane displacement components
u (a, b) and v (c, d) near the hole drilled in a thin-walled cylindrical shell of length 100
mm under tension

Interferograms which are analogous to those presented in Figure 9.5 but
obtained for the shell of length $L = 180$ mm are shown in Figure 9.6. The
obvious distinctions in the form of fringe patterns are revealed as a result of a
comparison of corresponding interferograms in Figures 9.5 and 9.6 which have
been recorded at the same stress level. These distinctions are due to the
different influences of boundary conditions on the deformation of the central
section of the shell. This influence for thin-walled structures mainly results in
the redistribution of the normal to the object surface displacement component

Figure 9.6. Interferograms for determination of the in-plane displacement components u (a, b) and v (c, d) near the hole drilled in a thin-walled cylindrical shell of length 180 mm under tension

w, but, as will be shown below, the values of the hole diameter change in the tangential directions Δu and Δv are almost the same for the two cases. This emphasizes the principal advantage of the tangential displacement components implementation for residual stress determination instead of the normal displacement component.

The quantitative results of the interpretation of the fringe patterns shown in Figure 9.6 for determination of the displacement component u have the following magnitudes:

$$\Delta N_{12}^1 = 10 \text{ (see Figure 9.6(a))},$$
$$\Delta N_{12}^2 = 10 \text{ (see Figure 9.6(b))}.$$

The analogous magnitudes for the displacement component v are:

$$\Delta N^1_{34} = 3.5 \text{ (see Figure 9.6(c))},$$
$$\Delta N^2_{34} = 3.7 \text{ (see Figure 9.6(d))}.$$

In this case relation (9.20), taking into account the physical sign of the tangential displacement components derived from the corresponding fringe pattern, results in the following values:

$$\Delta u = 7.3 \,\mu\text{m}, \qquad \Delta v = -2.6 \,\mu\text{m}. \tag{9.25}$$

Now to receive the stress values in accordance with expression (9.13) we have to establish the model describing state I of the surface under study (see Section 9.1) in terms of stress concentration factors α_1 and α_2 for the tension of a thin-walled shell with a small circular hole. As such a model, the classical solution of Lurie can be effectively used [18]:

$$\frac{\tilde{\varepsilon}_\varphi}{\varepsilon_1} = 1 - \cos 2\varphi \left[2 + \sqrt{\frac{3(m^2 - 1)}{m^2} \frac{\pi r_0^2}{4Rh}} \right] \tag{9.26}$$

where $m = 1/\mu$, μ is a material Poisson's ratio; $R = 30\,\text{mm}$ is the external radius of the shell; $h = 1.5\,\text{mm}$ is the thickness of the shell; and $r_0 = 1.6\,\text{mm}$ is the radius of the hole to be drilled in the shell. For the geometrical parameters of the shell presented above, the values of the stress concentration factor α_1 and α_2, which can be obtained from expression (9.26) for $\varphi = 90°$ and $\varphi = 0$, respectively, almost coincide with those used for the derivation of relation (9.14). Therefore, this relation can be used for residual stress determination in thin-walled cylindrical shells.

Substituting expression (9.24) and then (9.25) into (9.14) gives us respectively $\sigma_1 = 89\,\text{MPa}$, $\sigma_2 = 17\,\text{MPa}$ for the shell of length $L = 100\,\text{mm}$ and $\sigma_1 = 85\,\text{MPa}$, $\sigma_2 = 0$ for the shell of length $L = 180\,\text{mm}$. The stress values $\sigma_1 = 70\,\text{MPa}$ and $\sigma_2 = 0$ were held in lateral cross-sections of both shells investigated. These magnitudes are in a good agreement with experimental data to within the error (9.23).

PURE BENDING OF A BEAM OF RECTANGULAR CROSS-SECTION

All the above examples are related to stress determination in thin-walled structures where drilling a through hole is required. The other broad class of applications of the technique deals with stress determination in thick-walled or three-dimensional objects that have to be carried out by means of drilling a partially-through hole. To establish accuracy of the hole-drilling method in such a case and to clarify the questions concerning the choice of a correct model for a quantitative description of the stress concentration problem (state I

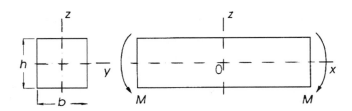

Figure 9.7. Schematic of a beam bending

in Figure 9.1), a pure bending of the long beam of rectangular cross-section is used as an example.

The beam was made of aluminium alloy (elasticity modulus $E = 72\,000$ MPa, Poisson's ratio $\mu = 0.32$). A diagram of the beam and scheme of its loading are shown in Figure 9.7. The dimensions of the beam's cross-section were: width $b = 30$ mm and height $h = 18$ mm. Under pure bending conditions according to the loading scheme shown in Figure 9.7, the surface layers of the beam were subjected to one-axis compression when $z = -h/2$, and to tension when $z = h/2$. Interferograms of the beam, which were obtained as a result of drilling the partially-through hole of diameter $2r_0 = 3$ mm and depth being equal to the diameter, are presented in Figure 9.8. Evidently, the fringe patterns in Figures 9.8(a) and (b) which correspond to the displacement component u (the x_1-axis is directed along the axis of the beam) can be identified as the case of the hole diameter decreasing in the direction x_1 in accordance with the criterion described in Section 9.2. The opposite sign of the hole diameter change in direction x_2 results from the analysis of the fringe patterns presented in Figures 9.8(c) and (d).

The quantitative interpretation of the fringe pattern shown in Figures 9.8(a) and (b) gives us the following values of absolute fringe orders needed for determination of the displacement component u:

$$\Delta N_{12}^1 = 8.4 \text{ (see Figure 9.8(a))},$$
$$\Delta N_{12}^2 = 8.6 \text{ (see Figure 9.8(b))}.$$

The analogous results for the displacement component v are:

$$\Delta N_{34}^1 = 1 \text{ (see Figure 9.8(c))},$$
$$\Delta N_{34}^2 = 1 \text{ (see Figure 9.8(d))}.$$

Substituting these values into expression (9.20), taking into account the physical signs of the displacement components indicated above, yields

$$\Delta u = -6.2\,\mu\text{m}, \ \Delta v = +0.7\,\mu\text{m}. \tag{9.27}$$

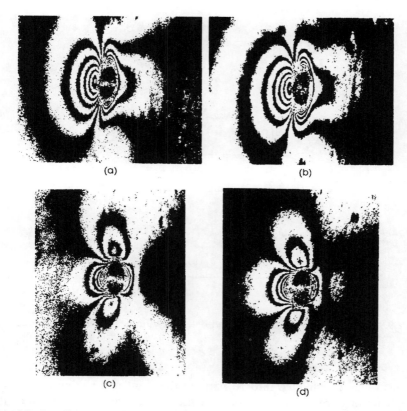

Figure 9.8. Interferograms for determination of the in-plane displacement components u (a, b) and v (c, d) near the partially through hole drilled in a beam under bending

In order to derive the required stress distribution from the interferogram presented above, it is necessary to establish the model describing the corresponding stress concentration problem. As such a model, the finite element solution of the problem about tension or compression of the parallelepiped with dimensions $20r_0 \times 20r_0 \times 8r_0$ with a partially-through hole of diameter $2r_0$ and depth $h = 2r_0$ located at the centre of the face $20r_0 \times 20r_0$ (see Figure 9.9) is used. The solution of this problem was performed by the software package FITING [16].

Let us briefly consider the main parameters of the numerical models used for the calculation of the stress concentration factors. Taking advantage of the symmetry, the quarter part of the parallelepiped is represented as a discrete model. The fragment of the structure representing by itself the cylinder of radius $4r_0$ surrounding the hole is idealized as a separate subconstruction. This

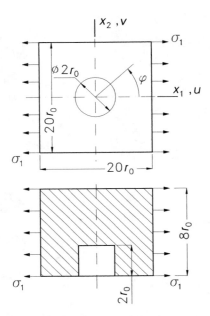

Figure 9.9. Model for stress concentration determination near a partially through hole

subconstruction is divided into 10 parts in the angular direction φ, into nine parts (counting from the hole edge) in the radial direction r, and into nine layers through the object thickness, five of which divide the hole depth. As practical experience proves, such a finite element mesh in the hole vicinity ensures the high accuracy of local strains and stresses determination that is comparable with the results of the analytical solutions of analogous elasticity problems. It should also be noted that the relation between the hole radius and linear dimensions of the object numerically investigated results in a uniform stress distribution near the external edges of structure.

The following procedure is implemented to define the stress concentration factors α_1 and α_2 contained in expressions (9.13) in the case involved. As known, the finite element method allows us to calculate the displacement components in the nodes of the mesh with a high accuracy. These nodes are located, in particular, on the hole edge. Then, strains and stresses are calculated at the points situated inside each finite element by means of various numerical differentiation procedures that may exert a negative influence on the accuracy of the stress concentration calculation. Therefore, the distributions of the tangential displacement components along the hole edge are used in our case to determine the stress concentration factors α_1 and α_2 (9.3). These dependencies

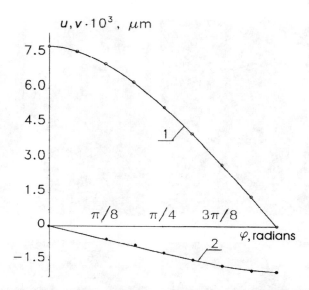

Figure 9.10. Distributions of the in-plane displacement conponents along the edge of a partially through hole corresponding to one-axis tension

are approximated with the Fourier series to establish the spectrum of corresponding decomposition.

The distributions of the tangential displacement components u (1) and v (2) obtained as a result of the solution to the above-described numerical problem are presented in Figure 9.10. These graphs reveal the displacement component values for structures made of aluminium alloy (the elasticity modulus $E = 72\,000$ MPa, Poisson's ratio $\mu = 0.32$) and correspond to one-axis stress field $\sigma_1 = 140$ MPa. It has been established that the above-presented distributions can be represented in the form (9.7) accurate to within one per cent. This not only confirms the validity of expression (9.10) in the case of small partially-through hole drilling on the surface of the three-dimensional object, but also allows us to immediately obtain the stress concentration factors α_1 and α_2 from the equation of the form (9.9). The values of these coefficients $\alpha_1 = 2.60$ and $\alpha_2 = 0.67$ were used in Section 9.1 to derive relations (9.15).

Substituting values (9.27) into (9.15) results in

$$\sigma_1 = 99 \text{ MPa}, \ \sigma_2 = -7 \text{ MPa}.$$

These magnitudes are in excellent agreement with the prediction of the elasticity theory resulting from the known conditions of the specimen loading

$$\sigma_1 = 100 \text{ MPa}, \ \sigma_2 = 0.$$

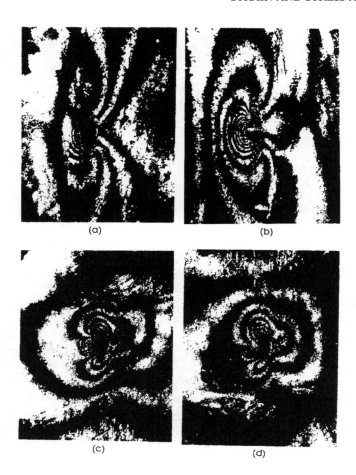

Figure 9.11. Typical interferograms for determination of the in-plane displacement components u (a, b) and v (c, d) along the edge of a partially through hole drilled near a welding seam

In conclusion to this chapter, we have represented, only as an example, the fringe patterns resulting from residual stress redistributions after hole drilling in a real structure. Figure 9.11 demonstrates the interferograms obtained for a partially-through hole made near the welding seam between two steel sheets. It is evident that the quality of these fringe patterns suffices in order to obtain all the information needed for the determination of residual stresses, namely, the directions of the principals residual strains, the signs of the corresponding diameter change, and the number of fringes which finish at half of the hole boundary.

References

1. Birger, I. A. (1963) *Residual stresses*. Mashgiz., Moscow.
2. Mathar, J. (1934) Determination of initial stress by measurement the deformations around drilled holes. *Trans. ASME* **56**, 249–254.
3. Flaman, M., Mills, B. and Boag, J. (1987) Analysis of stress-variation-with-depth measurement procedures for the center-hole method of residual stress measurement. *Exp. Techn.* **11**, 34–37.
4. Flaman, M. and Herring, J. (1987) Comparison of four hole-producing-techniques for the center hole residual stress measurement method. *Exp. Techn.* **11**, 477–479.
5. Kobayashi, A. S., (ed.) (1987) *Handbook of Experimental Mechanics*. Prentice Hall, Englewood Cliffs, New Jersey.
6. Antonov, A. A., Bobrik, A. I., Morozov, V. K. and Chernyshev, G. N. (1980) Residual stresses determination by making a holes and holographic interferometry. *Proc. AN SSSR: Mekh. Tverd. Tela* N2, 182–189.
7. Lobanov, L. M., Kasatkin, B. S., Pivtorak, V. A. and Andrushenko, S. G. (1983) Residual stresses determination by holographic interferometry using a single hologram. *Dokl. AN SSSR* N3, 557–561.
8. Rassokha A.A. (1983) Technologic residual stresses investigation by holographic and speckle interferometry. Probl.Prochn. N1, 111-115.
9. Antonov A.A., Kozintsev V.M. (1989) Reflection hologram application for residual stresses measurement. Zavod.Labor. 55, 84-87.
10. Antonov, A. A., Morozov, V. K. and Chernyshev, G. N. (1988) Stresses determination through holography of disturbed solid surface. *Proc. AN SSSR. Mekh. Tverd. Tela* N3, 185–189.
11. Kurnosov, D. G. and Yakutovich, M. V. (1946) Residual stresses measurement by hole-drilling method. *Zavod. Labor.* **12**, 960–967.
12. Shtanko, A. E. and Guzikov, M. N. (1985) Compact holographic interferometer for residual stresses estimation. In *Proc. II Sympos.: Residual technologic stresses*, pp. 366–370, Izd. IPM AN SSSR, Moscow.
13. Nelson, D. and McCrickerd, J. (1986) Residual-stress determination through combined use of holographic interferometry and blind-hole drilling. *Exp. Mech.*, **26**, 371–378.
14. Lingli Wang and Jingtang Ke (1988) The measurement of residual stresses by sandwich holographic interferometry. *Opt. Lasers Engng* **9**, 111–119.
15. Timoshenko, S. P. and Goodier, J. N. (1970) *Theory of Elasticity*. McGraw-Hill, New York.
16. Baryshnicov, V. I., Grishin, V. I., Donchenko, V. Yu. and Tikhonov, Yu. V. (1983) Implementation of the finite element method to local strength investigation of aircraft structural elements. *Proc. TsAGI.* **14**, 66–73.
17. Furse, I. E. (1981) Kinematic design of fine mechanism in instruments. *J. Phys. E.: Sci. Instrum.* **14**, 164–217.
18. Savin, G. N. (1951) *Stress Concentration Near the Holes*. Gostekhizdat, Moscow, Leningrad.

10 CORRELATION HOLOGRAPHIC INTERFEROMETRY AND SPECKLE PHOTOGRAPHY IN CONTACT INTERACTION MECHANICS

During the discussion on methods of double-exposure holographic interferometry in Section 2.1 and on double-exposure speckle photography in Sections 3.2 and 3.3, it was pointed out that high-contrast interference fringes may be observed if the surface microrelief does not change between exposures. The changes in the object surface microrelief between hologram exposures are followed by a decorrelation of the reconstructed light waves. As a result, the fringe contrast (visibility) decreases or completely disappears. Also, the changes in the object surface roughness between exposures during the specklegram recording procedure result in a decorrelation of the speckle patterns. In this case, Young's fringes contrast decreases too.

These phenomena, on the one hand, restrict the application of these two methods for the displacements measurements in high elasto-plastic strain concentration zones. On the other hand, new methods to investigate the mechanical processes, associated with the object surface microrelief changes, can be developed on this basis. The interferometric fringe visibility reduction was used to study the corrosion process by Ashton *et al.* [1] and Petrov and Presnyakov [2], cavitation erosion by Dmitriev *et al.* [3], and solid bodies contact interaction by Atkinson and Labor [4] and Shchepinov *et al.* [5]. The methods based on the contrast changes of the fringes due to any process, having destroyed the object surface microrelief, will be hereinafter called correlation holographic interferometry and correlation speckle photography.

Irreversible (plastic) deformation results from any macroelastic contact between objects with rough surfaces. In this chapter the methods of correlation holographic interferometry and speckle photography are discussed and applied to solving mechanics contact interaction problems. Some contact surface measurements and determination of the contact stress (pressure) are among these problems.

10.1   Fringe contrast in holographic interferometry and speckle photography for random surface microrelief changes

Let's once again consider the recording of a holographic interferogram of an object with a rough surface by the double-exposure technique. We shall assume that after the first exposure the microrelief was destroyed on the fragment of the object surface due to contact interaction with another body. Complex amplitudes of the reconstructed light waves, scattered by the object surface before A_1 and after A_2 with a change in its microrelief are expressed as

$$A_1 = a_1 \exp(-i\varphi_1),$$
$$A_2 = a_2 \exp[-i(\varphi_1 - \psi + \varphi_0)] \qquad (10.1)$$

where a_1, a_2 are the amplitudes of the reconstructed light waves; φ_1 is the object wave phase during the first exposure; $\varphi_1 - \varphi_2 = \psi$ is the phase variation of the object wave, caused by microrelief change; φ_0 is the regular phase shift manually introduced into the object wave to produce carrier interference fringes located on the object surface; and φ_2 is the object wave phase after the change of the surface roughness.

The interference pattern observed under simultaneous reconstruction of the object waves (10.1) recorded on a hologram is described by an intensity I_H distribution:

$$I_H = < |A_1 + A_2|^2 > \qquad (10.2)$$

where $< \ldots >$ denote averaging over an object surface area, corresponding to the limited resolution of the optical system used to observe the object. Let us consider that the size of this area substantially exceeds the characteristic transverse size of the surface roughness, i.e the optical system is not capable of resolving microrelief details. Subsequently, let's assume that the amplitudes a_1, a_2 and phases φ_1, φ_2 are statistically independent quantities. Thus equation (10.2) can be modified as follows:

$$I_H = < a_1^2 > + < a_2^2 > + 2 < a_1 a_2 > < \cos(\varphi_0 + \psi) > . \qquad (10.3)$$

In the following we shall assume that the random roughness nature of the opaque surfaces is transformed into the scattered wave phase only, i.e. $a_1 = a_2$. Taking this fact into account, equation (10.3) may be simplified as

$$I_H \propto 1 + < \cos(\varphi_0 + \psi) > . \tag{10.4}$$

Expression (10.4) can simply be transformed to the form

$$I_H \propto 1 + \sqrt{< \cos \psi >^2 + < \sin \psi >^2} \cos\left[\varphi_0 + tg^{-1}\left(\frac{< \sin \psi >}{< \cos \psi >}\right)\right]. \tag{10.5}$$

Contrast γ_H of the carrier interference fringes can be derived from the previous expression, thus

$$\gamma_H = \frac{I_{Hmax} - I_{Hmin}}{I_{Hmax} + I_{Hmin}} = \sqrt{< \cos \psi >^2 + < \sin \psi >^2}. \tag{10.6}$$

Furthermore, an additional phase shift caused by the surface microrelief change and influencing the carrier interference fringes shape will be represented as follows:

$$\Delta \varphi_0 = tg^{-1}\left(\frac{< \sin \psi >}{< \cos \psi >}\right).$$

The terms $< \cos \psi >$ and $< \sin \psi >$ in equation (10.6) were treated analytically by Osintsev et al. [6], using the principal relation of holographic interferometry (2.24) and distribution function, which was determined by the character of the irreversible changes of surface microrelief elements. Thus, according to equation (10.6) the changes in surface microrelief result in a contrast decrease and in a shape distortion of the carrier interference fringes.

A physical description of specklegram recording by the double-exposure technique in the case of a random microrelief variation is quite complex. This problem was considered in detail by Ostrovsky and Shchepinov [7,8] and the final results of their investigations will be used below. This way, the intensity I_S in the screen plane, obtained by pointwise filtering of a double-exposed specklegram (see Figure 3.6) is given:

$$I_S \sim 1 + | < \exp(i\psi) > |^2 \cos(\vec{\omega} \, \vec{d}_0) \tag{10.7}$$

where $\vec{\omega}$ is the radius-vector in the screen plane; $\vec{d}_0$ is the photoplate displacement between the exposures, which causes formation of carrier Young's fringes; and ψ is the random function describing the phase variation in the light wave. It should be clear from formula (10.7) that contrast γ_S of the carrier fringes is

$$\gamma_S = \frac{I_{Smax} - I_{Smin}}{I_{Smax} + I_{Smin}} = | < \exp(i\psi) > |^2.$$

Considering the probability density function $\rho(\psi)$ for random phase changes, Young's fringes contrast becomes

$$\gamma_S = < | \int_{-\infty}^{\infty} [\exp(i\psi)]\rho(\psi)d\psi|^2 > = < \cos \psi >^2 + < \sin \psi >^2. \qquad (10.8)$$

Comparing equations (10.6) and (10.8), the fundamental result is

$$\gamma_S = \gamma_H^2. \qquad (10.9)$$

Thus correlation speckle photography has a higher sensitivity according to the surface microrelief changes compared to correlation hologram interferometry.

10.2 Optical set-ups for visualization of object surface areas with changed microrelief

Approaches to determine surface microrelief destruction can be developed based on the carrier fringes or image subtraction techniques both for correlation holographic interferometry and for correlation speckle photography. Carrier fringes is the simplest way to visualize areas on the object surface subjected to irreversible microrelief changes. When a holographic interferogram is being recorded, an irreversible (plastic) deformation of the surface microrelief elements between exposures results in scattered light waves decorrelation. As consequence, the contrast of the carrier interference fringes decreases down to their complete disappearance.

Holographic interferograms with carrier fringes are best recorded by the double-exposure technique, yielding high-contrast interference fringe patterns. The illuminating beam displacement between hologram exposures produces the required carrier fringe pattern. Object rigid displacements or plastic deformations due to contact interaction are also accompanied by the formation of carrier fringes.

We can also find it very useful to apply speckle photography to visualize areas with changed surface microrelief by means of formation of carrier fringes. Remember that plastic deformation of surface microrelief elements will cause decorrelation of two displaced speckle patterns which are recorded on the double-exposed specklegram.

Let's consider an optical arrangement for recording specklegrams which is shown in Figure 3.4. During the first exposure a speckle pattern corresponding to the initial object state is recorded on the photoplate SF. Then, the contact interaction should be accomplished and photoplate SF should be displaced in its plane by $|\vec{d_0}|$. After this the second specklegram exposure should be made. Thus, two displaced speckle patterns with respect to one another are recorded on the specklegram. Note that the shift magnitude $|\vec{d_0}|$ should be somewhat greater than the average speckle size.

Pointwise filtering of any area of a double-exposed specklegram thus obtained results in the appearance of Young's fringes with the same spacing and orientation. These fringes can be observed on a screen by conventional methods (see Section 3.3). Depending on the changed microrelief on the object surface, the specklegram can reproduce fringes with a lower contrast or no fringes at all. By scanning the specklegram fragments with a narrow laser beam in small increments the contact zone on the object surface can be identified.

Speckle photography allows us to visualize the contact area on the object surface completely against the background of the carrier fringes. In this case the double-exposed specklegram SF should be inserted into a spatial filtering set-up proposed by Klimenko *et al.* [9] and shown in Figure 10.1. A circular aperture with a small diameter situated in plane A is located at a distance Δl from the lens L focal plane (Fourier plane). Parallel straight fringes are observed through the aperture. The spacing and orientation of the fringes depend on the aperture location in the filtration plane and the distance Δl from the Fourier plane.

Correlation holographic interferometry is capable of visualizing areas of contrast reduction in the carrier fringes corresponding to surface fragments with changed microrelief. The optical set-ups discussed above allow us to visualize the boundary of the area where microrelief changes by way of a set of dashed lines or dots. To receive a continuous boundary of the contact interaction area, an optical set-up that is capable of performing an image subtraction should be implemented, as suggested by Debrus *et al.* [10, 11]. The main principle of this approach is to add the phase shift π in order to compare the object waves between two exposures. Consequently, dark field

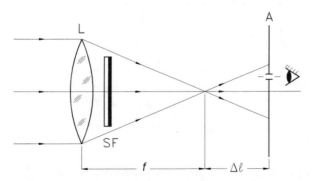

Figure 10.1. Optical arrangement for spatial filtering outside the Fourier plane

interferograms may be observed with light spots of various intensity, corresponding to the surface areas with destroyed microrelief.

Let us briefly discuss the image subtraction hologram technique, based on Fourier type holograms, and suggested by Klimenko and Ryabuho [12]. Figure 10.2(a) illustrates an optical set-up required to make a Fourier hologram H, on which the spatial spectrum of the object wave is recorded instead of the object image. The Fourier transform of the object light field is carried out by the lens L_1.

The step-by-step procedure to make the double-exposure Fourier hologram should be as follows. The first exposure is made for the initial object field spectrum. Then, a linear phase shift is introduced by way of a reference beam

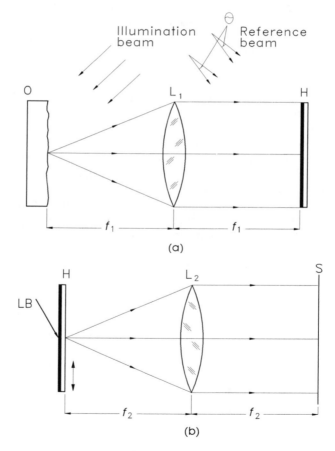

Figure 10.2. Optical arrangements for image subtraction by using Fourier holograms: Fourier hologram recording (a) and image subtraction (b)

inclination through the angle θ and the object surface undergoes a contact interaction. The phase shift made between exposures results in the fact that the spatial spectrum of the scattered object light field will be modulated by a set of uniformly spaced straight interference fringes. To perform an image subtraction, the hologram should be reconstructed by the unexpended laser beam LB through the dark fringe centre (Figure 10.2(b)). Thus the areas of changed microrelief will be visualized as light spots on the dark background. To direct the laser beam to the centre of the dark fringe, hologram H must be displaced in its plane smoothly and carefully. In the optical set-up shown in Figure 10.2(b) the lens L_2 performs an inverse Fourier transform of the light field reconstructed by the hologram.

For bodies with a flat surface, the image subtraction technique based on speckle photography can be applied [13]. When recording the specklegram the first exposure should be made in the initial state of the investigated surface. Then the photoplate should be displaced in its own plane by $|\vec{d_0}|$ and contact interaction carried out. The second exposure should be made to obtain the resultant speckle pattern. Thus the double-exposed specklegram obtained should be spatially filtered by means of an optical arrangement similar to that shown in the right-hand side of Figure 3.2. Subsequent to the photoplate displacement $|\vec{d_0}|$ between exposures, the intensity distribution in the lens focal plane is modulated by a set of uniformly spaced interference fringes. If the mask with a narrow slit, coinciding with the centre of a dark fringe, is located in the Fourier plane then an image subtraction can be performed. When observing through the slit we can see a fully dark surface image, in addition to the light areas where the microrelief changed. The latter is caused by the partial or complete decorrelation of speckle structures according to these surface fragments.

10.3 Contact area determination

Atkinson and Labor [4] first proposed use of hologram interferometry to measure the plastic component of the contact area. They proposed a hypothesis whereas in the first approximation the contrast variation γ_H can be expressed through the area squares ratio corresponding to the plastic component of the actual contact surface F and contour contact surface F_0 in the following way:

$$\gamma_H = 1 - \frac{F}{F_0}. \tag{10.10}$$

Speckle photography may also be applied to determine the plastic component of the contact area. In reference [14] it is shown that Young's fringes contrast variation is expressed as follows:

$$\gamma_s = \left(1 - \frac{F}{F_0}\right)^2. \qquad (10.11)$$

We can see that formulae (10.10) and (10.11) are similar. Therefore, both for correlation hologram interferometry and for correlation speckle photography, the contrast variations in the carrier fringes appear in the plastic component of the actual contact area. However, in the case of speckle photography the contrast decrease is described by quadratic dependence while in the case of hologram interferometry this dependence is linear. Therefore, the above-derived equation (10.9) is valid.

Now let's consider the contact contour area recording by correlation holographic interferometry. If between hologram exposures we submit the object surface to contact interaction with another body and, in addition, displace the illuminating beam, then we can reconstruct the image change by way of the carrier fringe patterns. The contrast of the carrier fringes will be lower in the contact area compared to the undisturbed surface. In this way the contact contour surface is determined by the condition

$$\gamma_H < \gamma_{H0}$$

where γ_{H0} is fringe contrast outside the contact area.

The problem becomes more simple if the light waves, scattered by the object surface in the contact area before and after loading, are completely decorrelated. In this case the contact surface will be bounded by discontinuities in the carrier fringes pattern.

To illustrate the technique under discussion let us consider the problem of contact interaction under loading between a plate and sphere made of steel. The plate had a hardness of 30 HRC, and the sphere with a radius of 300 mm of 49 HRC. The initial roughness of both contacting surfaces was characterized by the parameter $R_a = 1.5$.

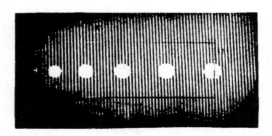

Figure 10.3. Holographic interferogram with six visualized indentations of sphere into plate surface corresponding to the increasing load

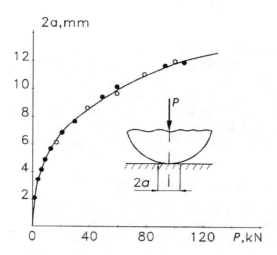

Figure 10.4. Dependence of contact area diameter 2a against load P. (Reproduced by permission from Ostrovsky, Shchepinov and Yakovlev, Holographic Interferometry in Experimental Mechanics. Fig 6.25 Vol 60, copyright Springer Verlag GmbH & Co. KG)

A typical interferogram with six contact spots is shown in Figure 10.3. The discontinuities in the carrier fringes pattern show the contact area boundary. The experimental dependence of the contact area diameter 2a versus the load P, obtained for two series of indentations, are indicated by the light and dark points in Figure 10.4. These series of indentations were produced under different loads and carrier fringes. For comparison, an available theoretical Hertz's solution, contained in reference [15], is given and shown by the curve. The experimental data and theory are comparably good.

Correlation hologram interferometry also permits determination of the character of contact interaction between the bodies. For example, if an elastic type contact interaction has occurred, then there are no residual deformations of the object in the neighbourhood of the contact area. In an opposite case, the interaction had an elasto-plastic character. Moreover, as shown by Shchepinov *et al.* [5], residual deformations result in carrier fringe pattern distortion.

For small contact indentations the application of the approach discussed above is impeded because of the limited density of the carrier fringes. This problem can be overcome by using the image subtraction holographic technique discussed in Section 10.2.

The carrier interference fringes or image subtraction approaches can be applied to determine the contour contact surface by speckle photography. For the first case, the contour contact surface is determined based on Young's fringe contrast according to

$$\gamma_s < \gamma_{s0}$$

where γ_{s0} is Young's fringe contrast outside the contact area. The second case is preferable, because the sensitivity of the correlation speckle photography is considerably higher than that of correlation hologram interferometry. Therefore in the majority of cases speckle patterns in the contact areas before and after the contact interaction are decorrelated completely. Many examples are presented by Ostrovsky and Shchepinov [8], where correlation speckle photography was used to visualize solid contact surfaces.

10.4   Contact pressure measurement

The contact pressure influence on carrier fringe contrast variation can be discussed by using, as an example, a rigid cylinder end indentation in a plate surface (Figure 10.5). Contact pressure values are available in an analytical form [15, 16]:

$$Q = \frac{P}{2\pi a \sqrt{a^2 - r^2}} \tag{10.12}$$

where P is the load; a is the cylinder radius; and r is the coordinate measured from the indentation centre.

Subsequent to equation (10.12), there is a wide region with a constant pressure approximately Q_0 in the centre of the contact area which is equal (see Figure 10.5):

$$Q_0 = \frac{P}{2F} \tag{10.13}$$

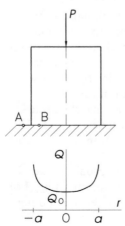

Figure 10.5. Disrtibution of contact pressure Q for contact interaction of cylinder end with a thick plate

Figure 10.6. Holographic interferogram of plate surface area after contact interaction with a cylinder end under load P = 350 kN

where $F = \pi a^2$ is the area of the cylinder end. Thus, by varying the contact pressure Q_0 and measuring the carrier fringes contrast in the central zone of the contact indentation, the experimental dependence for γ versus Q_0 may be obtained. A typical holographic interferogram of the thick steel plate surface with two indentations produced by a steel cylinder with diameter 30 mm is shown in Figure 10.6. We can observe that the carrier fringes contrast in the central areas of these indentations is practically constant, but some are lower compared to the fringes contrast on the surface areas which underwent a contact interaction. Along the contact area boundary, the fringe contrast is even lower than at the indentation centre. This means that the contrast variations in the contact areas on the holographic interferogram qualitatively agree with the actual contact pressure distribution.

The problem under discussion can also be treated by using speckle photography. Two typical Young's fringe patterns, obtained by scanning a double-exposed specklegram with a narrow laser beam, are shown in Figure 10.7. For the surface point A, located outside the contact area (see Figure 10.5), Young's fringes have the highest contrast (Figure 10.7(a)). On the other hand, for the surface point B located in the indentation area with the greatest

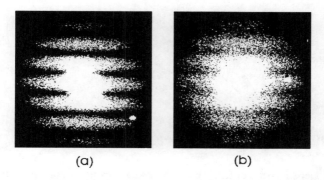

(a)   (b)

Figure 10.7. Young's fringe patterns obtained by pointwise specklegram filtering in points A (a) and B (b)

contact pressure, the fringes contrast is significantly lower (Figure 10.7(b)). Thus, it may be concluded that Young's fringes contrast corresponds to the contact pressures variations.

Quantitative dependence for fringe contrast variations $\Delta\gamma/\gamma_0 = \gamma_0 - \gamma/\gamma_0$, where γ_0 is the reference fringe contrast outside the contact area, against contact pressure Q_0, was experimentally obtained. To achieve this result, two thick plates, one made of steel with hardness 39 HRC and the second of aluminium alloy were loaded by using a steel cylinder with radius 30 mm and hardness 45 HRC. The cylinder end roughness was prepared and was $R_a = 0.32\,\mu$m. The surface roughness of the plates was prepared in the range from $R_a = 0.65\,\mu$m to $R_a = 2.50\,\mu$m. This range corresponds to the surface microrelief parameters of most structural elements operating under contact interaction conditions.

For each loading, a contact indentation was placed in the free undisturbed plate surface. To compare the results obtained by using two different experimental techniques, an optical set-up was created to allow simultaneous recording of the holograms and specklegrams [6].

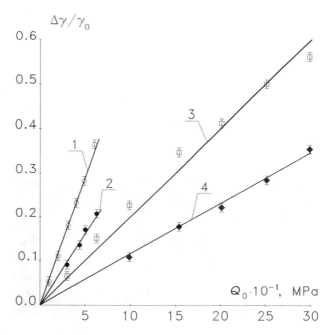

Figure 10.8. Dependencies of fringe contrast change against contact pressure Q between cylinder end and a plate. (Reproduced from Sirohi: Speckle Metrology, Vol 38, by courtesy of Marcel Dekker Inc.)

Some experimental results are presented in Figure 10.8. Dependencies 1, 2 correspond to the plate made of aluminium alloy ($R_a = 0.63\,\mu m$) and 3, 4 correspond to the steel plate with the same roughness. Curves 1 and 3 are results from a specklegram interpretation, while curves 2 and 4 are derived from a hologram. The first conclusion can be drawn here is that all experimentally obtained dependencies between $\Delta\gamma/\gamma_0$ and Q_0 are linearly proportional. Considering these results, we can conclude that the sensitivity of speckle photography defined by the inclinations of curves 2 and 4 is higher than the sensitivity for correlation holographic interferometry (see curves 1 and 3 in Figure 10.8). This fact qualitatively substantiates equation (10.9). Further investigations showed that the dependence between quantities $\Delta\gamma/\gamma_0$ and Q_0 is described by a linear function for a reasonably wide range of roughness, hardness and mechanical properties of materials.

Thus, it may be proposed that, in a general case, contact pressure may be approximated by the function

$$Q = C\frac{\Delta\gamma}{\gamma_0} \qquad (10.14)$$

where C is an unknown constant. From the static equilibrium condition it follows that

$$P = \int_F Q\,dF = C\int_F \frac{\Delta\gamma}{\gamma_0}\,dF \qquad (10.15)$$

where F is the area of contact interaction and P is the load component normal to the contact surface. In most cases the load P is known or can be determined accurately. Using expression (10.15) the constant C may be calculated:

$$C = \frac{P}{\displaystyle\int_F \frac{\Delta\gamma}{\gamma_0}\,dF}$$

and substituting it into expression (10.14) the formula to determine contact pressure takes the form:

$$Q = \frac{P}{\displaystyle\int_F \frac{\Delta\gamma}{\gamma_0}\,dF}\,\frac{\Delta\gamma}{\gamma_0}. \qquad (10.16)$$

The problem of variable contact pressures determination by the holographic technique is illustrated in the example of contact interaction between a thick steel plate and a steel sphere with radius 500 mm, presented by Ostrovsky *et al.* [17]. Two indentations were made under the same load $P = 140\,kN$. Two different carrier fringes systems having the same direction but different

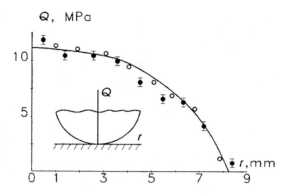

Figure 10.9. Distribution of contact pressure Q in the interaction area between sphere and a plate corresponding to radial direction. (Reproduced by permission from Ostrovsky, Shchepinov and Yakovlev, Holographic Interferometry in Experimental Mechanics. Fig 6.31 Vol 60, copyright Springer Verlag GmbH & Co. KG)

spacings were used. Each contact action was produced at different locations on the plate surface. The distribution of the contact pressure along the radial direction on the interaction surface area was determined by equation (10.16). These results are shown in Figure 10.9 by light and dark dots. In the same figure the theoretical Hertz solution is presented which coincides with experimental data.

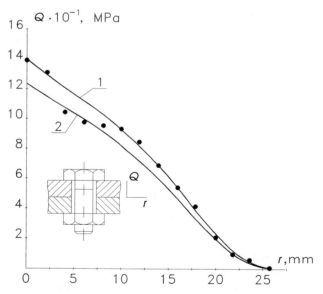

Figure 10.10. Distribution of contact pressure between two plates in a bolt joint. (Reproduced from Sirohi: Speckle Metrology, Vol 38 by courtesy of Marcel Dekker Inc.)

In the work of Ostrovsky and Shchepinov [7], correlation speckle photography was applied to determine variable contact pressure between two plates made of aluminium alloy in a bolt joint fragment. The thickness of both plates was $h = 15$ mm, and the hole diameter was 15 mm.

The surfaces of the plate in connection have a roughness characterized by the parameter $R_a = 1.5\,\mu$m. Contact pressure distribution $Q(r)$, corresponding to the load $P = 180$ kN, was calculated using formula (10.16) and is shown in Figure 10.10 by dots.

To verify the experimental results, the problem discussed was solved by a finite element method, which took into account the surface friction with coefficients 0.0, 0.1 and 0.2. Curve 1 in Figure 10.10 represents one of the numerical results, corresponding to friction coefficient 0.2, and most distinctly coincides with experimental data. For comparison, curve 2 illustrates a numerical solution for friction coefficient 0.1. Thus, the correlation methods presented in this section are capable of enhancing the numerical solutions of contact problems in cases when the value of the friction coefficient is unknown.

References

1. Ashton, R. A., Slovin, D. and Gerritsen, H. I. (1971) Interferometric holography applied to elastic stress and surface corrosion. *Appl. Opt.* **10**, 440–441.
2. Petrov, K. N. and Presnyakov, Yu. P. (1978) Holographic interferometry of corrosion process. *Opt. Spectrosk.* **44**, 309–311.
3. Dmitriev, A. P., Dreiden, G. V., Osintsev, A. V. *et al.* (1989) Cavitation errosion study by correlation holographic interferometry. *Zhur. Tekhn. Fiz.* **59**, 192–197.
4. Atkinson, I. T. and Labor, M. I. (1976) Measurement of the area of real contact between, and wear, articulating surface using hologram interferometry. *Proc. Conf. Application of Holography and Optical Data Processing*, Jerusalim, pp. 289–298.
5. Shchepinov, V. P., Morozov, B. A., Novikov, S. A. and Aistov, V. S. (1980) Contact surface determination by holographic interferometry. *Zhur. Tekhn. Fiz.* **50**, 1926–1928.
6. Osintsev, A. V., Ostrovsky, Yu. I., Presnyakov, Yu. P. and Shchepinov, V. P. (1992) Fringes contrast in methods of correlation speckle photography and correlation holographic interferometry. *Zhur. Tekhn. Fiz.* **62**, 128–137.
7. Ostrovsky, Yu. I. and Shchepinov, V. P. (1992) Correlation holographic and speckle interferometry. In Wolf, E. (ed.), *Progress in Optics 30*, pp. 87–135, Elsevier Science Publishers BV.
8. Ostrovsky, Yu. I. and Shchepinov, V. P. (1993) Correlation speckle interferometry in the mechanics of contact interaction. In Sirohy, R. S. (ed.), *Speckle metrology*, pp. 507–538, Marcel Dekker, New York.
9. Klimenko, I. S., Ryabukho, V. P. and Feduleev, B. V. (1983) On the separation of information concerning different motions in holographic interferometry based on spatial filtration. *Opt. Spektrosk.* **55**, 140–147.
10. Debrus, S., Francon, M. and Grover, C. P. (1971) Detection of differences between two images. *Opt. Commun.* **4**, 172–174.
11. Debrus, S., Francon, M. and Grover, C. P. (1972) Detection of differences between two images: an improved method. *Opt. Commun.* **6**, 15–17.

12. Klimenko, I. S. and Ryabukho, V. P. (1985) Application of holographic image subtraction basing on space filtration for visualization of surface microrelief destruction. *Opt. Spectrosk.* **59**, 398–403.
13. Osintsev, A. V., Ostrovsky, Yu. I., Shchepinov, V. P. and Yakovlev, V. V. (1988) Contact surface determination by speckle photography. *Zhur. Tekhn. Fiz.* **58**, 1420–1423.
14. Osintsev, A. V., Presnyakov, Yu. P. and Shchepinov, V. P. (1990) Natural contact surface determination by speckle photography. In *Proc. Sympos. Meth. Primen. Golograph. Interf.*, pp. 62–69, KUAT.
15. Timoshenko, S. P. and Goodier, J. N. (1970) *Theory of elasticity.* McGraw-Hill, New York.
16. Osintsev, A. V., Ostrovsky, Yu. I., Shchepinov, V. P. and Yakovlev, V. V. (1991) Fringe contrast changes in holographic interferometry and speckle photography for solids contact interaction. *Zhur. Tekhn. Fiz.* **61**, 134–139.
17. Ostrovsky, Yu. I., Shchepinov, V. P. and Yakovlev, V. V. (1991) *Holographic Interferometry in Experimental Mechanics.* Springer Series in Optical Science, 60. Springer-Verlag, Berlin.

11 SOME PRACTICAL APPLICATIONS

Although some practical applications have already been discussed, particularly in Chapters 5 to 8, in this chapter we shall consider a few examples which show the efficiency of holographic interferometry and speckle photography techniques for solving specific strength problems.

One of the pioneering works in holographic interferometry by Powell and Stetson [1] is devoted to vibration analysis. So far, the visualization of vibration modes for objects with diffusely reflected surfaces having a complex shape has been possible within the most important holographic interferometry applications. The new holographic vibrometry techniques now being developed allow us to investigate the influence of real structure irregularities on the resonance frequency spectrum and vibration modes [2–9]. For instance, the problem concerning oscillations of thin plates with holes is of methodological and practical interest. In real structures plate elements can be densely perforated by small holes or can contain large cut-outs.

The model of the inner mounting shaft of a VVER type reactor was investigated in order to determine the vibration resonance frequencies and modes influenced by different boundary conditions. Determination of the displacement fields for structures with a complex three-dimensional geometry is given for a reduced model of a general circular pump body of a VVER type reactor. The experimental data obtained were used to verify software packages for three-dimensional stress–strain analysis by finite element methods.

As demonstrated in Chapters 7 and 8, determination of elasto-plastic deformations of materials and structure elements is a principal advantage of the holographic interferometry and speckle photography techniques. Bearing capacity estimation for a high-power turbogenerator rotor conductor was obtained by a combined application of holographic interferometry and speckle photography to study the elastic-plastic deformation process. All the practical problems mentioned undoubtedly enhance the book while adding to its contents.

11.1   Vibration of plates with holes

Plates with holes are widely spread structural elements which undergo vibration loading. Consequently, the investigation of vibration resonance frequencies and modes is of principal interest. The variety of plates with holes may be classified conventionally into two important groups:

- Plates perforated with numerous small holes
- Plates with large cut-outs comparable with overall structure dimensions.

First, let's consider the results obtained by holographic interferometry for circular perforated plate vibration. The investigated objects were made of aluminium alloy and had a diameter $D = 120$ mm and thickness $h = 1.5$ mm [10]. The plates were supported rigidly along the outer contour. Three types of perforations were studied:

- 85 holes of diameter $d = 5$ mm located in a triangular grid.
- 37 holes of diameter $d = 5$ mm and 48 holes of diameter $d = 7$ mm located in alternate rows in a triangular grid.
- 97 holes of diameter $d = 7$ mm located in a square grid.

Hologram recording was carried out by means of an optical set-up in which the illumination and viewing directions coincided with the plate surface normal (see Figure 2.19(a)). Real-time interferometry and the carrier fringe method were used to obtain resonance frequencies [11].

The vibration modes, which were visualized using the Stetson–Powell method, are denoted by two parameters $m{:}n$. The first corresponds to the number of nodal circles, including the fixed outer contour, and the second is equal to the number of nodal diameters. Holographic interferograms of the plate with the first perforation type for the initial eight vibration modes are represented on Figure 11.1. Similar fringe patterns were obtained for circular plates with two different perforation types. The corresponding vibration modes coincide with the same modes of regular circular plates. In addition, due to the structural anisotropy caused by perforation, the locations of the nodes and maximum-amplitude peaks are predetermined. Evidently, the nodal diameters and circles are located on lines with a minimum number of holes, and the interference fringes representing maximum amplitude peaks coincide with the conditional curves on the plate surface containing the greatest number of holes.

The resonance frequencies for all three types of plate perforations are listed in Table 11.1. The experimental frequency values are normalized by the magnitude of the resonance frequency value corresponding to the vibration mode 1:0 for the regular circular plate of the same radius. From this data list

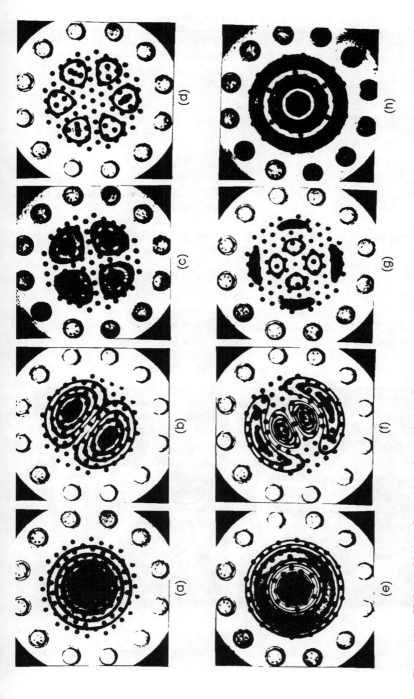

Figure 11.1. Holographic interferograms of vibrating perforated plates corresponding to resonance modes: 1:0-(a); 1:1-(b); 1:2-(c); 1:3-(d); 2:0-(e); 2:1-(f); 2:2-(g); 3:0-(h)

Table 11.1 Normalized resonance vibration frequencies

Vibration mode	Regular plate	Perforated plates		
		Type 1	Type 2	Type 3
1:0	1	0.948	0.997	0.935
1:1	2.075	1.948	2.084	1.936
1:2	3.411	3.172	3.287	3.133
1:3	5.011	4.632	4.686	4.631
2:0	3.581	3.611	3.758	3.579
2:1	5.939	5.476	5.541	5.479
2:2	7.554	7.591	7.569	7.593
3:0	8.611	7.991	8.072	7.933

analysis we can conclude that the resonance frequency spectrums related to different types of perforations almost coincide with the corresponding values for the regular plate.

Now we shall consider the experimental results of the vibration analysis for short cylindrical shells with plane bottoms [12]. The cylindrical shells were $l = 20$ mm high with a wall thickness $t = 1.5$ mm. The bottom had a diameter $D = 240$ mm and the same thickness $h = 1.5$ mm. Experiments were carried out for those objects having one, two, three and four circular cut-outs in the

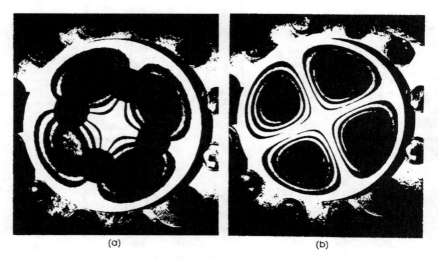

(a)   (b)

Figure 11.2. Holographic interferogram of vibrating circular plate containing four cut-outs corresponding to the cases when nodal diameters pass through the holes (a) and between the holes (b)

Table 11.2 Resonance frequencies v, Hz for plates, corresponding to modes with $m = 1$

Plate	$n = 00$	$n = 1$	$n = 2$	$n = 3$	$n = 4$
Regular	990	2070	3402	5054	6926
1 cutout	993	2096	3479	5035	6920
		2015	3281	4812	6626
2 cutouts	998	2123	3542	5092	6900
		1974	3186	4641	6380
3 cutouts	996	2047	3445	4465	—
		2038	3432	4434	—
4 cutouts	992	2038	3793	4774	6629
		2032	3622	—	5922

bottom. The cutouts were located along a circle of diameter $d = 60$ mm and having the same diameters $d = 24$ mm. For comparison, the bottom without holes was also investigated.

As the experiments showed, the vibration modes with node diameters have a split spectrum of resonance frequencies, i.e. similar vibration modes appear for two close frequencies. For example, in Figure 11.2 the fringe patterns corresponding to vibration modes 1:2 for the bottom with four cut-outs are shown. In the former case, the node diameters run through the holes in the direction of minimum stiffness (Figure 11.2(a)) and in the latter case, they run between the holes in the direction of maximum stiffness (Figure 11.2(b)).

Resonance frequencies, obtained for the bottom without holes (see Table 11.2), accurately correlate with the theoretical solution corresponding to circular plates rigidly supported along their contour (a gap in the range of 5 per cent). Also, by analysing the list of experimental data in Table 11.2 we can conclude that the splitting of resonance frequencies corresponding to the same vibration modes increases for a large number of node diameters. By using the obtained experimental data, the design-basis for the vibration of plates with perforations or large cut-outs can be improved.

11.2 Determination of vibration resonance frequencies and modes of the inner mounting reactor shaft model

Dynamic loads applied to inner-mounted equipment of water-cooling power-generating reactors may sometimes have faults or cause accidents in nuclear power stations. The stringent requirements as to reliability and safety of nuclear power stations force us to extend the spectrum of dynamic loads under consideration. Heat transfer medium vibration is one of the principal loads which must be taken into account.

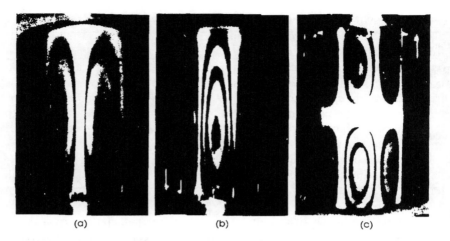

Figure 11.3. Holographic interferograms of shaft model with flat bottom resonance modes: 1:2, 2151 Hz-(a); 1:3, 2517 Hz-(b); 2:2, 4075 Hz-(c)

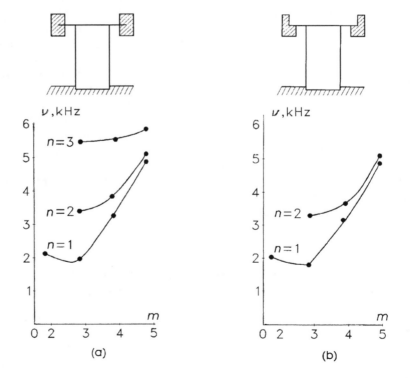

Figure 11.4. (a) and (b) *(Caption opposite)*

In order to determine the spectrum of forced vibrations of the equipment placed inside the reactor vessel, complementary information on the resonance frequencies and modes for the reactor shaft is required for designers. The shaft itself represents a cylindrical shell with varying wall thickness and an elliptic bottom. The main difficulty for calculations of the vibration resonance frequencies of the shaft is confined to obtaining an adequate account of various boundary conditions. Some results of experimental investigations of the influence of the boundary conditions on resonance vibration frequencies [13] are discussed below.

The objects investigated were models of the reactor shaft of scale 1:44, having a flat and elliptic bottom, made of aluminium alloy. Oscillations of the models were simulated by a piezoceramic device in the range of 400 Hz to 5 KHz.

The vibration modes of cylindrical shells can be characterized by a pair of numbers $m{:}n$, where m is the number of half-waves in the circumferential

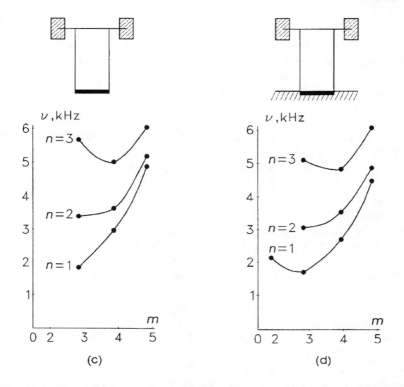

Figure 11.4. Dependencies of resonance frequencies against wave-numbers corresponding to various support conditions of shaft model with flat bottom: var.1-(a); var.2-(b); var.3-(c) and var.4-(d)

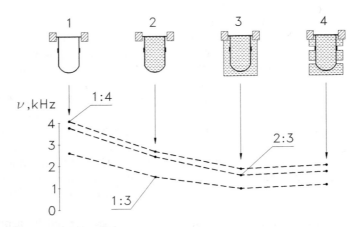

Figure 11.5. Resonance frequencies of shaft model with elliptical bottom variation due to change of support conditions and adding of inner and outer medium

direction and n is the number of half-waves in the axis (longitudinal) direction. The holographic interferograms showed some typical vibration modes of the shaft model which are represented in Figure 11.3.

A model with flat bottom was investigated under different support conditions:

(1) Rigidly clamped upper model end and lower end without bottom mounted with interference fit (Figure 11.4(a))
(2) Simply supported upper end of the model and lower end without bottom mounted with interference fit (Figure 11.4 (b))
(3) Rigidly clamped upper model end and free lower end with flat bottom (Figure 11.4(c))
(4) Rigidly clamped upper model end and the lower end with flat bottom mounted with interference fit (Figure 11.4(d)).

The obtained dependencies of resonance frequencies versus the wave numbers m and n are characteristic for cylindrical shells. It should be pointed out that for the third variant of boundary conditions only the lowest mode with fractional half-waves number in the longitudinal direction appeared ($m = 1$; $n = 0.5$). The free support of the upper end (variant 2) and the rigid clamping of the upper end (variant 1) did not differ with the resonance frequencies and vibrations modes. The gap of corresponding resonance frequencies in the same modes did not exceed 10 per cent.

The resonance frequency variations due to the different support conditions for the shaft model with an elliptic bottom, are illustrated in Figure 11.5.

The following variants were investigated experimentally and are illustrated in Figure 11.5:

(1) Model vibrations in air
(2) Model vibrations with water inside
(3) Model vibrations with water inside and outside
(4) Model vibrations with water inside and outside and, also, with modelling of reactor shaft fixture in lower nozzle and in flow dissector.

The added fluid masses (variants 2 and 3) caused a decrease in the resonance frequencies compared to the same ones obtained in the air medium (variant 1) by 1.5 to 3.0 times. By adding the supports, the resonance frequencies increased by approximately 16 per cent.

These investigations allow us to estimate the resonance frequencies spectrum of a real shell structure, and also improve the theoretical predictions particularly with regard to a presence of the inner and outer working medium and support conditions.

11.3 Displacement fields determination for the main circulator of a nuclear reactor

Currently, the verification of numerical codes developed to compute the stress–strain state of various structural elements is a problem. This is associated with the need for a more detailed description of the overall structure mechanics which account for real boundary conditions and material properties. Usually, numerical codes are tested by simulation of typical objects with an elementary shape for which accurate theoretical solutions are available. However, for designing structural elements with a complex three-dimensional geometry this approach is evidently inadequate.

One of the ways to solve this is by experimental determination of three-dimensional displacement fields on a deformed object surface and subsequent comparison of these displacement fields with numerical results. This approach was used to verify software packages for solid modelling by a finite element method [14].

The object under investigation was the frame of a main circulation pump (MCP) of a nuclear reactor VVER, made of stainless steel, with a scale 1:10 (see Figure 11.6). The model was manufactured in the form of a welded frame and included rigid end 1, shell 2, having ellipsoid shape, straight 3 and thyroidal 4 connections, and the base 5. The model was loaded by internal pressure.

To determine the displacement fields on the model surface, the holographic interferometry was used. The complexity of the MCP frame did not permit us to fully observe the object surface by using a typical optical set-up.

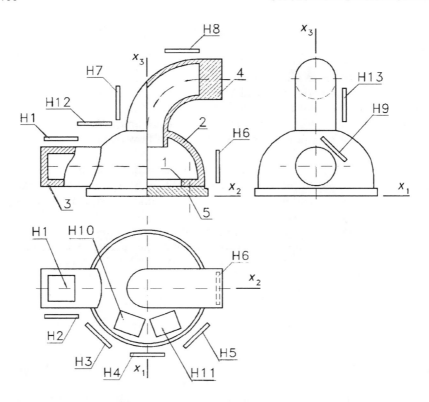

Figure 11.6. Drawing of main circulation pump model and scheme of photoplates location and orientation used to record reflection holograms

Consequently, an approach was developed which joined a few hologram interferometers based on reflection holograms, recorded near the main parts of model surface. The locations of thirteen photoplates, denoted as H1, H2, H3,..., H13, to record the reflection holograms are shown in Figure 11.7. The orientation of each hologram was defined in a global coordinate system (x_1, x_2, x_3).

In order to choose a loading increment which provided an optimal interference fringe spacing over the model surface in full, a few doube-exposed holograms were made by using the optical set-up of Leith and Upatniek. A typical holographic interferogram, recorded at internal pressure increment $q = 3.33$ MPa, is shown in Figure 11.7.

To identify the absolute fringe number for each reflection interferometer, an elastic rubber tape was attached between the model surface and the rigid base. The holographic interferogram shown in Figure 11.7 was used as a guide to choose the locations of these tapes.

Figure 11.7. Holographic interferogram of MCP model obtained by Leith–Upatnieks optical arrangement

Typical holographic interferograms, obtained by viewing the double-exposed reflection holograms, for straight connection (H2), frame shell (H10) and thyroidal connection (H13) are shown in Figures 11.8 (a), (b) and (c), respectively. In the fringe patterns represented we can see the elastic tape used for absolute determination of fringe numbers. A quantitative interpretation of the interferograms was carried out using the method described in Section 2.4. In this way, for each double-exposed reflection hologram, twelve fringe patterns were obtained corresponding to the viewing angle $\psi = 30°$. Absolute fringe numbers were determined in the nodes of the grid, specially painted on the object surface. Linear interpolation of the fringe order distributions was used. In the first step, three displacement components were determined in different local coordinate systems corresponding to the initial photoplate orientation. Then, all experimental data were transformed into a common global coordinate system (x_1, x_2, x_3). It should be pointed out that the displacements calculated on the intersected boundaries between the adjacent reflection holograms coincide with each other. This is good indirect evidence of the experimental data reliability.

As an example, in Figure 11.9(a) the diagram of displacement component d_3 in the plane of model symmetry is shown, which corresponds to internal pressure increment $q = 10\,\text{MPa}$. This displacement component is dominant by an absolute value in the problem under consideration. The distribution character of component d_3 in the connections demonstrates the bending deformation of these structural elements.

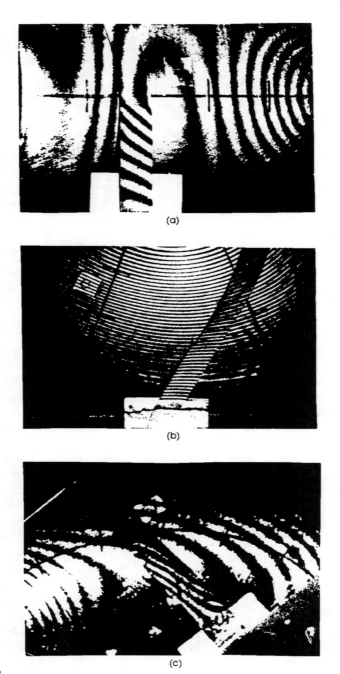

Figure 11.8

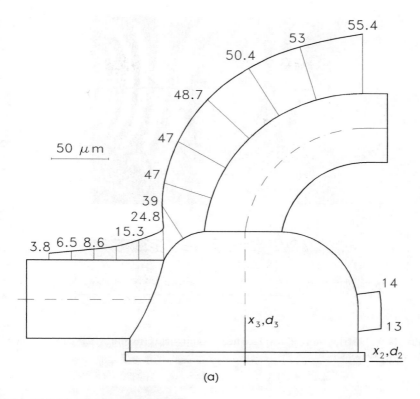

(a)

Figure 11.9 (a). *Caption on page 472*

A diagram of the displacement component d_3 in the same section is shown in Figure 11.9(b), which also refers to an internal pressure of 10 MPa. As can be concluded from this diagram, the straight connection undergoes a tension deformation.

Three-dimensional displacement fields for the MCP model were numerically computed by finite element simulation. The numerically obtained distribution of the displacement component d_3 differs from the experimental one on a constant magnitude of the order $10\,\mu$m. This may be due to an inadequate stiffness assignment of the model-base joint. But on the whole, the experimental and numerical results are in good correlation along the whole model surface.

Figure 11.8. *(opposite)* Holographic interferograms of MCP model obtained by using doubly-exposed reflection holograms corresponding to: straight nozzle H2-(a); frame shell H10-(b); toroidal nozzle H13-(c)

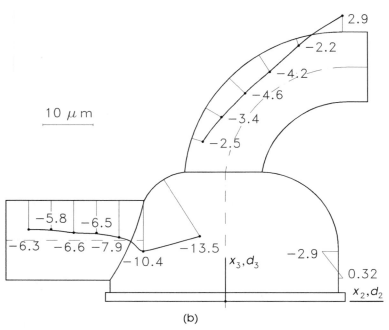

(b)

Figure 11.9. Distribution of displacement components in the MCP model plane of symmetry: d_3-(a), d_2-(b)

11.4 Deformations of a rotor winding conductor of a powerful turbo-generator

The winding conductor cross-section was square with a circular hole (waterway) in the centre. The conductor was made of a highly expensive copper–silver alloy. Under a centrifugal force which arises in the upper turn, the stress level may overcome the yield limit of the copper–silver alloy, and as consequence, the bearing capacity is lost and the centre waterway shape irreversibly changes. With regard to the plastic deformations of this alloy, the kinetic deformation study of rotor winding conductors is very significant.

The elasto-plastic strain in the conductor cross-section was investigated on samples with dimensions 24×24 mm and an axial channel of 15 mm in diameter. The samples were cut from a real winding conductor. The set-up for the samples loading is shown in Figure 11.10. The lower sample end was supported on a steel plate and the upper end was connected with a special piece, which was a model of a neighbouring conductor turn, and made of actual alloy.

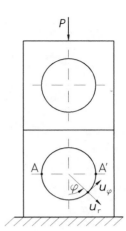

Figure 11.10. Rotor winding conductor loading arrangement

Investigations combining the holographic interferometry and speckle photography techniques were carried out to analyse sample surface deformations. The initiation and propagation of residual plastic deformations over a conductor cross-section were studied by the holographic technique, and measurements of the tangent displacement components were carried out by speckle photography. The applied technique, based on aperture function variations, was discussed in Section 3.7 in detail using, as example, a compression of a steel plate with a cut-out.

Samples, cut from a piece of a winding conductor, were tested in the range of loads P from 3.9 to 18.35 kN. Thus, the maximum load under testing exceeded the operating load by two times.

In Figure 11.11(a) the measurement results obtained by speckle photography for radial u_r (curve 1) and circumferential u_t (curve 2) displacement components along the hole contour are shown, which correspond to sample load $P = 7.25$ kN. Using the one-axis stress–strain diagram σ-ε for copper–silver alloy, a circumferential stress σ_φ along the axial channel contour in the conductor sample can be determined (see Figure 11.11(b)). As can be expected, the maximum compression stresses appeared at points A and A' on the horizontal hole diameter. Different boundary conditions corresponding to the upper and lower sample ends caused asymmetric stresses distribution with respect to the axis passing through points A and A'. The stresses in the contour points with coordinates $\varphi = 0$ and $\varphi = \pi$ differ by approximately 40 per cent.

The support surfaces of a real winding conductor are characterized by an initial non-flatness resulting from the manufacturing technology. In the first stage, a square extrusion is produced and then the waterway channel is drifted.

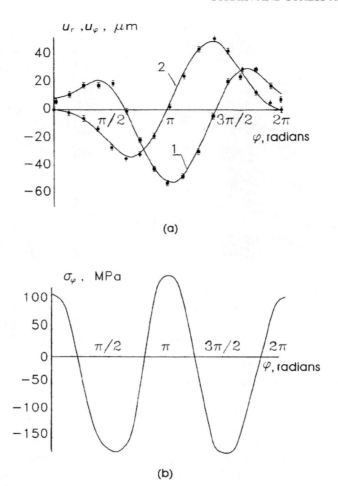

Figure 11.11. Distributions of radial 1 and circumferential 2 displacement components - (a) and circumferential stresses- (b) in the channel contour

Hence, the support surfaces of the samples have an average curvature radius of 500 mm. As experiments showed for samples with flat support surfaces, the maximum stresses decreased by 20 per cent.

Experimental information obtained by holographic interferometry proved our previous conclusions. The initiation moment of plastic deformation was observed under load $P \approx 7\,kN$. Circular interference fringes, corresponding to residual irreversible displacements, formed in zones with maximum compression stresses in the axial channel boundary (points A and A' in Figure 11.10). According to the load increase, the plastic zones grow in size until the sample

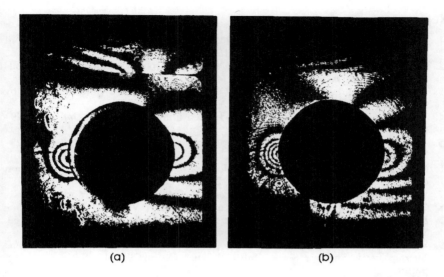

(a) (b)

Figure 11.12. Holographic interferograms corresponding to initial plastic deformations (a) and completely plastically deformed cross section (b)

cross-section becomes completely plastically deformed. By way of illustration, Figure 11.12(a) shows a holographic interferogram corresponding to the moment of the plastic deformation initiation, and Figure 11.12(b) shows another interferogram which corresponds to a plastically deformed conductor cross-section under a load substantially exceeding the operating conditions.

The flatness deviations of the sample surfaces are clearly observed in the holographic interferograms. For instance, in the upper contact zone of Figure 11.12(a), we can see a circular fringe which initiated due to plastic deformation at this point of the sample surface. The axial channel shape variations for the load range used were insignificant, therefore, for the operating loads the deviation from the initial circular shape will not exceed 1 per cent.

In spite of the acting stresses exceeding the material yield limit, the experiments reveal that while the turbo-generator is in use, the development of plastic strain in the conductor turns occur locally and, thus, the bearing capacity of the rotor winding conductor does not reach the limit. Also, it was revealed that the non-flatness of the conductor support surfaces mainly influences the plastic strain values. By improving the flatness level of the winding conductor it is possible to make available the bearing capacity of the conductor under lower-strength material properties and, therefore, the silver content in the alloy can be decreased.

References

1. Powell, R. L. and Stetson, K. A. (1965) Interferometric analysis by wavefront reconstruction. *J. Opt. Soc. Amer.* **55**, 1593–1598.
2. Listovets, V. S. and Ostrovsky, Yu. I. (1974) Interferometric-holographic methods of vibration analysis. *Zhur. Tekhn. Fiz.* **44**, 1345–1373.
3. Archbold, E. and Ennos A. E. (1968) Observation of surface vibration modes by stroboscopic hologram interferometry. *Nature* **217**, 842–843.
4. Aleksoff, C. C. (1971) Temporally modulated holography. *Appl. Opt.* **10**, 1329–1342.
5. Aleksoff, C. C. (1969) Time average holography extended. *Appl. Phys. Lett.* **14**, 23–25.
6. Mottier, F. M. (1969) Time-average holography with triangular phase modulation of reference wave. *Appl. Phys. Lett.* **15**, 285–287.
7. Vest, C. M. (1979) *Holographic Interferometry.* Interscience, New York.
8. Ostrovsky, Yu. I., Shchepinov, V. P. and Yakovlev, V. V. (1991) *Holographic Interferometry in Experimental Mechanics.* Springer Series in Optical Science 60. Springer-Verlag, Berlin.
9. Vicram, C. S. (1994) Study of vibrations. In Rastogi, P. K., (ed.) *Holographic Interferometry.* Springer Series Optical Science 68, pp. 293–318, Springer-Verlag, Berlin.
10. Osintsev, A. V., Yakovlev, V. V. and Shchepinov, V. P. (1981) Investigation of perforated plates vibration modes by holographic interferometry. In *Physics and Mechanics of Deformation and Fracture*, pp. 68–71, Energoatomizdat, Moscow.
11. Stetson, K. A. and Powell, R. L. (1965) Interferometric hologram evaluation and real-time vibration analysis of diffuse objects. *J. Opt. Soc. Amer.* **55**, 1694–1695.
12. Osintsev, A. V., Shchepinov, V. P. and Yakovlev, V. V. (1983) Experimental investigation of flat bottoms with cut-outs vibration. In *Deformation and Fracture of Materials and Structures of Nuclear Equipment*, pp. 95–99, Energoatomizdat, Moscow.
13. Leonov, M. A., Osintsev, A. V., Khairetdinov, V. Y. and Shchepinov, V. P. (1991) Boundary conditions influence on the dynamical characteristics of inner mounting reactor shuft. In *Bearing Capacity of Materials and Structures Elements of NPR*, pp. 67–76, Energoatomizdat, Moscow.
14. Aistov, V. S., Balalov, V. V., Kiselev, A. S., Tutnov, A. A. and Shchepinov, V. P. Investigation of deformation for main circulator of nuclear power reactor by holographic interferometry and finite elements methods. *Vopros. Atomn. Nauk. Tekhn.*, to be published.

SUBJECT INDEX